Theory of Probability

Volume 2

Theory of Probability

A critical introductory treatment

Volume 2

BRUNO DE FINETTI

(1905 – 1985)

Translated by

ANTONIO MACHI

Assistant Professor of Mathematics at the University of Rome, Italy

and

ADRIAN SMITH

Professor of Mathematics at the University of Nottingham, UK

Wiley Classics Library Edition Published 1990

INTERSCIENCE PUBLISHERS

JOHN WILEY & SONS

Chichester · New York · Brisbane · Toronto · Singapore

First published 1970 © Giulio Einaudi editore s.p.a.,
Torino, under the title of *Teoria Delle Probabilità,* by Bruno de Finetti.
Copyright © 1975 by John Wiley & Sons Ltd.
Reprinted August 1978

Wiley Classics Library edition copyright © 1990 by John Wiley & Sons Ltd.
Baffins Lane, Chichester
West Sussex PO19 1UD,
England

Other Wiley Editorial Offices

John Wiley & Sons, Inc., 605 Third Avenue,
New York, NY 10158-0012, USA

Jacaranda Wiley Ltd, G.P.O. Box 859, Brisbane,
Queensland 4001, Australia

John Wiley & Sons (Canada) Ltd, 22 Worcester Road,
Rexdale, Ontario M9W 1L1, Canada

John Wiley & Sons (SEA) Pte Ltd, 37 Jalan Pemimpin 05-04,
Block B, Union Industrial Building, Singapore 2057

British Library Cataloguing in Publication Data available

Printed in Great Britain by Biddles Ltd, Guildford

This work is dedicated to my colleague Beniamino Segre who about twenty years ago pressed me to write it as a necessary document for clarifying one point of view in its entirety

Foreword

It is an honour to be asked to write a foreword to this book, for I believe that it is a book destined ultimately to be recognized as one of the great books of the world.

The subject of probability is over two hundred years old and for the whole period of its existence there has been dispute about its meaning. At one time these arguments mattered little outside academia, but as the use of probability ideas has spread to so many human activities, and as probabilists have produced more and more sophisticated results, so the arguments have increased in practical importance. Nowhere is this more noticeable than in statistics, where the basic practices of the subject are being revised as a result of disputes about the meaning of probability. When a question has proved to be difficult to answer, one possibility may be that the question itself was wrongly posed and consequently unanswerable. This is de Finetti's way out of the impasse. Probability does not exist.

Does not exist, that is, outside of a person: does not exist, objectively. Probability is a description of your (the reader of these words) uncertainty about the world. So this book is about uncertainty, about a feature of life that is so essential to life that we cannot imagine life without it. This book is about life: about a way of thinking that embraces all human activities.

So, in a sense, this book is for everyone; but necessarily it will be of immediate appeal to restricted classes of readers.

Philosophers have recently increased their interest in probability and will therefore appreciate the challenging ideas that the author puts forward. For example, those of the relationships between possibility and tautology. They will notice the continual concern with reality, with the use of the ideas in practical situations. This is a philosophy intended to be operational and to express the individual's appreciation of the external world.

Psychologists are much concerned with the manner of this appreciation, and experiments have been performed which show that individuals do not reason about uncertainty in the way described in these volumes. The experiments provide a descriptive view of man's attitudes: de Finetti's approach is normative. To spend too much time on description is unwise when a normative approach exists, for it is like asking people's opinion of 2 + 2, obtaining an average of 4·31 and announcing this to be the sum. It would be better to teach them arithmetic. I hope that this book will divert psychologists' attentions away from descriptions to the important problem, ably discussed in this book, of how to teach people to assess probabilities.

Mathematicians will find much of interest. (Let me hasten to add that some people may approach the book with fear because of the amount of mathematics it contains. They need not worry. Much of the material is accessible with no mathematical skill: yet more needs only a sympathetic appreciation of notation. Even the more mathematical passages use mathematics in a sparse and yet highly efficient way. Mathematics is always the servant—never the master (see Section 1.9.1).) Nevertheless, the mathematician will appreciate the power and elegance of the notation and, in particular, the discussion of finite additivity. He will be challenged by the observation that 'mathematics is an instrument which should conform itself strictly to the exigencies of the field in which it is to be applied'. He will enjoy the new light shed on the calculus of probabilities.

Physicists have long used probabilistic notions in their understanding of the world, especially at the basic, elementary-particle level. Here we have a serious attempt to connect their use of uncertainty with the idea as used outside physics.

Statisticians are the group I can speak about with greatest confidence. They have tended to adopt a view of probability which is based on frequency considerations and is too narrow for many applications. They have therefore been compelled to introduce artificial ideas, like confidence intervals, to describe the uncertainties they need to use. The so-called Bayesian approach has recently made some significant impression, but de Finetti's ideas go further still in replacing frequency concepts entirely—using his notion of exchangeability—and presenting an integrated view of statistics based on a single concept of uncertainty. A consequence of this is that the range of possible applications of statistics is enormously widened so that we can deal with phenomena other than those of a repeatable nature.

There are many other groups of people one would like to see reading these volumes. Operational research workers are continually trying to express ideas to management that involve uncertainty: they should do it using the concepts contained therein. One would like (is it a vain hope?) to see politicians with a sensible approach to uncertainty—what a blessing it would be if they could appreciate the difference between prediction and prevision (p. 98).

The book should therefore be of interest to many people. As the author says (p. 14) 'it is . . . an attempt to view, in a unified fashion, a group of topics which are in general considered separately, each by specialists in a single field, paying little or no attention to what is being done in other fields.'

The book is not a text on probability in the ordinary sense and would probably not be useful as a basis for a course of lectures. It would, however, be suitable for a graduate seminar wherein sections of it were discussed and analysed. Which sections were used would depend on the type of graduates, but with the continuing emphasis on unity, it would be valuable in bringing different disciplines together. No university should ignore the book.

It would be presumptious of *me* to say how *you* should read the two volumes but a few words may help your appreciation. Firstly, do not approach it with preconceived ideas about probability. I address this remark particularly to statisticians, who can so easily interpret a formula or a phrase in a way that they have been used to, when de Finetti means something different. Let the author speak for himself. Secondly, the book does not yield to a superficial reading. The author has words of wisdom to say about many things and the wisdom often only appears after reflection. Rather, dip into parts of the book and read those carefully. Hopefully you will be stimulated to read the whole. Thirdly, the style is refreshing—the translators have cleverly used the phrase 'a whimsical fashion' (Section 1.3.3)—so that every now and again delightful ideas spring to view; the idea that we shall all be Bayesian by 2020, or how to play the football pools. But, as I said, this is a book about life.

University College London,
November 1973

D. V. Lindley

Preface

Is it possible that in just a few lines I can achieve what I failed to achieve in my many books and articles? Surely not. Nevertheless, this preface affords me the opportunity, and I shall make the attempt. It may be that misunderstandings which persist in the face of refutations dispersed or scattered over some hundreds of pages can be resolved once and for all if all the arguments are pre-emptively piled up against them.

My thesis, paradoxically, and a little provocatively, but nonetheless genuinely, is simply this:

PROBABILITY DOES NOT EXIST.

The abandonment of superstitious beliefs about the existence of Phlogiston, the Cosmic Ether, Absolute Space and Time, . . . , or Fairies and Witches, was an essential step along the road to scientific thinking. Probability, too, if regarded as something endowed with some kind of objective existence, is no less a misleading misconception, an illusory attempt to exteriorize or materialize our true probabilistic beliefs.

In investigating the reasonableness of our own modes of thought and behaviour under uncertainty, all we require, and all that we are reasonably entitled to, is consistency among these beliefs, and their reasonable relation to any kind of relevant objective data ('relevant' in as much as subjectively deemed to be so). This is Probability Theory. In its mathematical formulation we have the Calculus of Probability, with all its important off-shoots and related theories like Statistics, Decision Theory, Games Theory, Operations Research and so on.

This point of view is not bound up with any particular philosophical position, nor is it incompatible with any such. It is strictly *reductionist* in a methodological sense, in order to avoid becoming embroiled in philosophical controversy.

Probabilistic reasoning—always to be understood as subjective—merely stems from our being uncertain about something. It makes no difference whether the uncertainty relates to an unforseeable future, or to an unnoticed past, or to a past doubtfully reported or forgotten; it may even relate to something more or less knowable (by means of a computation, a logical deduction, etc.) but for which we are not willing or able to make the effort; and so on.

Moreover, probabilistic reasoning is completely unrelated to general philosophical controversies, such as Determinism versus Indeterminism, Realism versus Solipsism—including the question of whether the world 'exists', or is simply the scenery of 'my' solipsistic dream. As far as Determinism and Indeterminism are concerned, we note that, in the context of gas theory or heat diffusion and transmission, whether one interprets the underlying process as being random or strictly deterministic makes no difference to one's probabilistic opinion. A similar situation would arise if one were faced with forecasting the digits in a table of numbers; it makes no difference whether the numbers are random, or are some segment—for example, the 2001st to the 3000th digits—of the decimal expansion of π (which is not 'random' at all, but certain; possibly available in tables and, in principle, computable by you).

The only relevant thing is uncertainty—the extent of our own knowledge and ignorance. The actual fact of whether or not the events considered are in some sense *determined*, or known by other people, and so on, is of no consequence.

The numerous, different, opposed attempts to put forward particular points of view which, in the opinion of their supporters, would endow Probability Theory with a 'nobler' status, or a 'more scientific' character, or 'firmer' philosophical or logical foundations, have only served to generate confusion and obscurity, and to provoke well-known polemics and disagreements—even between supporters of essentially the same framework.

The main points of view that have been put forward are as follows.

The *classical* view, based on physical considerations of symmetry, in which one should be *obliged* to give the same probability to such 'symmetric' cases. But which symmetry? And, in any case, why? The original sentence becomes meaningful if reversed: the symmetry is probabilistically significant, in someone's opinion, if it leads him to assign the same probabilities to such events.

The *logical* view is similar, but much more superficial and irresponsible inasmuch as it is based on similarities or symmetries which no longer derive from the facts and their actual properties, but merely from the sentences which describe them, and from their formal structure or language.

The *frequentist* (or *statistical*) view presupposes that one accepts the classical view, in that it considers an *event* as a class of *individual events*, the latter being 'trials' of the former. The individual events not only have to be 'equally probable', but also 'stochastically independent'... (these notions when applied to individual events are virtually impossible to define or explain in terms of the frequentist interpretation). In this case, also, it is straightforward, by means of the subjective approach, to obtain, under the appropriate conditions, in a perfectly valid manner, the result aimed at (but unattainable) in the statistical formulation. It suffices to make use of the notion of exchangeability. The result, which acts as a bridge connecting this new approach with the old, has been referred to by the objectivists as 'de Finetti's representation theorem'.

It follows that all the three proposed definitions of 'objective' probability, although useless *per se*, turn out to be useful and good as valid auxiliary devices when included as such in the subjectivistic theory.

The above-mentioned 'representation theorem', together with every other more or less original result in my conception of probability theory, should not be considered as a discovery (in the sense of being the outcome of advanced research). Everything is essentially the fruit of a thorough examination of the subject matter, carried out in an unprejudiced manner, with the aim of rooting out nonsense.

And probably there is nothing new; apart, perhaps, from the systematic and constant concentration on the unity of the whole, avoiding piecemeal tinkering about, which is inconsistent with the whole; this yields, in itself, something new.

Something that may strike the reader as new is the radical nature of certain of my theses, and of the form in which they are presented. This does not stem from any deliberate attempt at radicalism, but is a natural consequence of my abandoning the reverential awe which sometimes survives in people who at one time embraced the objectivistic theories prior to their conversion (which hardly ever leaves them free of some residual).

It would be impossible, even if space permitted, to trace back the possible development of my ideas, and their relationships with more or less similar positions held by other authors, both past and present. A brief survey is better than nothing, however (even though there is an inevitable arbitrariness in the selection of names to be mentioned).

I am convinced that my basic ideas go back to the years of High School as a result of my preference for the British philosophers Locke, Berkeley and, above all, Hume! I do not know to what extent the Italian school textbooks and my own interpretations were valid: I believe that my work based on exchangeability corresponds to Hume's ideas, but some other scholars do not agree. I was also favourably impressed, a few years later, by the ideas of Pragmatism, and the related notions of operational definitions in Physics. I particularly liked the Pragmatism of Giovanni Vailati—who somehow 'Italianized' James and Peirce—and, as for operationalism, I was very much struck by Einstein's relativity of 'simultaneity', and by Mach and (later) Bridgman.

As far as Probability is concerned, the first book I encountered was that of Czuber. (Before 1950—my first visit to the USA—I did not know any English, but only German and French.) For two or three years (before and after the 'Laurea' in Mathematics, and some application of probability to research on Mendelian heredity), I attempted to find valid foundations for all the theories mentioned, and I reached the conclusion that the classical and frequentist theories admitted no sensible foundation, whereas the subjectivistic one was fully justified on a normative–behaviouristic basis. I had some indirect knowledge of De Morgan, and found that some of Keynes' ideas were in partial agreement with mine; some years later I was informed of the similar approach that had been adopted by F. P. Ramsey.

Independent ideas, which were more or less similar, were put forward later by Harold Jeffreys, B. O. Koopman, and I. J. Good (with some beautiful new discussion which illustrated the totally illusory nature of the so-called *objective* definitions of probability). I could add to this list the name of Rudolf Carnap, but this would be not altogether proper in the light of his (to me strange) superposition of the idea of a logical framework onto his own vivid, subjective behaviouristic interpretation. (Richard Jeffreys, in publishing Carnap's posthumous works, seems convinced of his underlying subjectivism.) A singular position is occupied by Robert Schlaifer, who arrived at the subjectivistic approach directly and with impressive freshness and originality, with little knowledge of previous work in the field. A similar thing, although in a different sense, may be said of George Pólya, who discussed *plausible reasoning* in mathematics in the sense of the probability (subjective, of course) of a supposed theorem being true, given the state of mind of the mathematician, and its (Bayesian) modification when new information or ideas appear. The following statement of his is most remarkable: 'It seems to me more philosophical to consider the general idea of *plausible reasoning* instead of its isolated particular cases' like *inductive* (and analogical) *reasoning*. (There have been so many vain attempts to build a theory of induction without beliefs—like a theory of elasticity without matter.)

A very special mention must be reserved, however, for Leonard J. Savage and Dennis V. Lindley, who escaped from the objectivistic school, after having grown up in it, by a gradual discovery of its inconsistencies, and through a comparison of its ambiguities with the clarity of the subjectivistic theory, and the latter's suitability for every kind of practical or theoretical problem. I have often had the opportunity of profitable exchanges of ideas with them, and, in the case of Savage, of actual collaboration, I wrote briefly of Savage's invaluable contributions as a dedication to my book *Probability, Induction and Statistics*, which appeared a few months after his sudden and premature death.

One should note, however, that, even with such close colleagues, agreement ought not to be absolute, on every detail. For example, not all agree with the rejection of countable-additivity.

Finally, having mentioned several of the authors who are more or less connected with the subjectivistic (and Bayesian) point of view, I feel an obligation to recall three great men—the first two, unfortunately, no longer with us—who, although they all shared an opposed view about our common subject, were always willing to discuss, and were extraordinarily friendly and helpful on every occasion. I refer to Guido Castelnuovo, Maurice Fréchet and Jerzy Neyman.

Rome, 16 July 1973 — Bruno de Finetti

Translators' Preface

In preparing this English translation, we were concerned to achieve two things: first of all, and most importantly, to translate as accurately as possible the closely argued *content* of the book; secondly, to convey something of the flavour of the author's idiosyncratic *style*; the sense of the painstaking struggle for understanding that runs through the Italian original.

Certain of Professor de Finetti's works have already appeared in English, the principal references being Kyburg and Smokler's *Studies in Subjective Probability* (Wiley, 1964), and the author's *Probability, Induction and Statistics* (Wiley, 1972). For the purpose of comparison—and to avoid any possible confusion—we include the following preliminary notes on the terminological and notational usage that we have adopted.

In common with the above-mentioned translations, we use the word *coherent* when referring to degrees of belief which satisfy certain 'consistency' conditions, *random quantity* in place of the more usual 'random variable', and *exchangeable*, rather than 'equivalent' or 'symmetric'.

We part company with previous translations, however, in our treatment of the concept corresponding to what is usually called 'mathematical expectation'. In Kyburg's translation of de Finetti's monograph 'La Prévision: ses lois logiques, ses sources subjectives' (see Kyburg and Smokler, pp. 93–158), the corresponding word becomes 'foresight'. We shall use the word *prevision*. A discussion of the reasons for this choice is given more fully at the appropriate place in the text (Chapter 1, 1.10.3) but let us note straightaway that the symbol **P** now very conveniently represents both *probability* and *prevision*, and greatly facilitates their unified treatment as linear operators.

Readers who are familiar with the Italian original will realize that on occasions we have opted for a rather free style of translation; we did so, in fact, whenever we felt this to be the best way of achieving our stated aims. Throughout, however, we have been mindful of the 'misunderstandings' referred to by the author in his Preface, and we can but hope that our translation does nothing to add to these.

Finally, we should like to express our gratitude to Professor de Finetti, who read through our translation and made many helpful suggestions; to the editor at John Wiley & Sons for getting the project under way; and to Mrs Jennifer Etheridge for her care in typing our manuscript.

A. Machí
A. F. M. Smith

Contents

CHAPTER 7

A Preliminary Survey

7.1 WHY A SURVEY AT THIS STAGE?

7.1.1. Our discussion of the requirements of the conceptual formulation of the theory of probability has already revealed its wide range of application. It applies, in fact, whenever the factor of uncertainty is present. The range of problems encountered is also extensive. Diverse in nature and in complexity, these problems require a corresponding range of mathematical techniques for their formulation and analysis, techniques which are provided by the calculus of probability. For a number of reasons, it is useful to give a preliminary survey, illustrating these various aspects. In setting out our reasons, and by inviting the reader to take note of certain things, we shall be able to draw attention to those points which merit and require the greatest emphasis.

First of all, we note that individual topics acquire their true status and meaning only in relation to the subject as a whole. This is probably true of every subject, but it is particularly important in the case of probability theory. In order to explore a particular area, it pays to get to know it in outline before starting to cover it in great detail (although this will be necessary eventually), so that the information from the detailed study can be slotted into its rightful place. If we were to proceed in a linear fashion, we would not only give an incomplete treatment, but also a misleading one, in that it would be difficult to see the connections between the various aspects of the subject. The same would be true of even the most straightforward problems if we had to deal with them without, at any given point, referring to any feature whose systematic treatment only came later (e.g. we would not be able to mention the connections with 'laws of large numbers', 'random processes' or 'inductive inference'). Nor would it be reasonable (either in general or in this particular case) to assume that individual chapters are approached only after all the preceding ones have been read, and their contents committed to memory. For an initial appreciation, it is necessary and sufficient to be clear about basic problems and notions rather than attempting to acquire a detailed knowledge. Moreover, the difficulties associated with this approach are easily avoided. It is sufficient to learn from the outset how to *understand* what these problems and concepts are about by concentrating on a small number of simple but meaningful examples. Although elementary and summary in nature, the approach is then both clear and concrete, and can be further developed by various additional comments and information.

7.1.2. In this preliminary survey, we shall, for this reason, concentrate on the case of *Heads and Tails*. This example, examined from all possible angles, will serve as a basic model, although other variants will be introduced from time to time (more for the sake of comparison and variety than from necessity). These simple examples will shed light on certain important ideas which crop up over and over again in a great many problems, even complex and advanced ones. This, in turn, facilitates the task of analysing the latter in greater depth. In fact, it often turns out that the result of such an analysis is simply the extension of known and intuitive results to those more complex cases. This also reveals that the detailed complications which distinguish such cases from the simple ones are essentially irrelevant.

Another reason for providing a survey is the following. Everybody finds problems in the calculus of probability difficult (non-mathematicians, mathematicians who are unfamiliar with the subject, even those who specialize in it if they are not careful†). The main difficulty stems, perhaps, from the danger of opting for the apparently obvious, but wrong, conclusion, whereas the correct conclusion is usually easily established, provided one looks at the problem in the right way (which is not—until one spots it—the most obvious). In this respect, the elementary examples provide a good basis for discussion and advice (which, although useful, is inadequate unless one learns how to proceed by oneself for each new case). Many of the comments we shall make, however, are not intended solely for the purpose of avoiding erroneous or cumbersome arguments when dealing with simple cases. More generally, they are made with the intention of clarifying the conceptual aspects themselves, and of underlining their importance, in order to avoid any misunderstandings or ambiguities arising in other contexts going beyond those of the examples actually used. In other words, we

† Feller, for example, repeatedly remarks on the way in which certain results seem to be surprising and even paradoxical (even simple results concerning coin tossing, such as those relating to the periods during which one gambler has an advantage over the other; Chapter 8, 8.7.6). He remarks on 'conclusions that play havoc with our intuition' (p. 68), and that 'few people will believe that a perfect coin will produce preposterous sequences in which no change of lead occurs for millions of trials in succession, and yet this is what a good coin will do rather regularly' (p. 81). Moreover, he can attest to the fact that 'sampling of expert opinion has revealed that even trained statisticians feel that' certain data are really surprising (p. 85).

All this is from W. Feller, *An Introduction to Probability Theory and its Applications*, 2nd ed., Wiley, New York (1957). Many of the topics mentioned in the present chapter are discussed in detail by Feller (in Chapter 3, in particular), who includes a number of original contributions of his own.

shall be dealing with matters which, as far as the present author is concerned, have to be treated as an integral part of the formulation of the foundations of the subject, and which could have been systematically treated as such were it not for the risk that one might lose sight of the direct nature of the actual results and impart to the whole enterprise a suggestion of argument for argument's sake, or of literary–philosophical speculation.

7.1.3. It turns out that the aims which we have outlined above are best achieved by concentrating mainly on examples of the 'classical' type—i.e. those based on combinatorial considerations. In fact, even leaving aside the need to mention such problems anyway, many of these combinatorial problems and results are particularly instructive and intuitive by virtue of their interpretations in the context of problems in probability. Indeed, it was once thought that the entire calculus of probability could be reduced essentially to combinatorial considerations by reducing everything down to the level of 'equiprobable cases'. Although this idea has now been abandoned, it remains true that combinatorial considerations do play an important rôle, even in cases where such considerations are not directly involved (cf. the examples that were discussed in Chapter 5, 5.7.4).

Given our stated purpose in this 'preliminary survey', it will naturally consist more of descriptive comments and explanations than of mathematical formulations and proofs (although in cases where the latter are appropriate we shall provide them). In the first place, we shall deal with those basic, straightforward schemes and analyses which provide the best means of obtaining the required 'insights'. Secondly, we shall take the opportunity of introducing (albeit in the simplest possible form) ideas and results that will be required in later chapters, and of subjecting them to preliminary scrutiny (although without providing a systematic treatment). Finally, we shall consider certain rather special results which will be used later (here they link up rather naturally with one of the examples, whereas introduced later they would appear as a tiresome digression).

7.2 HEADS AND TAILS: PRELIMINARY CONSIDERATIONS

7.2.1. Unless we specifically state otherwise, we shall, from now on, be considering events which You judge to have probability $\frac{1}{2}$, and to be stochastically independent. It follows that each of the 2^n possible results for n such events all have the same probability, $(\frac{1}{2})^n$. Conversely, to judge these 2^n results to be all equally probable implies that You are attributing probability $\frac{1}{2}$ to each event, and judging the events to be stochastically independent.

The events $E_1, E_2, \ldots, E_m, \ldots$ will consist in obtaining *Heads* on a given toss of a coin (we could think in terms of some preassigned number of tosses, n, or of a random number—e.g. 'until some specified outcome is obtained', 'those tosses which are made today', etc.—or of a potentially infinite number). We shall usually take it that we are dealing with successive tosses of the same coin (in the order $E_1, E_2, \ldots$), but nothing is altered if one thinks of the coin being changed every now and then, or even after every toss. In the latter case, we could be dealing with the simultaneous tossing of n coins, rather than n successive tosses (providing we establish some criterion other than the chronological one—which no longer exists—for indexing the E_i). We could, in fact, consider situations other than that of coin tossing. For example: obtaining an even number on a roll of a die, or at Bingo; or drawing a red card from a full pack, or a red number at Roulette (excluding the zero) and so on. We shall soon encounter further examples, and others will be considered later.

7.2.2. In order to represent the outcomes of n tosses (i.e. a sequence of n outcomes resulting in either *Heads* or *Tails*), we can either write $HHTHTTTHT$, or, alternatively, 110100010 (where Head = 1, Tail = 0).†

A. How many Heads appear in the n tosses? This is the most common question. We know already that out of the 2^n possibilities the number in which Heads appear h times is given by $\binom{n}{h}$. The probability, $\omega_h^{(n)}$, of h successes out of n events is therefore $\binom{n}{h}/2^n$. We shall return to this question later, and develop it further.

B. How many runs of consecutive, identical outcomes are there? In the sequence given above, there were six runs: $HH/T/H/TTT/H/T$. It is clear that after the initial run we obtain a new run each time an outcome differs from the preceding one. The probability of obtaining $h + 1$ runs is therefore simply that of obtaining h *change-overs*, and so we consider:

C. How many change-overs are there? In other words, how many times do we obtain an outcome which differs from the preceding one? For each toss, excluding the initial one, asking whether or not the toss gives the same outcome as the previous one is precisely the same as asking whether it gives Heads or Tails. The question reduces, therefore, to (A), and the probability that there are h change-overs in the n tosses is equal to $\binom{n-1}{h}/2^{n-1}$.

D. Suppose we know that out of $n = r + s$ tosses, r are to be made by Peter and s by Paul. *What is the probability that they obtain the same number*

† It should be clear that expressions like HHT (denoting that three consecutive outcomes—e.g. the first, second and third, or those labelled n, $n + 1$, $n + 2$—are Head–Head–Tail) are merely suggestive 'shorthand' representations. The actual logical notation would be $E_n E_{n+1} \tilde{E}_{n+2}$ (or $H_n H_{n+1} T_{n+2}$, if one sets $H_i = E_i$ and $T_i = \tilde{E}_i$). Let everyone be clear about this, so that no one inadvertently performs operations on HHT as though it were simply a product (it would be as though one thought that the year 1967, like *abcd*, being the product of four 'factors', i.e. 1, 9, 6, 7, were equal to 378).

of successes? Arguing systematically, we note that the probability of Peter obtaining h successes and Paul obtaining k is equal to $\binom{r}{h}\binom{s}{k}/2^n$. It follows that the probability of each obtaining the same number of successes is given by $(\frac{1}{2})^n \sum_h \binom{r}{h}\binom{s}{h}$ (the sum running from 0 to the minimum of r and s). As is well-known, however (and can be verified directly by equating coefficients in $(1 + x)^r . (1 + x)^s = (1 + x)^{r+s}$), this sum is equal to $\binom{n}{r} = \binom{n}{s}$, and the probability that we are looking for is identical to that of obtaining r (or s) successes out of n tosses.

This result could have been obtained in an intuitive manner, and without calculation, by means of a similar device to that adopted in the previous case. We simply note that the problem is unchanged if 'success' for Paul is redefined to be the outcome Tails rather than Heads. To obtain the same number of successes (h say) now reduces to obtaining s Heads and r Tails overall; $s = h + (s - h)$, $r = (r - h) + h$. Without any question, this is the most direct, natural and instructive *proof* of the combinatorial identity given above.

E. *What is the probability that the number of successes is odd*? There would be no difficulty in showing this to be $\frac{1}{2}$, by plodding through the summation of the binomial coefficients involved (the sums of those corresponding to evens and odds are equal!). If n were odd, it would be sufficient to observe that an odd number of Heads entails an even number of Tails, and so on.

A more direct and intuitive argument follows from noting that we need only concern ourselves with the final toss. The probability of a success is $\frac{1}{2}$ (no matter what happened on the preceding tosses), and hence the required probability is $\frac{1}{2}$. The advantage of this argument is that we see, with no further effort, that the same conclusion holds under much weaker conditions. It holds, in fact, for any events whatsoever, logically or stochastically independent, and with arbitrary probabilities, provided that one of them has probability $\frac{1}{2}$, and is independent of all combinations of the others (or, at least, of the fact of whether an odd or an even number of them occur).†

We shall return to this topic again in Section 7.6.9.

7.2.3. *Some comments.* The main lesson to be learned from these examples is the following. *In the calculus of probability, just as in mathematics in general, to be able to recognize the essential identity of apparently different problems is not only of great practical value, but also of profound conceptual importance.*

---

† Pairwise independence (which we consider here in order to show how much weaker a restriction it is) would not entitle us to draw these conclusions. We can obtain a counterexample by taking just three events, A, B, C, and supposing them all to be possible, with the four events 'only A', 'only B', 'only C' and 'all three' (ABC) equally probable ($p = \frac{1}{4}$). It is easily seen that A, B, C each have probability $\frac{1}{2}$ and are pairwise independent, but that the number of successes is certainly odd (either 1 or 3). If we had argued in terms of the complements, it would certainly be even (either 0 or 2).

In particular, arguments of this kind often enable us to avoid long and tedious combinatorial calculations; indeed, *they constitute the most intuitive and 'natural' approach to establishing combinatorial identities.*† Moreover, they should serve, from the very beginning, to dispel any idea that there might be some truth in *certain of the specious arguments one so often hears repeated.* For example: that there is some special reason (in general, that it is advantageous) to either always bet on Heads, or always on Tails; or, so far as the lottery is concerned, to always bet on the same number, perhaps one which has not come up for several weeks! All this, despite the fact that, by assumption, all the sequences are equally probable. It is certainly true that the probability of no Heads in ten successive tosses is about one in a thousand ($2^{-10} = 1/1024$), and in twenty tosses about one in a million ($2^{-20} = 1/1048576$), but the fact of the matter is that the probability of not winning in ten (or twenty) tosses if one always sticks to either Heads or Tails is always exactly the same (that given above). This is the case no matter whether or not the tosses are consecutive, or whether or not one always bets on the same face of the coin, or whether one alternates in a regular fashion, or decides randomly at every toss. To insist on sticking to one side of the coin, or to take the consecutive nature of the tosses into account, is totally irrelevant.

7.2.4. *F. What is the probability that the first* (*or, in general, the* rth) *success* (*or failure*) *occurs on the* hth *toss*? The probability of the first success occurring on the hth toss is clearly given by $(\frac{1}{2})^h$ (the only favourable outcome out of all the 2^h is given by 000 . . . 0001). Note that this probability, $(\frac{1}{2})^h$, is the same as that of obtaining *no successes in h tosses*; i.e. of having to perform more than h tosses before obtaining the first success.‡ The probability of the rth success occurring at the hth toss is given by $(\frac{1}{2})^h\binom{h-1}{r-1}$, because this is the probability of exactly $r - 1$ successes in the first $h - 1$ tosses multiplied by the probability ($\frac{1}{2}$) of a further success on the final (hth) toss.

G. A coin is alternatively tossed, first by Peter and then by Paul, and so on. *If the winner is the one who first obtains a Head, what are their respective*

---

† My 'philosophy' in this respect is to consider as a *natural proof* that which is based on a combinatorial argument, and as a more or less *dull verification* that which involves algebraic manipulation. In other fields, too, certain things strike me as mere 'verifications'. For example: proofs of vectorial results which are based upon components; properties of determinants established by means of expansions (rather than using the ideas of alternating products, volume or, as in Bourbaki, smoothly generated by means of an exterior power). Indeed, this applies to anything which can be proved in a synthetic, direct and (meaningfully) instructive manner, but which is proved instead by means of formal machinery (useful for the bulk of the theory, but not for sorting out the basic ideas).

‡ We observe that the probability of no successes in n tosses tends to zero as n increases. This is obvious, but it is necessary to draw attention to it, and to make use of it, if certain arguments are to be carried through correctly (cf. the *Comments* following (*G*)).

probabilities of winning? A dull, long-winded approach would be to sum the probabilities $(\frac{1}{2})^h$ for h odd (to obtain Peter's probability of winning), or h even (for Paul's probability), and this would present no difficulties. The following argument is more direct (although its real advantage shows up better in less trivial examples). If Peter has probability p, Paul must have probability $\tilde{p} = \frac{1}{2}p$, because he will find himself in Peter's shoes if the latter fails to win on the first toss; we therefore have $p = \frac{2}{3}$, $\tilde{p} = \frac{1}{3}$.

Comments. We have tacitly assumed that one or other of them certainly ends up by winning. In actual fact, we should have stated beforehand that, as we pointed out in the footnote to (*F*), the probability of the game not ending within n tosses tends to zero as n increases. In examples where this is not the case, the argument would be wrong.

7.2.5. *H. What is the probability of obtaining, in n tosses, at least one run of h successes* (i.e. at least h consecutive successes)? Let A_n denote the number of possible sequences of n outcomes which do not contain any run of h Heads. By considering one further trial, one obtains a set of $2A_n$ sequences, which contains all the A_{n+1} sequences with no run of h Heads in $n+1$ trials, plus those sequences where the last outcome—which is therefore necessarily a Head—forms the first such sequence. There are A_{n-h} of these, because they must be obtained by taking any sequence of $n-h$ trials with no run of h Heads, and then following on with a Tail and then h Heads. We therefore obtain the recurrence relation $A_{n+1} = 2A_n - A_{n-h}$, in addition to which we know that $A_0 = 1$, $A_n = 2^n$ for $n < h$, $A_h = 2^h - 1$, etc.

We are dealing here with a difference equation. It is well-known (and easy to see) that it is satisfied by x^n, where x is a root of the (characteristic) equation $x^{h+1} - 2x^h + 1 = 0$. The general solution is given by

$$A_n = a_0 + a_1x_1^n + a_2x_2^n + \ldots + a_hx_h^n,$$

where $1, x_1, \ldots, x_h$ are the $h+1$ roots, and the constants are determined by the initial conditions.

We shall confine attention to the case $h = 2$. The recurrence relation $A_{n+1} = 2A_n - A_{n-2}$ can be simplified† so that it reduces to that of the Fibonacci numbers (each of which is the sum of the two preceding ones); i.e. $A_{n+1} = A_n + A_{n-1}$. In fact, however, a direct approach is both simpler and more meaningful. Those of the A_{n+1} sequences ending in Tails are the A_n followed by a Tail; those ending in Heads are the A_{n-1} followed by Tail–Head; the formula then follows immediately.

Using the fact that $A_0 = 1$, $A_1 = 2$, $A_2 = 3$, we find that $A_3 = 5$, $A_4 = 8$, $A_5 = 13$, $A_6 = 21$, $A_7 = 34$, $A_8 = 55$, and so on, and hence that the required

† By writing it in the form $A_{n+1} - A_n - A_{n-1} = A_n - A_{n-1} - A_{n-2}$, we see that the expression is independent of n; for $n = 2$, we have $A_2 - A_1 - A_0 = 3 - 2 - 1 = 0$.

probability is $1 - A_n/2^n$. For four trials this gives $1 - \frac{8}{16} = \frac{1}{2}$, for eight trials $1 - \frac{55}{256} = 0{\cdot}785$, and so on. To obtain the analytic expression, we find the roots of $x^2 - x - 1 = 0$ (noting that

$$x^3 - 2x^2 + 1 = (x - 1)(x^2 - x - 1) = 0),$$

obtaining $x_{1,2} = (1 \pm \sqrt{5})/2$, and hence

$$A_n = [(1 + \sqrt{5})^{n+1} - (1 - \sqrt{5})^{n+1}]/2^{n+1}\sqrt{5}.$$

A similar argument will work for any $h > 1$. The A_{n+1} are of the form $A_n T$, $A_{n-1}TH$, $A_{n-2}THH, \ldots, A_{n-h+1}THHH\ldots H$ (with $h - 1$ Heads), where the notation conveys that an A_{n-k} is followed by a Tail and then by k Heads. It follows that A_{n+1} = the sum of the h preceding terms (and this is clearly a kind of generalization of the Fibonacci condition).

The equation one arrives at $(1 + x + x^2 + x^3 + \ldots + x^h = x^{h+1})$ is the same as the one above, divided by $x - 1$ (i.e. without the root $x = 1$).

Comments. We have proceeded by *induction*; i.e. with a *recursive* method. This is a technique which is often useful in probability problems—keep it in mind!

Note that in considering the A_{n-h} we have discovered in passing the probability that a run of h successes is completed for the first time at the $(n + 1)$th trial. This is given by $A_{n-h}/2^{n+1}$. It is always useful to examine the results that become available as byproducts. Even if they do not seem to be of any immediate interest, they may throw light on novel features of the problem, suggest other problems and subsequently prove valuable (You never know!).

On the other hand, this probability (i.e. that something or other occurs for the first time on the $(n + 1)$th trial) can always be obtained by subtracting the probability that it occurs at least once in the first $n + 1$ trials from the probability that it occurs at least once in the first n (or by subtracting the complementary probabilities).

This remark, too, is obvious, but it is important, nonetheless. It often happens that the idea is not used, either because it is not obviously applicable, or because it simply does not occur to one to use it.

7.2.6. *I. What is the probability that a particular trial (the* n*th say) is preceded by exactly* h *outcomes identical to it, and followed by exactly* k? In other words, what is the probability that it forms part of a run of (exactly) $h + k + 1$ identical outcomes (either all Heads or all Tails) of which it is the $(h + 1)$th (we assume that $h < n - 1$).† The probability is, in fact, equal to $(\frac{1}{2})^{h+k+2}$. We simply require the h previous outcomes and the k following

† We exclude the (possible) case $h = n - 1$, which would give a different answer ($(\frac{1}{2})^{h+k+1}$: why?).

to be identical, and the outcome of the trial preceding this run, and the one following it, to be different (in order to enclose the run).

J. What is the probability that the nth *trial forms part of a run of* (*exactly*) *m trials having identical outcomes*? This, of course, reduces to the previous problem with h and k chosen such that $h + k + 1 = m$ (naturally, we assume $m \leqslant n - 1$). For each individual possible position, the probability is $(\frac{1}{2})^{m+1}$, and there are m such cases since the nth trial could either occupy the 1st, 2nd, . . . , or the mth position in the run (i.e. we must have one of

$$h = 0, 1, 2, \ldots, m - 1).$$

The required probability is therefore $m/2^{m+1}$.

In particular, the probability of a particular trial being *isolated* (i.e. a Tail sandwiched between two Heads, or vice-versa) is equal to $\frac{1}{4}$; the same is true for a run of length two, and we have $\frac{3}{16}$ for a run of length three, $\frac{4}{32} = \frac{1}{8}$ for a run of length four, $\frac{5}{64}$ for a run of length five, and so on.

K. What is the probability that some 'given' run, the nth *say, has* (*exactly*) *length m*? The nth run commences with the $(n - 1)$th change-over, and has length m if the following $m - 1$ outcomes are identical and the mth is different. The probability of this is given by $(\frac{1}{2})^m$. For lengths 1, 2, 3, 4, 5, we therefore have probabilities $\frac{1}{2}, \frac{1}{4}, \frac{1}{8}, \frac{1}{16}, \frac{1}{32}$, and so on.

Comments. It might appear that (J) and (K) are asking the same question: i.e. in both cases one requires to know the probability that a run (or 'a run *chosen at random*') has length m. The problem is not well-defined, however, until *a particular* run has been specified. The two methods of doing so—on the one hand demanding that the run contain some given element, on the other that it be the run with some given label—lead to different results, yet both methods could claim to be 'choosing a run at random'. We shall often encounter 'paradoxes' of this kind, and this example (together with the developments given under (L)) serves precisely to draw attention to such possibilities.

Warning. All the *relevant* circumstances (*and, at first sight, many of them often do not seem relevant*!) must be *set out very clearly* indeed, in order to avoid essentially different problems becoming confused. The phrase 'chosen at random' (and any similar expression) *does not, as it stands, have any precise meaning.* On the contrary, assuming it to have some uniquely determined intrinsic meaning (which it does not possess) is a common source of error. Its use is acceptable, however, provided it is always understood as indicating something which subsequently has to be made precise in any particular case. (For example, it may be that at some given instant a person decides that 'choosing a run at random' will have the meaning implicit in (J), or it may be that he decides on that of (K), or neither, preferring instead some other interpretation ...). In order not to get led astray by the over-familiar form of words, one might substitute in its place the more neutral and accurate form 'chosen in some quite natural and systematic way (which will be made precise later)'.†

7.2.7. *L. What is the prevision of the length of a run (under the conditions given in (J) and (K), respectively)*? Let us begin with (*K*). It might appear that the random quantity L = 'length of the run' can take on possible values $1, 2, \ldots, m, \ldots$ with probabilities $\frac{1}{2}, \frac{1}{4}, \ldots, (\frac{1}{2})^m, \ldots$ and that, therefore, its prevision is given by $\sum m/2^m$. But are we permitting ourselves the use of the series, considering the sequence as infinitely long (a possibility we previously excluded, for the time being anyway), or should we take the boundedness of the sequence into account (by assuming, for instance, that the number of trials does not exceed some given N)? It seems to be a choice between the devil and the deep blue sea, but we can get over this by thinking of N as finite, but large enough for us to ignore the effect of the boundedness. In other words, we accept the series in an unobjectionable sense; i.e. as an asymptotic value as N increases. Although in this particular case the series $\sum mx^m = x/(1-x)^2$ presents no difficulties (we see immediately that it gives $\mathbf{P}(L) = 2$), there are often more useful ways of proceeding. Given that prevision is additive, and given that the length L is the sum of as many 1s as there are consecutive identical outcomes (the first of which is certain, the others having probabilities $\frac{1}{2}, \frac{1}{4}, \ldots$, etc., as we saw above), we obtain

$$\mathbf{P}(L) = 1 + \tfrac{1}{2} + \tfrac{1}{4} + \ldots + (\tfrac{1}{2})^m + \ldots = 2.$$

Alternatively, and even more directly (using an argument like that in (*G*)), we note that we must have $\mathbf{P}(L) = 1 + \frac{1}{2}\mathbf{P}(L)$ (and hence $\mathbf{P}(L) = 2$), since, if the second outcome is the same as the first (with probability $\frac{1}{2}$), the length of the run starting with it has the same prevision $\mathbf{P}(L)$.

Going through the same process in case (*J*), we would have

$$\mathbf{P}(L') = \sum m(m/2^{m+1}) = \sum m^2/2^{m+1} = x(1+x)/2(1-x)^3 = 3 \text{ (for } x = \tfrac{1}{2}),$$

but the argument could be simplified even more by recalling that L had the value 1 + the prevision of the identical outcomes to the right (which was also 1), and that here we should add the same value 1 as a similar prevision to the left, so that we obtain $1 + 1 + 1 = 3$. Of course, we must also assume n large, but, in any case, it would be easy to evaluate the rest of the series in order to put bounds on the error of the asymptotic formula were the accuracy to be of interest.

† See, for instance, the remarks in Chapter 5, 5.10.2 concerning the notion of 'equiprobable' in quantum physics, and those in Chapter 10, 10.4.5 concerning the 'random choice' of subdivision point of an interval, or in a Poisson process.

It turns out, therefore, that the previsions resulting from the two different methods of choosing the runs are different; 2 and 3, respectively (as one might have expected, given that the smaller values were more probable in the first case, and conversely for the others). If we ignore the certain value 1 for the initial outcome, the additional length turns out to be double in the second case (2 instead of 1), because the situation is the same on both sides (independently of the fact that there is no actual continuation to the left, since, by hypothesis, the initial term is the first one in the run; this in no way changes the situation on the right, however: '*the later outcomes neither know nor care about this fact*').

Comment. A sentence like the above, or, equivalently, 'the process has no memory' (as in the case of stochastic independence), is often all that is required to resolve a paradox, or to avoid mistakes (like those implicit in the *well-known specious arguments* which we mentioned in our Comments following (*E*)).

M. Suppose we toss a coin n times: what are the previsions of the number of

successes (*Heads*);
change-overs (*Head followed by Tail, or vice-versa*);
runs;
runs of length m;
tosses up to and including the hth *success*;
tosses up to and including the completion for the first time of a run of successes of length 2 (*or, in general, of length h*)?

Most questions of this kind are much easier than the corresponding questions involving probabilities (as one would expect, given the additivity of prevision).

So far as successes are concerned, at each toss the probability of success is $\frac{1}{2}$, and hence the prevision of the number of successes in n tosses is $\frac{1}{2}n$.

For the change-overs, the same argument applies (apart from the case of the first toss), and we have $\frac{1}{2}(n-1)$.

For runs, we always have 1 more than the number of change-overs, and hence the prevision is $\frac{1}{2}(n+1)$.

For runs of length m, we shall give, for simplicity, the asymptotic expression for n large in comparison to m (see (*L*)). For each toss (and, to be rigorous, we should modify this for the initial and final m tosses), we have probability $m/2^{m+1}$ of belonging to a run of length m, and so, in prevision, there are $nm/2^{m+1}$ such tosses out of n; there are, therefore, $n/2^{m+1}$ runs of this kind (since each consists of m tosses). In particular, we have, in prevision, $n/4$ isolated outcomes, $n/8$ runs of length two, and so on.

From (*F*), we can say that the prevision of the number of tosses required up to and including the hth success is given by $\sum k\binom{k-1}{h-1}/2^k$ (the sum being taken over 1 to ∞, and, as usual, thought of as an asymptotic value). It is sufficient, however, to restrict attention to the very simple case $h = 1$. The prevision of the number of tosses required for the first success is 2 (the probability of it occurring on the first toss is $\frac{1}{2}$, at the second $\frac{1}{4}$, etc.); i.e. it is given by $\sum m/2^m$ as in the first variant discussed under (L). It follows that the prevision of the number of tosses required for the hth success is $2h$ (and that this is therefore the value of the summation given above, a result which could be verified directly).

For the final problem, we note that the probability of the first run of two Heads being completed at the nth toss is given by $A_{n-2-1}/2^n$ (where the A denote Fibonacci numbers), and that the prevision is therefore given by $\sum nA_{n-3}/2^n$. There is a useful alternative method of approach, however. Let us denote this prevision by $\mathbf{P}(L)$, and note the following: if the first two tosses both yield Heads, we have $L = 2$ (and this is the end of the matter); if they yield Head–Tail, we have $\mathbf{P}(L|HT) = 2 + \mathbf{P}(L)$, because the situation after the first two tosses reverts back to what it was at the beginning; if the first toss results in a Tail, we have, similarly, $\mathbf{P}(L|T) = 1 + \mathbf{P}(L)$. Since the probabilities of these three cases are $\frac{1}{4}, \frac{1}{4}, \frac{1}{2}$, respectively, we obtain

$$\mathbf{P}(L) = (\tfrac{1}{4}) . 2 + (\tfrac{1}{4})(\mathbf{P}(L) + 2) + (\tfrac{1}{2})(\mathbf{P}(L) + 1) = \tfrac{3}{2} + \tfrac{3}{4}\mathbf{P}(L),$$

which implies that $(\frac{1}{4})\mathbf{P}(L) = \frac{3}{2}$, and hence that $\mathbf{P}(L) = 6$.

The argument for the first run of h Heads proceeds similarly. Let us briefly indicate how it goes for the case $h = 3$: first three tosses Head–Head–Head, probability $\frac{1}{8}$, $L = 3$; first three tosses Head–Head–Tail, probability $\frac{1}{8}$, $\mathbf{P}(L|HHT) = 3 + \mathbf{P}(L)$; first two tosses Head–Tail, probability $\frac{1}{4}$, $\mathbf{P}(L|HT) = 2 + \mathbf{P}(L)$; first toss Tail, probability $\frac{1}{2}$, $\mathbf{P}(L|T) = 1 + \mathbf{P}(L)$. Putting these together, we obtain $\mathbf{P}(L) = 14$. For $h = 4$, we obtain $\mathbf{P}(L) = 30$; the general result is given by $\mathbf{P}(L) = 2^{h+1} - 2$ (prove it!).

7.2.8. *Remarks*. Let us quickly run through some other possible interpretations, and, in so doing, draw attention to certain features of interest. Instead of simply dealing with the Head and Tail outcomes themselves, we could consider their 'matchings' with some given 'comparison sequence', $E_1^*, E_2^*, \ldots, E_n^*, \ldots$. For example, if the comparison sequence were chosen to be the alternating sequence $HTHTHT\ldots$, and we used 1 to denote a matching, 0 otherwise, then we obtain a 1 whenever a Head appears on an odd toss, or a Tail on an even. In this way, any problem concerning 'runs' can be reinterpreted directly as one concerning 'alternating runs'. The comparison sequence could be the sequence in which a gambler 'bets' on the outcome of the tosses: for example, $HHTHTTTHT$ = 'he bets on Heads at the 1st, 2nd, 4th and 8th tosses, and he bets on Tails at the 3rd, 5th, 6th, 7th and 9th tosses'. The outcomes 1 and 0 then denote that 'he wins' or 'he loses', respectively. It follows, in this case, with no distinction drawn between Heads or Tails, that 'runs' correspond to runs of wins or losses, whereas if 'runs' refer to Heads only, say (as in (H)), then they correspond to runs of wins only (and conversely, if 'runs' refer to Tails only). The comparison sequence could even be random. For example, it might have arisen as the result of another sequence of coin tosses, or from some other game or experiment (double six when rolling two dice; the room temperature being below 20 °C; whether or not the radio is broadcasting music; whether at least $\frac{1}{3}$ of those present have blond hair, etc.). This other experiment may or may not be performed simultaneously and could depend on the outcomes of the sequence itself (for example, if $E_n^* = \tilde{E}_{n-1}$ we obtain the case of 'change-overs' as in (C)). The only condition is that E_h^* be stochastically independent of E_h (for each h). In fact, if, conditional on any outcome for the $E_i (i \neq h)$, E_h has probability $\frac{1}{2}$, then the same holds for $(E_h^* = E_h)$, no matter what the event E_h^* is, provided that it is independent of E_h (this can be seen as a special case of (E) for $n = 2$, but there it is obvious anyway).

7.3 HEADS AND TAILS: THE RANDOM PROCESS

7.3.1. In Section 7.2, we confined ourselves to a few simple problems concerning the calculation of certain probabilities and previsions in the context of coin tossing. This provided a convenient starting point for our discussion, but now we wish to return to the topic in a more systematic manner. In doing so, we shall get to know many of the basic facts, or, at least, become acquainted with some of them, and we shall also encounter concepts and techniques that will later come to play a vital rôle. In particular, we shall see how second-order previsions often provide a fruitful way of getting at important results, and we shall encounter various distributions, random† processes, asymptotic properties and so on. Let us proceed straightaway to a consideration of why it is useful, even when not strictly necessary, to formulate and place these problems in a *dynamic* framework, as *random processes*, or as *random walks*.

An arbitrary sequence of events $E_1, E_2, \ldots, E_n, \ldots$ (which, unless we state otherwise, could be continued indefinitely) can, should one wish to do so, be considered as already constituting in itself a random function, $E_n = Y(n)$,‡ assigning either 1 or 0 to each positive integer n. To obtain more meaningful representations, one could, for instance, consider the *number of successes* $Y(n) = S_n = E_1 + E_2 + \ldots + E_n$, or the *excess of successes* over failures

$$Y(n) = 2S_n - n = (E_1 + E_2 + \ldots + E_n) - (\tilde{E}_1 + \tilde{E}_2 + \ldots + \tilde{E}_n).$$

---

† See Chapter 1, 1.10.2, for a discussion of the use of the words 'random' and 'stochastic'.

‡ We shall usually write $Y(n)$, $Y(t)$, only when the variable (e.g. time) is *continuous*. When it is *discrete* we shall simply write Y_n, Y_t, except, as here, when we wish to emphasize that we are thinking in terms of the *random process*, rather than of an *individual* Y_n.

The latter could also be considered as the *total gain*,

$$Y(n) = X_1 + X_2 + \ldots + X_n,$$

if $X_i = 2E_i - 1 = E_i - \tilde{E}_i$ is defined to be the gain at each event; i.e. $X_i = \pm 1$ (one gains 1 or loses 1 according as E_i is true or false; put in a different way, one always pays 1 and receives 2 if the event occurs, receives 1 and pays 2 if it does not).

If it should happen (or if we make the assumption) that events occur after equal (and unit) time intervals, then we can say that $Y(t)$ is the number (or we could take the excess) of successes up to time t (i.e. $Y(t) = Y(n)$ if $t = n$, and also if $n \leqslant t < n + 1$). For the time being, this merely serves to provide a more vivid way of expressing things (in terms of 'time' rather than 'number of events'), but later on it will provide a useful way of showing how one passes from processes in *discrete time* to those in *continuous time* (even here this step would not be without meaning if events occurred at arbitrary time instants $t_1 < t_2 < \ldots < t_n < \ldots$, especially if randomly, as, for instance, in the Poisson process; Chapter 8, 8.1.3).

7.3.2. The representation which turns out to be most useful, and which, in fact, we shall normally adopt, is that based on the *excess* of successes over failures, $Y(n) = 2S_n - n$. This is particularly true in the case of coin tossing, but to some extent holds true more generally.

The possible points for $(n, Y(n))$ in the (t, y)-plane are those of the lattice shown in Figure 7.1(a). They have integer coordinates, which are either both even or both odd ($t = n \geqslant 0, -t \leqslant y \leqslant t$). It is often necessary, however, to pick out the point which corresponds to the number of successes in the first n events, and, in order to avoid the notational inconvenience of $(n, 2h - n)$, we shall, by convention, denote it by $[n, h]$: in other words, $[t, z] = (t, 2z - t)$, $(t, y) = [t, \frac{1}{2}(y + t)]$. As can be seen from Figure 7.1(b), this entails referring to the coordinate system (t, z), with vertical lines ($t =$ constant), and downward sloping lines (making a 45° angle; $z =$ constant), in which the points of the lattice are those with integer coordinates (and the possible ones are those for which $t \geqslant 0, 0 \leqslant z \leqslant t$†).

† In Figure 7.1b, if we take the two bisecting lines (with respect, that is, to the axes of Figure 7.1a), and therefore take as coordinates

$$\tfrac{1}{2}(t + y) = \text{number of successes } (= z, \text{ in Figure 7.1b})$$

and

$$\tfrac{1}{2}(t - y) = \text{number of failures } (= t - z, \text{ in Figure 7.1b}),$$

we have a system often used in other contexts (for example, in batch testing: a horizontal dash for a 'good' item, a vertical dash for a 'defective' item). This is convenient in this particular case, but has the disadvantage (a serious one if one wishes to study the random process) of not showing up clearly the independent variable (e.g. time), which one would like to represent along the horizontal axis.

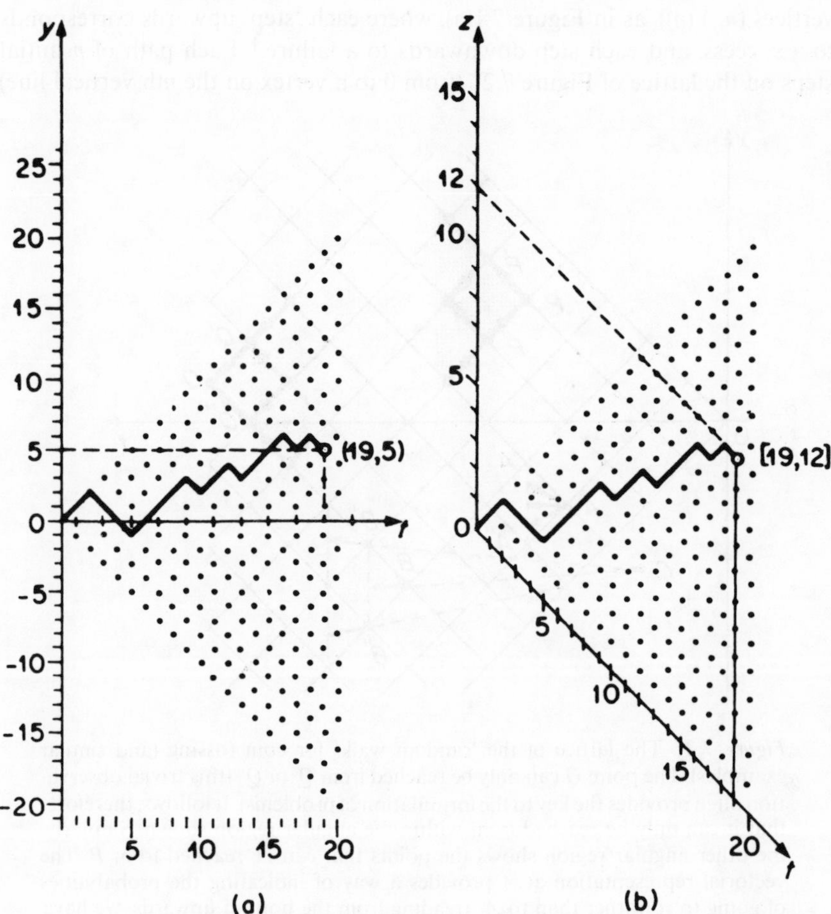

Figure 7.1 The lattice of possible points for the coin-tossing process. Coordinates: time $t = n$ = number of tosses (in both cases), together with: in (a): y = gain = $h - (n - h)$ = number of successes minus number of failures; in (b): $z = h$ = number of successes = $(n + y)/2$. The notations (t, y) and $[t, z]$ refer, respectively, to the coordinate systems of (a) and (b). For the final point of the path given in the diagrams, we have, for example,

$$(19, 5) \equiv [19, 12] \qquad (t = n = 19, h = 12, n - h = 7, y = 12 - 7 = 5)$$

The behaviour of the gain $Y(n)$ (and of the individual outcomes) can be represented visually by means of its *path*: i.e. the jagged line joining the vertices $(n, Y(n))$, as in Figure 7.1(a), where each 'step' upwards corresponds to a success, and each step downwards to a failure.† Each path of n initial steps on the lattice of Figure 7.2‡ (from 0 to a vertex on the nth vertical line)

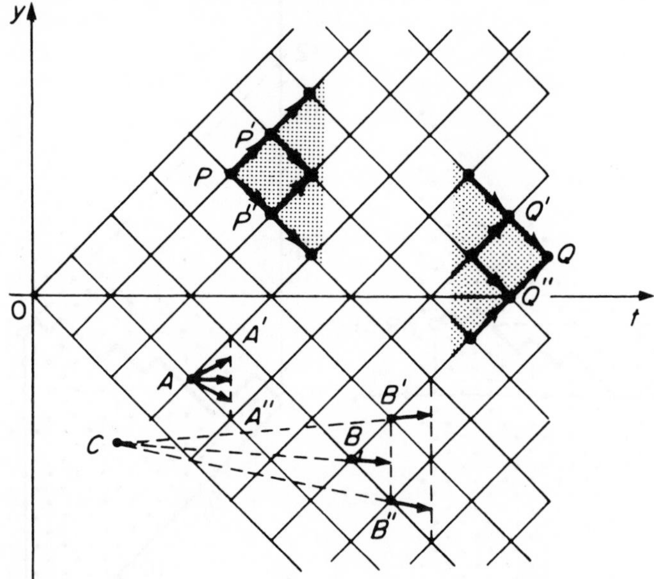

***Figure** 7.2* The lattice of the 'random walk' for coin tossing (and similar examples). The point Q can only be reached from Q' or Q'' (this trivial observation often provides the key to the formulation of problems). It follows, therefore, that it can only be reached from within the angular region shown. Similarly, the other angular region shows the points that can be reached from P. The vectorial representation at A provides a way of indicating the probabilities of going to A' rather than to A'' (reading from the bottom upwards, we have probabilities $p = 0{\cdot}2, 0{\cdot}5, 0{\cdot}7$). The other example of vectors at B, B' and B'' (meeting at C) will be of interest in Chapter 11, 11.4.1

† We observe, however, that this representation does not preserve the meaning of $Y(t)$, as given above, which requires it to change by a jump of ± 1 at the end of any interval $(n, n+1)$ (and not linearly). The use of the jagged line is convenient, however, not only visually, and for the random-walk interpretation (see below, in the main text), but also for drawing attention to interesting features of the process. It is convenient, for example, to be able to say that one is *in the lead* or *behind* when the path is in the positive or negative half-plane, respectively (in other words, not according to the sign of $Y(n)$, which could be zero, but according to the sign of $Y(n) + Y(n+1)$, where the two summands either have the same sign or one of them is zero).

‡ This representation also enables us to show clearly the probabilities at each step, and this is particularly useful when they vary from step to step (see, in particular, the end of Chapter 11, 11.4.1; the case of 'exchangeability'). All one has to do is the following: at each vertex, draw a vector, emanating from the vertex, and with components $(1, 2p-1)$, the prevision vector of the next step (downwards or upwards; i.e. $(1, -1)$ or $(1, 1)$ with probabilities $1-p$ and p).

corresponds to one of the 2^n possible sequences of outcomes of the first n events.

The interpretation as a *random walk* is now immediate. The process consists in starting from the origin 0, and then walking along the lattice, deciding at each vertex whether to step upwards or downwards, the decision being made on the basis of the outcomes of the successive events. The same interpretation could, in fact, be made on the y-axis (each step being one up or one down), and this would be more direct, although less clear visually than the representation in the plane. When we think, in the context of 'random walk', of Y_n as the distance of the moving point from the origin (on the positive or negative part of the y-axis) at time $t = n$, we are, in fact, using this representation: $Y_n = 0$ then corresponds to passage through the origin, and so on.

In fact, when one talks in terms of random walks, time is usually regarded as a parameter of the path (as for curves defined by parametric equations), so that, in general, we do not have an axis representing 'time' (and, if there is one, it is a waste of a dimension—even though visually useful). Normally, one uses the plane only for representing the random walk as a pair of (linearly independent) random functions of time (and the same holds in higher dimensions). An example of a random walk in two (or three) dimensions is given by considering the movement of a point whose coordinates at time n represent the gains of two (or three) individuals after n tosses (where, for example, each of them bets on Heads or Tails in any way he likes, with gains ± 1).

We have mentioned several additional points which, strictly speaking, had little to do with our particular example, but will save us repeating ourselves when we come to less trivial situations. Moreover, it should be clear by now that the specific set-up we have considered will be suitable for dealing with any events whatsoever, no matter what their probabilities are, and no matter what the probability distributions of the random functions are.

7.3.3. If we restrict ourselves to considering the first n steps (events), the 2^n probabilities, non-negative, with sum 1, of the 2^n paths (i.e. of the 2^n products formed by sequences like $E_1\tilde{E}_2\tilde{E}_3E_4 \ldots E_{n-1}\tilde{E}_n$) could be assigned in any way whatsoever. Thinking in terms of the random walk, the probabilities of the $(n + 1)$th step being upward or downward will be proportional to the probabilities of the two paths obtained by making E_{n+1} or $\tilde{E}_{n+1}$ follow that determined by the n steps already made.

The image of probability as mass might also prove useful. The unit mass, initially placed at the origin 0, spreads out over the lattice, subdividing at each vertex in the manner we have just described (i.e., in general, depending on the vertex in question and the path travelled in order to reach it). One could think of the distribution of traffic over a number of routes which split into two forks after each step (provided the number of vehicles N is assumed large enough for one to be able to ignore the rounding errors which derive from considering multiples of $1/N$). The mass passing through an arbitrary vertex of the lattice, (t, y), say, comes from the two adjacent vertices on the left, $(t-1, y-1)$ and $(t-1, y+1)$ (unless there is only one, as happens on the boundaries $y = \pm t$), and proceeds by distributing itself between the two adjacent vertices on the right, $(t+1, y-1)$ and $(t+1, y+1)$. Thinking of the random walk as represented on the y-axis, all the particles are initially at the point 0 (or in the zero position), and after each time interval they move to one or other of the two adjacent points (or positions); from y to $y-1$ or $y+1$. Consequently, whatever the probabilities of such movements are, the mass is alternatively all in the even positions or all in the odd: in any case, we have some kind of *diffusion* process (and this is more than just an image). In many problems, it happens, for example, that under certain conditions the process comes to a halt, and this might correspond to the mass being stopped by coming up against an absorbing barrier.

7.3.4. *Further problems concerning Heads and Tails.* There are many interesting problems that we shall encounter where the probabilities involved are of a special, simple form. The most straightforward case is, of course, that in which the events E_n are independent, and a further simplification results if they are equally probable. The simplest case of all is that of only two possible outcomes, each with probability $\frac{1}{2}$, and this is precisely the case of Heads and Tails that we have been considering.

The mass passing through a point *always divides itself equally* between the two adjacent points to the right. As a result of this, *each possible path of n steps always has the same probability*, $(\frac{1}{2})^n$, and *every problem concerning the probabilities of this process reduces to one of counting the favourable paths.*

Basing ourselves upon this simple fact, we can go back and give a systematic treatment of Heads and Tails, thinking of it now as a random process.†

† Figure 7.3 shows the results of successive subdivisions, $(\frac{1}{2}, \frac{1}{2})$, $(\frac{1}{4}, \frac{1}{2}, \frac{1}{4})$, $(\frac{1}{8}, \frac{3}{8}, \frac{3}{8}, \frac{1}{8})$, etc. Figure 7.4 shows a simple apparatus invented by Bittering. It is a box with two sets of divisions into compartments, one set being on top of the other, and shifted through half the width of a compartment. The middle section of the bottom half of the box is filled with sand, and then the box is turned upside down. The sand now divides itself between the two central compartments of what was the top half and is now the bottom. By repeatedly turning the box over (shaking it each time to ensure a uniform distribution of the sand within the compartments), one obtains successive subdivisions (the ones we have referred to above—those of Heads and Tails). By arranging the relative displacement of the overlapping compartments to be in the ratio $p:1-p$, one can obtain any required Bernoulli distribution (cf. Section 7.4.2).

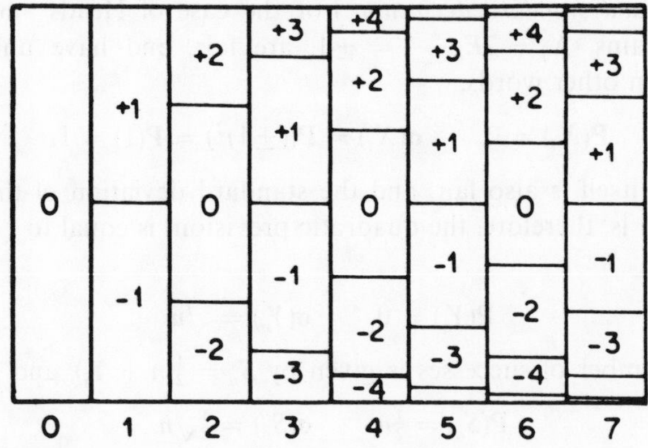

Figure 7.3 Subdivision of the probability for the game of Heads and Tails

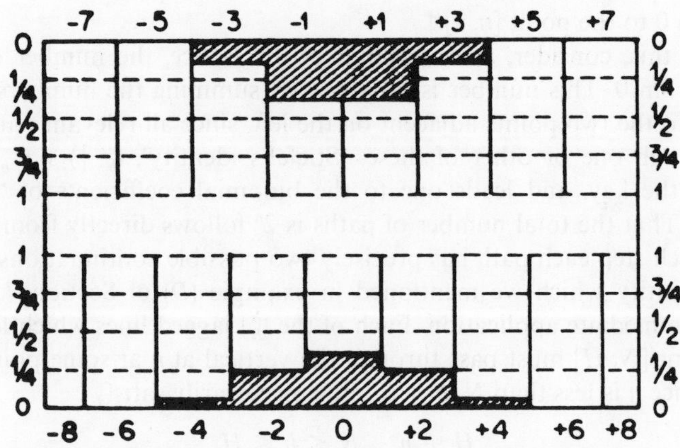

Figure 7.4 Bittering's apparatus. Probabilities of Heads and Tails: below, after four tosses; above, after three tosses

It is in this context that we shall again encounter well-known combinatorial ideas and results, this time in a form in which they are especially easy to remember, and which provides the most meaningful way of interpreting and representing them.

Prevision and standard deviation. For the case of Heads and Tails, the individual gains, $X_i = 2E_i - 1 = \pm 1$, are fair, and have unit standard deviations: in other words,

$$\mathbf{P}(X_i) = 0, \qquad \boldsymbol{\sigma}(X_i) = \mathbf{P}((\pm 1)^2) = \mathbf{P}(1) = 1.$$

The process itself is also fair, and the standard deviation of the gain in n tosses (which is, therefore, the quadratic prevision) is equal to $\sqrt{n}$: in other words,

$$\mathbf{P}(Y_n) = 0, \qquad \boldsymbol{\sigma}(Y_n) = \sqrt{n}.$$

The total number of successes is given by $S_n = \frac{1}{2}(n + Y_n)$, and so we have

$$\mathbf{P}(S_n) = \tfrac{1}{2}n, \qquad \boldsymbol{\sigma}(S_n) = \tfrac{1}{2}\sqrt{n}.$$

Successes in n tosses. We already know (case (A), Section 7.2.2) that the probability of h successes in n tosses is given by

$$(1) \qquad \omega_h^{(n)} = \tbinom{n}{h}/2^n \qquad (h = 0, 1, 2, \ldots, n).$$

This corresponds to the fact that $\binom{n}{h}$ is the number of paths which lead from the origin 0 to the point $[n, h]$.†

To see this, consider, at each point of the lattice, the number of paths coming from 0. This number is obtained by summing the numbers corresponding to the two points adjacent on the left, since all relevant paths must pass through one or other of these. 'Stiefel's identity', $\binom{n-1}{h-1} + \binom{n-1}{h} = \binom{n}{h}$, provides the key, and leads one to the binomial coefficients of 'Pascal's triangle'. That the total number of paths is 2^n follows directly from the fact that at each step each path has precisely two possible continuations.

The identity which we mentioned in example (*D*) of Section 7.2.2 also finds an immediate application. Each of the $\binom{N}{H}$ jagged lines which lead to a given point $[N, H]$ must pass through the vertical at n at some point $[n, h]$, where, since n is less than N, $h = S_n$ must necessarily satisfy

$$H + n - N \leqslant h \leqslant H$$

(because, between n and N, there can be no reduction in either the number of successes or of failures). There are $\binom{n}{h}$ paths from the origin arriving at $[n, h]$, and $\binom{N-n}{H-h}$ paths leading from this point to $[N, H]$. There are, therefore, $\binom{n}{h}\binom{N-n}{H-h}$‡ paths from the origin to $[N, H]$ which pass through the given intermediate point $[n, h]$; summing over the appropriate values of h, we

† Recall that, in our notation, $[n, h]$ represents the fact that $S_n = h$; in other words, it represents the point $(n, 2h - n)$, where $Y_n = 2S_n - n = 2h - n$.

‡ It is often not sufficiently emphasized that the *basic operation of combinatorial calculus is the product*; this should always be borne in mind, using this and many similar applications as examples.

must obtain $\binom{N}{H}$, thus establishing the identity. Further, we see that

$$\binom{n}{h}\binom{N-n}{H-h}\Big/\binom{N}{H}$$

is the probability of passing through the given intermediate point conditional on arriving at the given final destination (in other words, we obtain

$$\mathbf{P}(Y_n = 2h - n \mid Y_N = 2H - N) \qquad \text{or} \qquad \mathbf{P}(S_n = h \mid S_N = H);$$

we shall return to this later).

The rth *success.* Problem (F) of 7.2.4 can be tackled in a similar fashion, by reasoning in terms of crossings of sloping lines rather than vertical ones. In fact, the rth success is represented by the rth step upward; i.e., the step which takes one from the line $y = 2r - 2 - t$ to the line $y = 2r - t$ (i.e. from the rth to the $(r + 1)$th downward sloping line of the lattice, starting from $y = -t$). It is obvious that the rth failure can be dealt with by simply referring to upward sloping lines rather than downward ones. We have already shown the probability of the rth success at the hth toss to be $\binom{h-1}{r-1}/2^h$. We note that the favourable paths are those from 0 to $[h, r]$ whose final step is upward, i.e. passing through $[h - 1, r - 1]$, and that there are, in fact, $\binom{h-1}{r-1}$ of these.

If one is interested in considering the problem conditional on the path terminating at $[N, H]$, it is easily seen that there are $\binom{h-1}{r-1}\binom{N-h}{H-r}$ paths in which the rth success occurs at the hth toss (they must go from 0 to $[h - 1, r - 1]$, and then, with a compulsory step, to $[h, r]$, and finally on to $[N, H]$). As above, we can sum over all possibilities to obtain $\binom{N}{H}$† (the sum being taken over $r \leqslant h \leqslant N - H + r$, since r successes cannot occur until at least r trials have been made, nor can there be more than $H - r$ failures in the final $H - r$ trials).

Dividing the sum by the total, we obtain, in this case also, the conditional probabilities (of the rth success at the hth toss, given that out of N tosses there are H successes; we must have $r \leqslant h \leqslant H \wedge N - H + r$). These are given by

$$\begin{aligned}\binom{h-1}{r-1}\binom{N-h}{H-r}\Big/\binom{N}{H} &= \mathbf{P}(S_{h-1} + 1 = S_h = r \mid S_N = H)\\ &= \mathbf{P}(Y_{h-1} + 1 = Y_h = 2r - h \mid Y_N = 2H - N).\end{aligned}$$

This result will also be referred to again later.

† In this way, we arrive at meaningful interpretations of two well-known identities involving products of binomial coefficients:

$$\binom{N}{H} = \sum_{h=0\vee(H+n-N)}^{n\wedge H} \binom{n}{h}\binom{N-n}{H-h} \quad \text{(holding for each fixed } n\text{, with } 1 \leqslant n \leqslant N\text{)},$$

$$\binom{N}{H} = \sum_{h=r}^{N-H+r} \binom{h-1}{r-1}\binom{N-h}{H-r} \quad \text{(holding for each fixed } r\text{, with } 1 \leqslant r \leqslant H\text{)}.$$

These simply give the number of paths from 0 to $[N, H]$, expressed in terms of the points at which they cross vertical (1st identity) or sloping (2nd identity) lines.

Gambler's ruin. The problem of the crossings of horizontal lines is more complicated, since, unlike the previous cases, more than one crossing is possible (in general, an unlimited number). It is, however, a very meaningful and important topic, and, in particular, relates to the classical gambler's ruin problem.

If a gambler has initial fortune c, then his ruin corresponds to his gain reaching $-c$. Similar considerations apply if two gamblers with limited fortunes play against each other. Here, we confine ourselves to just this brief comment, but we note that, for this and for other similar problems (some of them important), arguments in terms of paths will turn out to be useful. In particular, we shall make use of appropriate *symmetries* of paths by means of Desiré André's celebrated *reflection principle* (in particular, Chapter 8 will deal with topics of this kind).

7.4 SOME PARTICULAR DISTRIBUTIONS

7.4.1. Before we actually begin our study of random processes, we shall, on the basis of our preliminary discussions, take the opportunity to examine a few simple problems in more detail, and to consider some particular distributions.

In order to avoid repetition later (and for greater effectiveness), we shall consider these distributions straightaway, both in the special forms that are appropriate for Heads and Tails, and in the more general forms. It should be noted that although the form of representation which we have adopted is a valid and useful one, the property of *fairness* (together with the principle of *reflection* and the *equal* probabilities of the paths) only holds for the special case of Heads and Tails.

In order to achieve some uniformity in notation, we shall always use X to denote the random quantity under consideration, and $p_h = \mathbf{P}(X = x_h)$ to denote the probability concentrated at the point x_h.† In the examples we shall consider, however, it turns out that the possible values of x_h are always integer (apart from changes in scale, $x_h = h$). For the particular case of the 'number of successes', we shall always use $\omega_h^{(n)} = \mathbf{P}(S_n = h)$ for the p_h.

7.4.2. *The Bernoulli (or binomial) distribution.* This is the distribution of $S_n = E_1 + E_2 + \ldots + E_n$ (or of $Y_n = 2S_n - n$, or of the frequency S_n/n— they are identical apart from an irrelevant change of scale), when the events E_h are *independent and have equal probabilities*, $\mathbf{P}(E_h) = p$. When $p = \frac{1}{2}$, as in the case of Heads and Tails, we have the *symmetric* Bernoulli distribution. The distributions are, of course, different for different n and p. Given n,

† We shall use *concentrated* rather than *adherent* (cf. Chapter 6) because, in these problems, the possible values can only be, by definition, the x_h themselves (finite in number, and, in any case, discrete).

the possible values are $x_h = h = 0, 1, 2, \ldots, n$ (or $x_h = a + hb = a, a + b, a + 2b, \ldots, a + nb$), and their probabilities are given by

$$p_h = \binom{n}{h} p^h \tilde{p}^{n-h} \qquad (\text{if } p = \tfrac{1}{2},\ p_h = \binom{n}{h}/2^n), \tag{2}$$

i.e. the $\omega_h^{(n)}$ of the process.

For the case $p = \frac{1}{2}$, we know that $\mathbf{P}(X) = n/2$, and $\boldsymbol{\sigma}(X) = \sqrt{n}/2$ (cf. Section 7.3.4). Similarly, for arbitrary p, we see that $\mathbf{P}(X) = np, \boldsymbol{\sigma}(X) = \sqrt{(np\tilde{p})}$, because for each summand we have

$$\mathbf{P}(E_i) = \mathbf{P}(E_i^2) = p, \qquad \boldsymbol{\sigma}^2(E_i) = \mathbf{P}(E_i^2) - \mathbf{P}^2(E_i) = p - p^2 = p\tilde{p}.$$

Hence, using the second-order properties, we obtain, without calculation,

$$\begin{gathered}\boldsymbol{\sigma}^2(X) = \sum_{h=0}^{n} \binom{n}{h} p^h \tilde{p}^{n-h}(h - np)^2 = np\tilde{p} \\ (\text{in addition to } \mathbf{P}(X) = \sum_{h=0}^{n} \binom{n}{h} p^h \tilde{p}^{n-h} h = np).\end{gathered} \tag{3}$$

The behaviour of the $p_h = \omega_h^{(n)}$ in the case $p = \frac{1}{2}$ is governed by that of the binomial coefficients $\binom{n}{h}$. These are largest for central values ($h \simeq n/2$) and decrease rapidly as one moves away on either side. Unless one looks more carefully† at ratios like p_{h+1}/p_h, however, it is difficult to get an idea of *how rapidly* they die away:

$$\frac{\omega_{h+1}^{(n)}}{\omega_h^{(n)}} = \frac{n-h}{h+1} \qquad \left(\text{in general, } \frac{n-h}{h+1} \cdot \frac{p}{\tilde{p}} \text{ for } p \neq \tfrac{1}{2}\right). \tag{4}$$

The same conclusions hold for general p, except that the maximum is attained for some $h \simeq np$ (instead of $\simeq \frac{1}{2}n$).

In fact, a consideration of Tchebychev's inequality suffices to show that the probability of obtaining values far away from the prevision is very small. Those h which differ from np by more than $n\varepsilon$ ($\varepsilon > 0$), i.e. corresponding to frequencies h/n not lying within $p \pm \varepsilon$, have, *in total*, a probability less than $\boldsymbol{\sigma}^2(X)/(n\varepsilon)^2 = np\tilde{p}/(n\varepsilon)^2 = p\tilde{p}/n\varepsilon^2$ (and this is far from being an accurate bound, as will be clear from the asymptotic evaluations which we shall come across shortly; equation (20) of Section 7.5.4).

Comments. The p_h can be obtained as the coefficients of the expansion of

$$(\tilde{p} + pt)^n = \sum_{h=0}^{n} p_h t^h = \sum_{h=0}^{n} \omega_h^{(n)} t^h$$

† As is done, for the purpose of providing an elementary exposition, in B. de Finetti and F. Minisola, *La matematica per le applicazioni economiche*, Chapter 4. See also a brief comment later (in Section 7.6.3).

(the alternative notation being chosen to avoid any ambiguity in the discussion which follows). It suffices to observe that the random quantity

$$\prod_{i=1}^{n} (\tilde{E}_i + tE_i)$$

is the sum of the constituents multiplied by t^h, where h is the number of positive outcomes (giving, therefore, $S_n = h$). Its value is thus given by

$$\sum_{h=0}^{n} (S_n = h)t^h = t^{S_n},$$

and its prevision, i.e. the characteristic function $\phi(u)$ with $t = e^{iu}$ (or $u = -\mathrm{i}\log t$), by

$$\text{(5)} \qquad \mathbf{P}(t^{S_n}) = \sum \mathbf{P}(S_n = h)t^h.$$

A generalization. In the same way, we observe that the result holds even if the E_i (always assumed stochastically independent) have different probabilities p_i. In this case, the $\omega_h^{(n)}$ are given by

$$\text{(6)} \qquad \sum_{h=0}^{n} \omega_h^{(n)} t^h = \prod_{i=1}^{n} (\tilde{p}_i + p_i t).$$

On the other hand, this only expresses the obvious fact that $\omega_h^{(n)}$ is the sum of the products of h factors involving the p_i, and $n - h$ involving the complements $\tilde{p}_i$.

In particular, we see that $\boldsymbol{\sigma}^2(S_n) = \sum_i p_i\tilde{p}_i$, and that this formula (like $np\tilde{p}$, of which it is an obvious generalization) continues to hold, even if we only have pairwise independence. (Recall that this is not sufficient for many other results concerning the distribution.)

7.4.3. *The hypergeometric distribution.* As in the previous case, we are interested in the distribution of $X = S_n$ (or $Y_n = 2S_n - n$, or S_n/n, which, as we have already remarked, only differ in scale). The difference is that we now *condition on the hypothesis that, for some given $N > n$, we have $S_N = H$.*

In deriving the required distribution, it suffices that the $\binom{N}{H}$ paths from 0 to $[N, H]$ appear equally probable to us. It does not matter, therefore, whether we choose to think in terms of Heads and Tails (where initially all

2^N paths were equally probable, and the paths compatible with the hypothesis remain such), or in terms of events which, prior to the hypothesis, were judged independent and equally probable, but with $p \neq \frac{1}{2}$ (because in the latter case all the remaining $\binom{N}{H}$ paths have the same probability, $p^H \tilde{p}^{N-H}$).

Instead of thinking in terms of these representations (whose main merit is that they show the links with what has gone before), it is useful to be able to refer to something rather more directly relevant. The following are suitable examples: drawings without replacement from an urn (containing N balls, H of which are white); counting votes (where a total of N have been cast, H of which are in favour of some given candidate); ordering N objects, H of which are of a given kind (N playing cards, H of which are 'Hearts'; N contestants, H of whom are female). In all these cases, the $N!$ possible permutations are all considered equally probable. (Or, at least, all $\binom{N}{H}$ possible ways of arranging the two different kinds of objects must be regarded as equally probable; it is these which correspond to the $\binom{N}{H}$ paths involving H upward steps and $N - H$ downward steps.)

Under the given assumptions (or information), each event $E_i (i \leqslant N)$ has probability $\mathbf{P}(E_i) = H/N$† (and, for convenience, we shall write $H/N = q$). These events are not independent; in fact, we shall see later that they are negatively correlated.

Observe that, as a result of changes in the state of information, problems which were initially distinct may come to be regarded as identical, and assumptions about equal probabilities, or independence, may cease to hold (or conversely in some cases). These and other considerations will appear obvious to those who have entered into the spirit of our approach. Those who have come to believe (either through ignorance or misunderstanding) that properties like stochastic independence have an objective and absolute meaning that is inherent in the phenomena themselves, will undoubtedly find these things rather strange and mystifying.

The distribution that concerns us (for example, that of the number of white balls appearing in the first n drawings—or something equivalent in one of the other examples) will be different for every triple n, N and H (or, equivalently, n, N, q). For $q = \frac{1}{2}$, i.e. for $H = N - H = \frac{1}{2}N$, the distribution is symmetric; $\mathbf{P}(S_n = h) = \mathbf{P}(S_n = n - h)$. The possible values are the integers $x_h = h$, where $0 \vee H - (N - n) \leqslant h \leqslant n \wedge H$ (or $x_h = a + hb$;

† Given that we assume the hypothesis $S_N = H$ to be already part of our knowledge or information, we take $\mathbf{P}(E)$ to mean $\mathbf{P}(E|S_N = H)$. In this situation, the E_i have probability $q = H/N$, but are not stochastically independent (even if they are the outcomes of coin tossing, or rolling a die, etc., where, prior to the information about the frequency of successes out of N tosses, they were judged independent and equally probable). In particular, $p_h = \omega_h^{(n)} = \mathbf{P}(S_n = h)$ *in this case* is what we would have written as $\mathbf{P}(S_n = h|S_N = H)$ in the previous case.

e.g. $= 2h - n$, or $= h/n$), and their probabilities are, as we saw already in Section 7.3.4,

$$p_h = \omega_h^{(n)} = \frac{\binom{n}{h}\binom{N-n}{H-h}}{\binom{N}{H}}$$

$$= \frac{\binom{H}{h}\binom{N-H}{n-h}}{\binom{N}{n}}$$

(7)

$$= \binom{n}{h} \frac{H(H-1)(H-2)\ldots(H-h+1)(N-H)(N-H-1)(N-H-2)\ldots(N-H-(n-h)+1)}{N(N-1)(N-2)\ldots(N-n+1)}$$

$$= \binom{n}{h} q^h \tilde{q}^{n-h} \frac{\left[\left(1-\frac{1}{H}\right)\left(1-\frac{2}{H}\right)\ldots\left(1-\frac{h-1}{H}\right)\right]\left[\left(1-\frac{1}{N-H}\right)\left(1-\frac{2}{N-H}\right)\ldots\left(1-\frac{n-h-1}{N-H}\right)\right]}{\left(1-\frac{1}{N}\right)\left(1-\frac{2}{N}\right)\ldots\left(1-\frac{n-1}{N}\right)}.$$

The interpretation of the four different forms is as follows.

The *first form* (as we already know) enumerates the paths.

The *second form* enumerates those n-tuples, out of the total of $\binom{N}{n}$ that can be drawn from N events, which contain h out of the H, and $n - h$ out of the $N - H$.

The *third form* (which can be derived from the previous two) can be interpreted directly, observing that the probability of first obtaining h successes, and then $n - h$ failures, is given by the product of the ratios (of white balls, and then of black balls) remaining prior to each drawing:

$$\frac{H}{N}\cdot\frac{H-1}{N-1}\cdot\frac{H-2}{N-2}\cdots\frac{H-h+1}{N-h+1}\cdot\frac{N-H}{N-h}\cdot\frac{N-H-1}{N-h-1}$$

$$\times\frac{N-H-2}{N-h-2}\cdots\frac{N-H-(n+h)+1}{N-n+1}.$$

But this is also the probability for any other of the $\binom{n}{h}$ orders of drawing this number of successes and failures (even if the ratios at each drawing vary, the result merely involves permuting the factors in the numerator, and is, therefore, always the same).

We see already that, provided n is small in comparison to N, H and $N - H$, all the ratios differ little from q (the drawings already made do not seriously alter the composition of the urn). The results will therefore not differ much from the Bernoulli case (drawings from the same urn *with* replacement).

The *fourth form* shows the relation between the two cases explicitly, by displaying the correction factor.

The behaviour of the p_h in this case is similar to that of the Bernoulli case, and can be studied in the same way (by considering ratios p_{h+1}/p_h). The maximum is obtained by the largest h which does not exceed

$$nq[1 - 2/(N+2)] - (H-3)/(N+2)$$

(the reader should verify this!), and as one moves further and further away on each side, the p_h decrease. Compared with the Bernoulli case, the terms around the maximum are larger, and those far away are smaller. Some insight into this can be obtained by looking at the final formula.†

For the prevision, we have, of course, $\mathbf{P}(X) = nq = nH/N$. The standard deviation $\boldsymbol{\sigma}(X)$, on the other hand, is a little smaller than $\sqrt{(np\tilde{p})}$ (the result for the case of independence), and is given by

$$\boldsymbol{\sigma}^2(X) = nq\tilde{q}[1 - (n-1)/(N-1)].$$

In fact, if we evaluate the correlation coefficient r between two events ($r = r(E_i, E_j)$, $i \neq j$) we obtain $r = -1/(N-1)$. More specifically,

$$\mathbf{P}(E_iE_j) = (H/N)((H-1)/(N-1)) = q^2(1 - 1/H)/(1 - 1/N),$$

from which it follows that

$$\begin{aligned} r &= [\mathbf{P}(E_iE_j) - \mathbf{P}(E_i)\mathbf{P}(E_j)]/\boldsymbol{\sigma}(E_i)\boldsymbol{\sigma}(E_j) \\ &= q^2[(N+H-1)/N(N-1)]/q\tilde{q} = -1/(N-1). \end{aligned}$$

It is possible, however, to avoid the rather tiresome details (which we have skipped over) by using the argument already encountered in Chapter 4, 4.17.5, and observing that

$$\boldsymbol{\sigma}^2(S_n) = nq\tilde{q} + 2\tbinom{n}{2}rq\tilde{q} = nq\tilde{q}[1 + (n-1)r].$$

For $n = N$, we have $\boldsymbol{\sigma}(S_N) = 0$, because $(S_N = H) =$ the certain event, and hence

$$(8) \qquad 1 + (N-1)r = 0, \qquad r = -1/(N-1).$$

---

† For this correction factor, the variant of Stirling's formula (equation (28)) which is given in (30) (cf. Section 7.6.4) yields the approximation (for $n \ll N$)

$$\exp\left\{-\frac{1}{2}\frac{n}{N}\left[1 + \frac{2(n-\frac{1}{2})}{\eta(1-\eta)}(\xi - \eta) - \frac{n}{\eta(1-\eta)}(\xi-\eta)^2\right]\right\},$$

where we have set $\eta = H/N$ and $\xi = h/n$ (i.e. the percentage of white balls in the urn, and the frequency of white balls drawn in a sample of n, respectively). In the special case $\eta = \frac{1}{2}$ ($H = \frac{1}{2}N$; half the balls in the urn are white, half black), the expression simplifies considerably to give

$$\exp\left\{-\frac{1}{2}\frac{n}{N}[1 - 4n(\xi - \tfrac{1}{2})^2]\right\}.$$

On the basis of this, we conclude that (approximately) the distribution gives higher probabilities than the Bernoulli distribution in the range where h lies between $n\eta \pm \sqrt{[n\eta(1-\eta)]}$ (i.e. between $m \pm \sigma$), with a maximum at $n\eta$, and lower values outside this interval.

Remarks. Note how useful it can be to bear in mind that apparently different problems may be identical, and how useful it can be to have derived different forms of expression for some given result, to have found their probabilistic interpretations, and to be in a position to recognize and use the simplest and most meaningful form in any given situation. Note, also, that one should be on the lookout for the possibility of reducing more complicated problems to simpler ones, both heuristically and, subsequently, by means of rigorous, detailed, analytical arguments, either exact or approximate.

In the case considered, we can go further and note that, by virtue of their interpretations within the problem itself, we have $\omega_h^{(n)} = \omega_{H-h}^{(N-n)}$. In other words, for given N and H, the distributions for complementary sample sizes n and $N - n$ are identical if we reverse $h = 0, 1, \ldots, H$ to $H, \ldots, 1, 0$ (and this can be seen immediately by glancing at the formula). It follows that, among other things, what is claimed to hold for 'small n' must also hold for 'large n' (i.e. n close to N). The approximation does not work for central values ($n \sim \frac{1}{2}N$), and we note, in particular, that, for n lying between N and $N - H$, not all the values $h = 0, 1, \ldots, n$ are possible (since they themselves must lie between H and $n - (N - H)$).

7.4.4. *The Pascal distribution.* This is the distribution of X = 'the number of tosses required before the rth Head is obtained' (more generally, it arises for any independent events with arbitrary, constant probability p). Alternatively, it is the distribution of X such that $S_X = r > S_{X-1}$. By changing the scale, we could, of course, consider $X' = a + bX$. One example of this which often crops up is $X' = X - r$ = 'the number of failures preceding the rth success', but many of the other forms, such as those considered in the previous case, do not make sense in this context.

A new feature is that the distribution is unbounded, the possible values being $x_h = h = r, r + 1, r + 2, \ldots$ (up to infinity, and, indeed, $+\infty$ must be included as a possible value, along with all the integers, since it corresponds to the case where the infinite set of trials result in less than r successes). In line with our previous policy, we shall avoid critical questions by deciding that if the rth success is not obtained within a maximum of N trials (where N is very large compared with the other numbers in question) we shall set $X = N$. (To be precise, we shall consider $X' = X \wedge N$ instead of X.) Were we to consider $X' = X - r$, the possible values would be $0, 1, 2, \ldots$ (and this is one of the reasons why this formulation is often preferred; another reason will be given in the discussion that follows; see equation (15)).

For each r and p, we have, of course, a different distribution:

$$(9) \qquad p_h = \binom{h-1}{r-1} p^r \tilde{p}^{h-r} \qquad \left(\text{if } p = \tfrac{1}{2},\ p_h = \binom{h-1}{r-1}\Big/2^h\right).$$

In fact (as we saw in (F), Section 7.2.4, for the special case $p = \frac{1}{2}$), in order to obtain $X = h$, we must have $r - 1$ successes in the first $h - 1$ trials, together with a success on the hth trial. In terms of the random process representation, we are dealing with the crossing of the line $y = 2r - x$ (or, if one prefers, with the mass that ends up there if the line acts as an absorbing barrier).

Note that the series $\sum p_h$ sums up to 1. It must, of course, be convergent with sum $\leqslant 1$ by its very meaning; the fact that it is $= 1$, and not < 1, ensures that, as n increases, the probability that $X > n$ tends to zero (and, in particular, the probability that $X = \infty$ is 0).

So far as the behaviour of the distribution is concerned, the p_h again increase until they reach a maximum (attained for the greatest $h \leqslant r/p$), and then decrease to zero (asymptotically, like a geometric progression with ratio $\tilde{p}$). A more intuitive explanation of the increase is that it continues so long as the prevision of the number of successes, $\mathbf{P}(S_h) = hp$, does not exceed the required number of successes, r.

The geometric distribution. For $r = 1$, the Pascal distribution reduces to the special case of the geometric distribution:

$$p_h = p\tilde{p}^{h-1}, \tag{10}$$

forming a geometric progression ($p_1 = p$, with ratio $\tilde{p} = 1 - p$). If, for example, the first failure corresponds to elimination from a competition, this gives the probability of being eliminated at the hth trial, when the probability of failure at each trial is p. (N.B. For the purposes of this particular example, we have, for the time being, interchanged 'success' and 'failure'.) In particular, it gives the probability that a machine first goes wrong the hth time it is used, or that a radioactive atom disintegrates in h years time, and so on, where the probability of occurrence is p on each separate occasion. (If the probability of death were assumed to be constant, rather than increasing with age, this would also apply to the death of an individual in h years time.)

The property of giving the same probability, irrespective of the passing of time, or of the outcomes of the phenomenon in the past, is known as the *lack of memory* property of the geometric distribution. The waiting time for a particular number to come up on the lottery† has, under the usual assumptions, a geometric distribution (the ratio is $\tilde{p} = \frac{17}{18} = 94{\cdot}44\%$, for a single city; $\tilde{p} = (\frac{17}{18})^{10} = 56\%$ for the whole set of ten cities). This provides further confirmation, if such were needed, of the absurdity of believing that numbers which have not come up for a long time are more likely to be drawn in future.

To put this more precisely: it is absurd to use the small probabilities of long waiting times, which are themselves evaluated on the basis of the usual assumptions, and are given by the geometric distribution (or to invoke

† *Translators' note.* Cf. the footnote on p. 61 of Volume I.

their comparative rarity, statistically determined in accordance with it), to argue, on the basis of independence, against the very assumptions with which one started—i.e. the *lack of memory* property. If, on the other hand, someone arrived at a coherent evaluation of the probabilities by a different route, we might not judge him to be reasonable, but this would simply be a matter of opinion.

Finally, let us give the explicit expression for the case $r = 2$ (again, this could be thought of as elimination from a competition, but this time at the second failure): it reduces to $p_h = (h-1)p^2\tilde{p}^{h-2}$.

Prevision and standard deviation. In order to calculate $\mathbf{P}(X)$ and $\boldsymbol{\sigma}(X)$, it turns out to be sufficient to do it for the case $r = 1$. We obtain

(11) $$\mathbf{P}(X) = p\sum_{h=1}^{\infty} h\tilde{p}^{h-1} = p/(1-\tilde{p})^2 = 1/p,$$

(12) $$\mathbf{P}(X^2) = p\sum_{h=1}^{\infty} h^2\tilde{p}^{h-1} = p\tilde{p}\sum_{h=1}^{\infty} h(h-1)\tilde{p}^{h-2} + p\sum_{h=1}^{\infty} h\tilde{p}^{h-1}$$
$$= 2p\tilde{p}/p^3 + 1/p = (2-p)/p^2$$

(verify this!), and hence,

$$\boldsymbol{\sigma}^2(X) = \mathbf{P}(X^2) - \mathbf{P}^2(X) = (1-p)/p^2.$$

For general r, it suffices to note that

(13) $$\mathbf{P}(X) = r/p,$$

(14) $$\boldsymbol{\sigma}^2(X) = r(1-p)/p^2,$$

because (as we already observed for $\mathbf{P}(X)$ in the case $p = \frac{1}{2}$; cf. (M), Section 7.2.7) we can consider X as the sum of r terms, $X_1 + X_2 + \ldots + X_r$, stochastically independent, and each corresponding to $r = 1$ (X_i = the number of trials required after the $(i-1)$th failure until the ith failure occurs).

Comments. This technique will also be useful in what follows: note that it can also be used for $X' = X - r$ if we consider r summands of the form $X_i' = X_i - 1$.

In this context (i.e. with h transformed into $h + r$), the p_h are given by

(15) $$p_h = \binom{h+r-1}{r-1}p^r\tilde{p}^h = \binom{h+r-1}{h}p^r\tilde{p}^h = (-1)^h\binom{-r}{h}p^r\tilde{p}^h,$$

where the definition $\binom{x}{h} = x(x-1)\ldots(x-h+1)/h!$ is extended to cover any real x (not necessarily integer, not necessarily positive).

If we do this, the distribution then makes sense for any real $r > 0$. This generalized form of the Pascal distribution (which has integer r) is called the *negative binomial distribution* (simply because it involves the notation $\binom{-r}{h}$).

For $r = 0$, the distribution is concentrated at the origin ($p_0 = 1$, $p_h = 0$, $h \neq 0$); for $r \sim 0$, we have $p_h \simeq r\tilde{p}^h/h$ ($h \neq 0$) (the logarithmic distribution; cf. Chapter 6, 6.11.2), and hence

(16) $$p_0 \simeq 1 - r\sum_{h=1}^{\infty} \tilde{p}^h/h = 1 - r\log(1/p).$$

We shall make use of this later on, and the significance of the extension to non-integer r will also be explained.

The prevision in this case is clearly given by

(17) $$\mathbf{P}(X') = \mathbf{P}(X) - r = r\tilde{p}/p,$$

whereas the standard deviation, $\sqrt{(r\tilde{p})}/p$, is unaltered; this also holds for non-integer r.

Another form. We have already seen (in Section 7.3) that, if $S_N = H$ is assumed to be known, the problem of the location, h, of the rth success leads to the distribution

(18) $$p_h = \binom{h-1}{r-1}\binom{N-h}{H-r}\Big/\binom{N}{H},$$

rather than to the Pascal distribution. For an example where this distribution occurs, consider an election in which N votes have been cast and are counted one at a time. Suppose further that a candidate is to be declared elected (or a thesis accepted) when r votes in favour have been counted. Given that a total of $H \geqslant r$ out of N were actually favourable, equation (18) gives the probability that success is assured by the counting of the hth vote. (Another example, with $N = 90$ and $H = r = 15$, is given by the probability of completing a 'full house' at Bingo with the hth number called.)

We shall restrict ourselves to considering the particularly simple case of $H = r = 1$, a case which is nonetheless important (and a study of the general case is left as an exercise for the reader). Clearly (even without going through the algebra), we have $p_h = 1/N$ for $h = 1, 2, \ldots, N$. If there has only been one success in N trials (or there is only one favourable vote in the ballot box, or only one white ball in the urn, or only one ball marked '90'), there is exactly the same probability of finding it on the first, second, ..., or Nth (final) trial.

7.4.5. *The discrete uniform distribution.* This is the name given to the distribution of an X which can only take on a finite number of equally spaced possible values, each with the same probability: e.g. $x_h = h$, $h = 1, 2, \ldots, n$ (or $x_h = a + bh$), with all the $p_h = 1/n$. As examples, we have a fair die ($n = 6$), a roulette wheel ($n = 37$) or the game of Bingo ($n = 90$).

It is easily seen that

$$\mathbf{P}(X) = \tfrac{1}{2}(n+1), \quad \mathbf{P}(X^2) = (1^2 + 2^2 + \ldots + n^2)/n = (4n^2 + 6n + 2)/12,$$

from which (subtracting $[\frac{1}{2}(n+1)]^2 = (3n^2+6n+3)/12$) we obtain

$$\boldsymbol{\sigma}^2(X) = (n^2+1)/12, \qquad \boldsymbol{\sigma}(X) = (n/\sqrt{12})\sqrt{(1+1/n^2)} \simeq n/\sqrt{12}.$$

A random process (*Bayes–Laplace, Pólya*). Using this distribution as our starting point, we can develop a random process similar to that which led to the hypergeometric distribution. In fact, we consider successive drawings (without replacement) from an urn containing N balls, with the possible number of white balls being any of $0, 1, 2, \ldots, N$, each with probability $1/(N+1)$. (This could arise, for example, if the urn were chosen from a set of $N+1$ urns, ranging over all possible compositions, and there were no grounds for attributing different probabilities to the different possible choices.)

Let us assume, therefore, that $\omega_H^{(N)} = 1/(N+1)(H = 0, 1, 2, \ldots, N)$, and that (as in the case of a known composition, H/N) all the permutations of the possible orders of drawing the balls are equally probable: in other words, that all the dispositions of H white and $N-H$ black balls (i.e. all the paths from 0 ending up at the same final point $[N, H]$) are equally probable. Each of these paths therefore has probability $1/\binom{N}{H}(N+1)$.

We shall now show that, under these conditions, the distribution for every S_n $(n < N)$ is uniform, just as we assumed it to be for S_N: i.e.

$$\omega_h^{(n)} = 1/(n+1) \qquad (h = 0, 1, 2, \ldots, n).$$

It can be verified, in a straightforward but tedious fashion, that

$$\sum_{H=0}^{N} \binom{n}{h}\binom{N-n}{H-h}\frac{1}{\binom{N}{H}(N+1)} = \frac{1}{n+1}$$

(the probabilities of the paths terminating at $[N, H]$ multiplied by the number of them that pass through $[n, h]$, the sum being taken over H). The proof by induction (from N to $N-1$, $N-2, \ldots$, etc.) is much simpler, however, and more instructive. It will be sufficient to establish the step from N to $N-1$. The probability $\omega_h^{(N-1)}$ that $S_{N-1} = h$ is obtained by observing that this can only take place if $H = h$ and the final ball is black, or if $H = h+1$ and the final ball is white; each of these two hypotheses has probability $1/(N+1)$, and the probabilities of a black ball under the first hypothesis and a white ball under the second are given by $(N-h)/N$ and $(h+1)/N$, respectively. It follows that

$$\omega_h^{(N-1)} = \frac{1}{N+1}\left(\frac{N-h}{N} + \frac{h+1}{N}\right) = \frac{1}{N}. \tag{19}$$

Expressed in words: if all compositions are equally probable, so are all the frequencies at any intermediate stage. This property ($\omega_h^{(n)} = 1/(n+1)$) can also hold for all n (without them being bounded above by some preassigned N), and leads to the important Bayes–Laplace process (which we shall meet in Chapter 11, 11.4.3) or, with a different interpretation, to the Pólya process (Chapter 11, 11.4.4) with which it will be compared.

7.5 LAWS OF 'LARGE NUMBERS'

7.5.1. We now return to our study of the random process of Heads and Tails (as well as some rather less special cases) in order to carry out a preliminary investigation of what happens when we have 'a large number' of trials. This preliminary investigation will confine itself to qualitative aspects of the order of magnitude of the deviations. In a certain sense, it reduces to simple but important corollaries of an earlier result, which showed that *the order of magnitude increases as the square root of n* (the number of trials).

In the case of Heads and Tails ($p = \frac{1}{2}$), the prevision of the *gain*, Y_n, is zero (the process is fair; $\mathbf{P}(Y_n) = 0$), and its standard deviation $\sigma(Y_n)$ (which, in a certain sense, measures 'the order of magnitude' of $|Y_n|$) is equal to $\sqrt{n}$. The *number of successes* (Heads) is denoted by S_n, and has prevision $\frac{1}{2}n$; its standard deviation (the order of magnitude, measured by σ) is equal to $\frac{1}{2}\sqrt{n}$. For the *frequency of successes*, S_n/n, the prevision and standard deviation are those we have just given, but now divided by n; i.e. $\frac{1}{2}$ and $\frac{1}{2}/\sqrt{n}$, respectively. In a similar way, one might be interested in the *average gain* (per toss), Y_n/n; this has prevision 0 and standard deviation $1/\sqrt{n}$.

The fact that

$$\mathbf{P}\left[\left(\frac{S_n}{n} - \frac{1}{2}\right)^2\right] = \frac{1}{4n} \to 0$$

is expressed by saying that *the frequency converges in mean-square to* $\frac{1}{2}$. This implies (cf. Chapter 6, 6.8.3) that it also converges *in probability*. Similarly, the average gain converges to 0 (both in mean-square and in probability).

We recall that convergence in probability means that, for positive ε and θ (however small), we have, for all n greater than some N,

$$\mathbf{P}\left(\left|\frac{S_n}{n} - \frac{1}{2}\right| > \varepsilon\right) < \theta.$$

A straightforward application of Tchebychev's inequality shows that the probability in our case is less than $\sigma^2/\varepsilon^2 = 1/4n\varepsilon^2$.

7.5.2. When referring to this result, one usually says, in an informal manner, that after a large number of trials it is *practically certain* that the frequency becomes *practically equal* to the probability. Alternatively, one might say that 'the fluctuations tend to cancel one another out'. One should be careful, however, to avoid exaggerated and manifestly absurd interpretations of this result (a common trap for the unwary). Do not imagine, for example, that convergence to the probability is to be expected because future discrepancies should occur in such a way as to 'compensate' for present discrepancies by being in the opposite direction. Nor should one imagine that this holds for the absolute deviations. It is less risky to gamble just a few times (e.g. ten plays at Heads and Tails at 1000 lire a time) than it is to repeat the same bet many times (e.g. a 1000 plays are 10 times more risky; $10 = \sqrt{(1000/10)}$). On the other hand, it would be less risky to bet 1000 times at 10 lire a time. Furthermore, if at a certain stage one is losing—let us say 7200 lire—the law of large numbers provides no grounds for supposing that one will 'get one's own back'.† In terms of prevision, this loss remains forever at the same level, 7200 lire. The future gain (positive or negative) has prevision zero, but, as one proceeds, the order of magnitude becomes larger and larger and, eventually, it makes the loss already suffered *appear negligible*. It is in this sense, and only in this sense, that the word 'compensate' might reasonably be used, since one would then avoid the misleading impression that it usually conveys. The fact remains, however, that the loss has already been incurred.

Observe once again how absurd it would be to imagine, *a priori*, some sort of correlation—which would be a consequence of laws and results derived on the basis of an assumption of independence!

7.5.3. In addition to this, one should note that the property we have established concerns the probability of a deviation $>\varepsilon$ between the probability and the frequency for *an individual* n (although it can be for any $n \geqslant N$). This clearly does not imply—although the fact that it does not can easily escape one's notice—that the probability of an 'exceptional' deviation occurring at least once for an n between some N and an $N + K$ greater than N is also small.

It is especially easy to overlook this if one gets into the habit of referring to events with small probability as 'impossible' (and even worse if one appears to legitimize this bad habit by giving it a name—like 'Cournot's principle', Chapter 5, 5.10.9). In fact, if an 'exception' were impossible for every individual case $n \geqslant N$, it would certainly be impossible to have even a single exception among the infinite number of cases from $n = N$ onwards.

If one wanted to use the word 'impossible' in this context without running into these problems, it would be necessary to spell out the fact that it should not be understood as meaning 'impossible', but rather 'very improbable'. However, anyone who states that 'horses are potatoes', making it clear that

† The illusory nature and pernicious influence of such assumptions are referred to in a popular, witty saying (possibly Sicilian in origin), in itself rather remarkable, given that popular prejudice seems on the whole to incline towards the opposite point of view. The saying concerns the answer given by a woman to a friend, who has asked whether it was true that her son had lost a large amount of money gambling: '*Yes, it's true*', she replies, '*But that's nothing*: *what is worse is that he wants to get his own back*!'.

when it refers to horses the meaning of 'potato' is not really that of 'potato' (but rather that of 'horse'), would probably do better not to create useless terminological complications in the first place (since, in order for it not to be misleading, it must be immediately followed by a qualification which takes away its meaning).

Now let us return to the topic of convergence. If the probability of a deviation $|S_n/n - \frac{1}{2}|$ at Heads and Tails being greater than ε were actually equal to $1/4n\varepsilon^2$, then, for any N, in some interval from N to a sufficiently large $N + K$ the prevision of the number of 'exceptions' (deviations $> \varepsilon$) would be arbitrarily large (approximately $(1/4\varepsilon^2)\log(1 + K/N)$). This follows from the fact that the series $\sum 1/n$ is divergent, and the sum between N and $N + K$ is approximately equal to $\log(1 + K/N)$. In fact, as we shall see later, the result we referred to at the beginning of Section 7.5.3 does hold. It just so happens that the Tchebychev inequality, although very powerful in relation to its simplicity, is not sufficient for this more delicate result. Stated mathematically, we have, for arbitrary positive ε and θ,

$$\mathbf{P}\left(\max_{N \leqslant n \leqslant N+K}\left|\frac{S_n}{n} - \frac{1}{2}\right| > \varepsilon\right) < \theta,$$

provided N is sufficiently large (K is arbitrary).† This form of stochastic convergence is referred to as *strong convergence*, and the result is known as the '*strong law of large numbers*'. By way of contrast, the word 'strong' is replaced by '*weak*' when we are referring to convergence in probability, or to the previous form of the law of large numbers.

7.5.4. In order to fix ideas, we have referred throughout to the case of Heads and Tails. Of course, the results also hold for $p \neq \frac{1}{2}$ (except that we then have to write $\sigma^2 = p\tilde{p}$, which is $< \frac{1}{4}$ unless $p = \frac{1}{2}$), and even in the case where the $p_i = \mathbf{P}(E_i)$ vary from event to event (provided $\sum p_i\tilde{p}_i$ diverges, which may not happen if the p_i get too close to the extreme values 0 and 1). In the latter case, our statement would, in general, assert that the difference between the frequency S_n/n in the first n trials and the arithmetic mean of the probabilities, $(p_1 + p_2 + \ldots + p_n)/n$, tends to zero (in mean-square and in probability). Only if the arithmetic mean tends to a limit p (or, as analysts would say, if the p_i are a sequence converging to *p in the Cesàro sense*) does the previous statement in terms of deviation from a fixed value hold (and the fixed value would then be the limit p).

We can, however, say a great deal more on the basis of what we have established so far. The only properties we have made use of are those of the

† Were it not for our finitistic scruples (cf. Chapter 6 and elsewhere), we could do as most people do, and write $\sup(n \geqslant N)$ in place of $\max(N \leqslant n \leqslant N + K)$, saying that it is 'almost certain' (i.e. the probability $= 1$) that $\lim(S_n/n) = \frac{1}{2}$ (in the sense given).

previsions and standard deviations of the gains of individual bets, $2E_i - 1$, and of their sums, Y_n. It is easy to convince oneself that for the conclusion (weak convergence) to hold we only require that the gains X_i $(i = 1, 2, \ldots)$ have certain properties. For example, it suffices that they have zero prevision and are *(pairwise) independent with constant, finite standard deviations.* More generally, we only require that they are (pairwise) uncorrelated and that the standard deviations $\boldsymbol{\sigma}(X_i) = \sigma_i$ are bounded, and such that $\sum \sigma_i^2$ diverges. Considering the case of zero prevision for convenience, we have

$$Y_n/n = (X_1 + X_2 + \ldots + X_n)/n \stackrel{\cdot}{\rightarrow} 0 \qquad \text{(and hence } \overset{<}{\rightarrow} 0).$$

Expressed in words: the (weak) law of large numbers holds for sums of uncorrelated random quantities under very general conditions, in the sense that *the arithmetic mean, Y_n/n, tends to* 0 *in quadratic prevision, and the probability of its having an absolute value* $>\varepsilon$ *(an arbitrary, preassigned positive value) also tends to* 0.

The *strong law of large numbers* also holds under very general conditions. The argument which ensures its validity if the sum of the probabilities p_h of 'exceptions' (deviations $|Y_h/h| > \varepsilon$) converges, turns out to be sufficient if these probabilities are evaluated on the basis of the normal distribution, and this will be the case if the X_h are assumed to be independent with standard unit normal distributions ($m = 0$, $\sigma = 1$). Asymptotically, however, this property also holds in the case of Heads and Tails, and for any other X_h which are identically distributed with finite variances (let us assume $\sigma = 1$).† We shall, as we mentioned above, restrict ourselves to the proof based on the convergence of $\sum p_h$. Afterwards, we shall mention the possibility of modifications which make the procedure much more powerful.

Since the distribution function of the (standard unit) normal cannot be expressed in a closed form, it is necessary, in problems of this kind, to have recourse to an asymptotic formula (which can easily be verified—by L'Hospital's rule, for example):

$$(20)\quad 1 - F(x) = \frac{1}{\sqrt{(2\pi)}}\int_x^{\infty} \mathrm{e}^{-\frac{1}{2}z^2}\,\mathrm{d}z \sim \frac{1}{\sqrt{(2\pi)}x}\mathrm{e}^{-\frac{1}{2}x^2} \qquad (\text{as } x \rightarrow +\infty).$$

It follows, since Y_h has standard deviation $\sqrt{h}$, that $|Y_h/h| > \varepsilon$ can be thought of as

$|Y_h/\sqrt{h}| > \varepsilon\sqrt{h} = \varepsilon\sqrt{h} \times$ the standard deviation of the standardized distribution,

† That the normal distribution frequently turns up is a fact which is well-known, even to the layman (though the explanation is often not properly understood). The case we are referring to here will be dealt with in Section 7.6; we shall not, therefore, enter into any detailed discussion at present.

and, therefore,

$$p_h = 1 - F(\varepsilon\sqrt{h}) - F(-\varepsilon\sqrt{h}) = 2[1 - F(\varepsilon\sqrt{h})]$$

$$\sim \frac{2}{\sqrt{(2\pi)}\varepsilon\sqrt{h}} e^{-\frac{1}{2}h\varepsilon^2} = \frac{K}{\sqrt{h}} e^{-ch}.$$

The geometric series $\sum e^{-ch}$ converges, however, and hence, *a fortiori*, so does the series $\sum p_h$, with the terms divided by $\sqrt{h}$. The remainder, from some appropriate N onwards, is less than some preassigned θ, and this implies that the probability of even a single 'exceptional deviation', $|Y_h/h| > \varepsilon$, for h lying between N and an arbitrary $N + M$, is less than θ. (If countable additivity were admitted, one would simply state the result 'for all $h \geqslant N$'.)

The conclusion can easily be strengthened by observing that convergence still holds even if the constant ε is replaced by some $\varepsilon(h)$ decreasing with h; for example,

$$\varepsilon(h) = \sqrt{(2a \log h)}/\sqrt{h}, \qquad \text{with } a > 1.$$

We then have $h\varepsilon^2 = 2a \log h$, and

$$p_h = \mathbf{P}(|Y_h/h| > \varepsilon(h)) = [K/\sqrt{(2a \log h)}]\, e^{-a \log h} = (\ldots)h^{-a}.$$

But the terms $(\ldots)$ tend to zero, the series $\sum h^{-a}$ $(a > 1)$ converges, and, *a fortiori*, $\sum p_h$ converges. Expressed informally, this implies that, from some N on, it is 'almost certain that Y_h will remain between $\pm c\sqrt{(2h \log h)}$', for $c > 1$.

The argument that follows exemplifies the methods that could be used to further strengthen the conclusions. Indeed, we shall see precisely how it is that one arrives at a conclusion which is, in a certain sense, the best possible ('we shall see', in the sense that we will sketch an outline of the proof without giving the details).

We note that were we to consider only the possible exceptions (Y_h lying outside the interval given above) at the points $h = 2^k$, instead of at each h, we could obtain the same convergent series by taking

$$\varepsilon(h)\sqrt{h} = \sqrt{(2a \log k)} \sim \sqrt{(2a \log \log h)}, \qquad \text{instead of } \sqrt{(2a \log h)}.$$

A conclusion which only applies to the values $h = 2^k$ is, of course, of little interest, but it is intuitively obvious that we certainly do not require a check on all the h. The graph of $y = Y(h)$ can scarcely go beyond the preassigned bounds if one checks that it has remained within them by scanning a sufficiently dense sequence of 'check points'. Well, one can show that the check points $h = 2^k$ (for example) are sufficiently dense for one to be able to conclude that—again expressed informally—it is almost certain that all the Y_h from some N on (an N which cannot be made precise), will, in fact, remain within much smaller bounds of the form $\pm c\sqrt{(2h \log \log h)}$, for $c > 1$.

What makes this result important is that, conversely, if $c < 1$, it is 'practically certain that these bounds will be exceeded, however far one continues'. This is the celebrated *law of the iterated logarithm*, due to Khintchin.

Note that, in order to prove the converse which we have just stated, the divergence of the series is not sufficient, unless the events are independent (Borel–Cantelli lemma). In the case under consideration, we do not have independence. We do, however, have the possibility of reducing ourselves to the latter case, because, if h'' is much larger than h', the contribution of the increment between h' and h'' (which is independent of $Y(h')$) is the dominating term in $Y(h'') = Y(h') + [Y(h'') - Y(h')]$.

All these problems can be viewed in a more intuitive light (and can be dealt with using other techniques, developed on the basis of other approaches) if we base ourselves on random processes on the real line (and, so far as the results we have just mentioned are concerned, in particular on the Wiener–Lévy process). It will be a question of studying the graph $y = Y(t)$ of a random function in relation to regions like $|y| \leqslant y(t)$ (a preassigned function), by studying the probability of the graph entering or leaving the region, either once, or several times, or indefinitely.

Finally, let us just mention the standard set of conditions which are sufficient for the validity of the strong law. The X_h are required to be independent, and such that $\sum \sigma_h^2/h^2$ converges (the Kolmogorov condition). The proof, which is based on an inequality due to Kolmogorov, one which, in a certain sense, strengthens the Tchebychev inequality, and on the truncation of the 'large values' of the X_h, goes beyond what we wanted to mention at this stage.

In the classical case (that of independent events with equal probabilities), the weak and strong laws of large numbers are also known as the Bernoulli and Cantelli laws, respectively.

7.5.5. *The meaning and value of such 'laws'*. In addition to their intrinsic meaning, both mathematically and probabilistically, the laws of large numbers, and other asymptotic results of this kind, are often assigned fundamental rôles in relation to questions concerning the foundations of statistics and the calculus of probability. It seems appropriate to provide some discussion of this fact, both in order to clarify the various positions, and, in particular, to clarify our own attitude.

For those who seek to connect the notion of probability with that of frequency, results which relate probability and frequency in some way (and especially those results like the 'laws of large numbers') play a pivotal rôle, providing support for the approach and for the identification of the concepts. Logically speaking, however, one cannot escape from the dilemma posed by the fact that the same thing cannot both be assumed first as a definition and then proved as a theorem; nor can one avoid the contradiction that arises from a definition which would assume as certain something that the theorem only states to be very probable. In general, this point is accepted, even by those who support a statistical-frequency concept of probability; the attempts to get around it usually take the form of singling out, separating off, and generally complicating, particular concepts and models.

An example of this is provided by the 'empirical law of chance'. A phrase created for the purpose of affirming the *actual occurrence* of something the 'law of large numbers' states to be *very probable* comes to be presented as an *experimental fact*. Another example is provided by 'Cournot's principle': this states, as we mentioned in Chapter 5, 5.10.9, that 'an event of small probability does not occur', and covers the above, implicitly, as a special case. Sometimes, the qualification 'never, or almost never' is added, but although this removes the absurdity, in doing so it also takes away any value that the original statement may have had.

In any case, this kind of thing does nothing to break the vicious circle. It only succeeds in moving it somewhere else, or disguising it, or hiding it. A veritable labour of Sisyphus! It always ends up as a struggle against *irresolvable* difficulties, which, in a well-chosen phrase of B. O. Koopman, '*always retreat but are never finally defeated*, unlike Napolean's Guard'.

In order for the results concerning frequencies to make sense, it is necessary that the concept of probability, and the concepts deriving from it which appear in the statements and proofs of these results, should have been defined and given a meaning beforehand. In particular, a result which depends on certain events being uncorrelated, or having equal probabilities, does not make sense unless one has defined in advance what one means by the probabilities of the individual events. This requires that probabilities are attributed to each of the given events (or 'trials'), that these all turn out to be equal, and that, in addition, probabilities are attributed to the products of pairs of events, such that these are all equal, and, moreover, equal to the square of the individual probabilities. In using the word 'attributed', we have, of course, used a word which fits in well with the subjectivistic point of view; in this context, however, it would make no difference if we were to think of such probabilities as 'existing', in accordance with the 'logical' or 'necessary' conception. In fact, the criticisms of the frequentistic interpretation made by Jeffreys, for instance, and the case against it which he puts forward (closely argued, and, I would say, unanswerable†), are in complete accord with the views we have outlined above. We acknowledge, of course, that there are differences between the necessary and subjectivistic positions

† Cf. Harold Jeffreys, *Scientific Inference*, Cambridge University Press; 1st ed. (1931), 2nd ed. (1957) and *Theory of Probability*, Oxford University Press; 1st ed. (1938), 2nd ed. (1948), 3rd ed. (1961). Particularly relevant are the following: Section 9.21 of the first work, entitled 'The frequency theories of probability', and Sections 7.03–7.05 of the second, in Chapter 7, 'Frequency definitions and direct methods'.

(the latter denies that there are logical grounds for picking out *one single* evaluation of probability as being *objectively* special and 'correct'), but we regard this as of secondary importance in comparison with the differences that exist between, on the one hand, conceptions in which probability is probability (and frequency is just one of the ingredients of the 'outside world' which might or might not influence the evaluation of a probability) and, on the other, conceptions in which probability is, to a greater or lesser extent, a derivative of frequency, or is an idealization or imitation of it.

7.5.6. From our point of view, the law of large numbers forms yet another link in the chain of properties which justify our making use of expected or observed frequencies in our (necessarily subjective) evaluations of probability. We now see how to make use of the prevision of a frequency in this connection. The law of large numbers says that, *under certain conditions*, the value of the probability is *not only equal to the prevision* $\mathbf{P}(X)$ of a frequency X, but, moreover, *we are almost certain that* X *will be very close to this value* (getting ever closer, in a way that can be made precise, as one thinks of an even larger number of events).

This really completes the picture for the special case we have considered. Rather than introducing new elements into the situation (something we shall come across when we deal with exchangeable events in Chapter 11, and in similar contexts), we shall use these results in order to consider rather more carefully the nature of this special case: i.e. independent events with equal probabilities. It is important to realize that these assumptions, so apparently innocuous and easily accepted, contain unsuspected implications. To judge a coin to be 'perfectly fair so far as a single toss is concerned', means that one considers the two sides to be equally probable on this (the first) toss, or on any other toss for which one does not know the outcomes of the previous tosses. To judge a coin to be 'perfectly fair so far as the random process of Heads and Tails is concerned' is a very different matter.

The latter is, in fact, an extremely rash judgement, which commits one to a great deal. It *commits* one, for example, to evaluating the probabilities as $\frac{1}{2}$ at each toss, *even if all the previous tosses* (a thousand, or a million, or $10^{1000}, \ldots$) *were all Heads*, or Heads and Tails alternately, and so on. Another consequence (although one which is well beyond the range of intuition) is that, for a sufficiently large number of tosses, one considers it advantageous to bet on the frequency lying between 0·49999 and 0·50001 rather than elsewhere in the interval $[0, 1]$ (and the same holds for $0{\cdot}5 \pm 10^{-1000}$, etc.). I once remarked that '*the main practical application of the law of large numbers consists in persuading people how unrealistic and unreasonable it is, in practice, to make rigid assumptions of stochastic independence and equal probabilities*'. The remark was taken up by L. J. Savage, to whom it was made, and given publicity in one of his papers. It was intended to be witty, in part facetious and paradoxical, but I think that it is basically an accurate observation.

Notwithstanding its great mathematical interest, there is clearly even less to be said from a realistic and meaningful point of view concerning the strong law of large numbers.

7.5.7. *Explanations based on 'homogeneity'*. First of all, it is necessary to draw attention to the upside-down nature of the very definitions of notions (or would-be notions) like those of *homogeneous* events, *perfect* coins and so on. Any definition that is framed in objective, physical terms, or whatever, is not suitable, because it cannot be used to *prove* that a given probabilistic opinion is a logical truth, nor can it *justify its imposition* as an article of faith.

If one wants to make use of these, or similar, notions, it is clear, therefore, that their meanings can only come about and be made precise as expressions of particular instances of probabilistic opinions (opinions which, had one already attributed to these notions a metaphysical meaning, preceding these personal opinions, one would have called a *consequence* of it).

I recall a remark, dating from about the time of my graduation, which has remained engraved upon my memory, having struck me at the time as being very accurate. A friend of mine used to say, half-jokingly, and in a friendly, mocking way, that it was never enough for me to define a concept, but that I needed to 'definettine' it. In actual fact, I had, by and large, adopted the mode of thinking advocated by authors like Vailati and Calderoni (or perhaps it would be more accurate to say that I found their approach to be close to my own). Papini used to say of Calderoni that 'what he wanted to do was to show what precautions one ought to take, and what procedures one ought to use, in order to arrive at statements which make sense'.† On the other hand, it was precisely this form of reasoning which, in successive waves, from Galileo to Einstein, from Heisenberg to Born, freed physics—and with it the whole of science and human thought—from those superstructures of absurd metaphysical dross which had condemned it to an endless round of quibbling about pretentious vacuities.

At the same time, and for the reasons we have just given, any attempt to define a coin as 'perfect' on the basis of there being no objective characteristics which prevent the probability of Heads from being $p = \frac{1}{2}$, or different tosses from being stochastically independent, is simply a rather tortuous way of making it appear that the above-mentioned objective circumstances play a decisive rôle. In fact, they are mere window-dressing. The real meaning only becomes clear when these circumstances are pushed on one side, and one simply proceeds as follows (and, in doing so, discovers that there are

† G. Papini, *Stroncature*, No. 14: 'Mario Calderoni'; G. Vailati, *Scritti* (in particular, see those works quoted in the footnotes to Chapter 11, 11.1.5).

two possible meanings of 'perfect'): we shall use the expression *perfect coin* in the *weak* sense as a shorthand statement of the fact that we attribute equal probabilities ($\frac{1}{2}$) to each of the two possible outcomes of a toss; we use the expression in the *strong* sense if we attribute equal probabilities ($(\frac{1}{2})^n$) to each of the 2^n possible outcomes of n tosses, for any n. This does not mean, of course, that in making such a judgment it is not appropriate (or, even less, that it is not admissible) to take into account all those objective circumstances that one considers relevant to the evaluation of probability. It merely implies that the evaluation (or, equivalently, the identification and listing of the circumstances that might 'reasonably' influence it) is not a matter for the theory itself, but for the individual applying it. From his knowledge of the theory, the individual will have at his disposal various auxiliary devices to aid him in sharpening his subjective analysis of individual cases; the standard schemes will serve as reference points for his idealized schemes. There is no way, however, in which the individual can avoid the burden of responsibility for his own evaluations. The key cannot be found that will unlock the enchanted garden wherein, among the fairy-rings and the shrubs of magic wands, beneath the trees laden with monads and noumena, blossom forth the flowers of *Probabilitas realis*. With these fabulous blooms safely in our button-holes we would be spared the necessity of forming opinions, and the heavy loads we bear upon our necks would be rendered superfluous once and for all.

7.5.8. Having dealt with the logical aspect, it remains to consider, in a more detailed fashion, the criticisms of those discussions based upon *homogeneity*, both from a practical point of view, and from the point of view of the 'realism' of such a notion in relation to actual applications. It is curious to observe that these kinds of properties (independence with equal probabilities) are even less realistic than usual in precisely those cases which correspond to the very empirical–statistical interpretation which claims to be the most 'realistic' (i.e. those attributing the 'stability of the frequency' to quasi-'physical' peculiarities of some phenomenon possessing 'statistical regularity').

Can we really believe that a coin—'perfect' so far as we can see—provides the perfect example of a phenomenon possessing these 'virtues'?. There appears to be room for doubt. Is it not indeed likely that 'suspicious' outcomes would lead us to re-evaluate the probability, somehow doubting its perfection, or the manner of tossing, or something else?

By way of contrast, we would have less reason for such suspicions and doubts if, from time to time, or even at each toss, the coin were changed. This would be even more true if coins of different denomination were used and the person doing the tossing were replaced, and more so again if the successive events considered were completely different in kind (for example:

whether we get an even or odd number with a die, or with two dice, or in drawing a number at Bingo, or for the number-plate of the first car passing by, or in the decimal place of the maximum temperature recorded today, etc.; whether or not the second number if greater than the first when we consider number-plates of cars passing by, ages of passers-by, telephone numbers of those who call us up, etc.; it is open to anyone to display their imagination by inventing other examples). Under these circumstances, it seems very unlikely that a 'suspicious' outcome, whatever it was, would lead one to expect similar strange behaviour from future events, which lack any similarity or connection with those that have gone before.†

This demonstrates that the homogeneity of the events (the fact of their being, in some sense, 'trials of the same phenomenon', endowed with would-be statistical virtues of a special kind) is by no means a necessary prerequisite for the possible acceptance of the properties of independence with equal probabilities. In fact, it is a positive obstacle to such an acceptance. If, in such a case, the above properties are accepted, it is not that they should be thought of as valid *because of homogeneity*, but, if at all, *in spite of homogeneity* (and it is much easier to accept them in other cases *because of heterogeneity*).‡ Despite all this, we continue to hear the exact opposite, repeated over and over, with the tiresome insistence of a silly catch-phrase.

The 'laws of chance' (although it is rather misleading to refer to them in this way) express, instead, precisely that which one can expect from a maximum of disorder, in which any kind of useful knowledge is lacking. Any increase in one's knowledge of the phenomena and of their 'properties' would, if it were to be at all useful, lead one to favour some subset of the 2^n possible outcomes, and hence would lead to an evaluation of probabilities which are *better in this specific case* (with respect to the judgment of the individual who makes his evaluation after taking it into account) than those which would be valid in the absence of any information of this kind. There exists no information, knowledge, or property, that can strengthen or give 'physical' (or philosophical, or any other) meaning to the situation which corresponds to a perfect symmetry of ignorance.§

† We have used the qualification 'suspicious' only after careful consideration (we have avoided, for example, 'exceptional', or 'strange', or 'unlikely'). Now is not the appropriate time for a detailed examination of this question, however. This will come later (in Chapter 11, 11.3.1), and will clarify the meaning of this term and the reasons why it was chosen.

‡ A fuller account of this may be found in 'Sulla "compensazione" tra rischi eterogenei', *Giorn. Ist. Ital. Attuari* (1954), 1–21.

§ The following point has been made many times, and should be unnecessary. We are not speaking of exterior symmetries (which could exist), nor of 'perfect ignorance' (which cannot exist—otherwise, we would not even know what we were talking about), but about symmetry of judgment as made by the individual (in relation to the notion of indifference which he had prior to obtaining information, whether a great deal or only a small amount).

7.6 THE 'CENTRAL LIMIT THEOREM'; THE NORMAL DISTRIBUTION

7.6.1. If one draws the histograms of the distribution of Heads and Tails (the binomial distribution with $p = \frac{1}{2}$) and compares them for various values of n (the number of tosses), one sees that the shape remains the same (apart from discontinuities and truncation of the tails, features which arise because of the discreteness, and tend to vanish as n increases). The shape, in fact, suggests that one is dealing with the familiar normal distribution (the Gaussian distribution, or 'distribution of errors', which we mentioned briefly in Chapter 6, 6.11.3, and will further treat in Section 7.6.6; cf. Figure 7.6). Figure 7.5 gives the histogram for $n = 9$ (which is, in fact, a very small n!), together with the density curve. The agreement is already quite good, and the curve and the boundary of the histogram would rapidly become indistinguishable if we took a larger n (not necessarily very large).

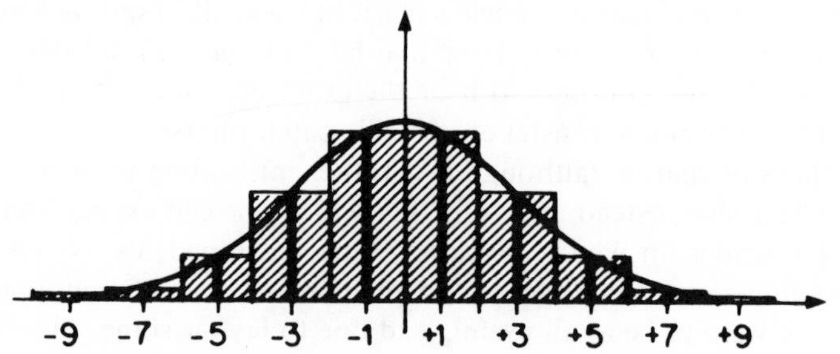

***Figure** 7.5* The binomial distribution: Heads and Tails ($p = \frac{1}{2}$) with n tosses, $n = 9$. The possible values for the gain run through the ten odd numbers from -9 to $+9$, and the height of the column indicates the probability concentrated on each of these numbers. To give a more expressive picture, the values are assigned uniformly within ± 1 of each point, and this makes much clearer the approach of the binomial to the normal distribution, which will, in fact, be shown to be the limit distribution (as $n \to \infty$)

In order to adjust the histograms to arrive at a unique curve, it is, of course, necessary to make an appropriate change of scale (we are concerned with convergence to a *type* of distribution; cf. Chapter 6, 6.7.1). The standard procedure of transforming to $m = 0$ and $\sigma = 1$ (Chapter 6, 6.6.6) is convenient and this is what we have done in the figure.†

If we were, in fact, to consider the representation in terms of the natural scale (the gain Y_n, or the number of successes S_n), it would flatten out more

† This holds for the actual distribution (discrete: the mass of every small rectangle concentrated at the centre). If one thinks of it as diffused, one must modify this slightly (an increase) as we shall see shortly (cf. 7.6.2, the case of F_n^I).

and more, since the deviation behaves like $\sqrt{n}$. (One could think in terms of Bittering's apparatus, in which less and less sand would remain in each section as one continued to overturn it; cf. Figure 7.4: on the other hand, a large number of sections would be needed if one were to continue for very long.) In contrast to this, if one were to represent it in relative terms (the mean gain per toss, Y_n/n, or the frequency, S_n/n) the curve would shrink like $1/\sqrt{n}$, and would rise up to a peak in the centre. The remainder outside this central interval would tend to zero by virtue of the Tchebychev inequality. An appropriate choice is somewhere in between; as we have seen, one can take $Y_n/\sqrt{n}$ (i.e. the standardized deviation, both of the gain, and of the frequency from its prevision, $\frac{1}{2}$).

A somewhat more detailed study of the distribution of Heads and Tails will show us straightaway that the convergence to the normal distribution which is suggested by a visual inspection does, in fact, take place. In this case, too, however, the conclusions are valid more generally. They are valid not only for any binomial process with $p \neq 0$ (the effect of any asymmetry tends to vanish as n increases†), but also for sums of arbitrary, independent random quantities, provided that certain conditions (which will be given at the end of the chapter) are satisfied.

7.6.2. The limit distribution F of a sequence of distributions is to be understood in the sense defined in Chapter 6, 6.7.1: $F_n \to F$ means that, in terms of the distribution functions, $F_n(x) \to F(x)$ at all but at most a countable set of points (more precisely, at all but the possible discontinuity points of $F(x)$). This does not imply, of course, that if the densities exist we must necessarily also have $f_n(x) \to f(x)$; even less does it imply that if the densities are themselves differentiable we must have $f'_n(x) \to f'(x)$. Conversely, however, it is true that these properties do imply the convergence of the distributions (indeed, in a stronger and stronger, and intuitively meaningful way; one only has to think in terms of the graph of the density function).

In our case, we can facilitate the argument by first reducing ourselves to the case just mentioned (albeit with a little trickery along the way). In fact, our distributions are discrete, standardized binomial with $p = \frac{1}{2}$, and hence with probabilities $p_h = \binom{n}{h}/2^n$ concentrated at the points $x_h = (2h - n)/\sqrt{n}$ (distance $2/\sqrt{n}$ apart, lying between $\pm\sqrt{n}$). In order to obtain a distribution admitting a density, it is necessary to distribute each mass, p_h, uniformly over the interval $x_h \pm 1/\sqrt{n}$; or, alternatively, with a triangular distribution on $x_h \pm 2/\sqrt{n}$. In this way, we obtain a continuous density (in the first case, the density is a step function; in the second, the derivative is a step function): we shall denote these two distributions by F_n^I and F_n^{II}, respectively.

† We mention this case explicitly since many people seem to doubt it (notwithstanding the fact that it is clearly covered by the general theorem). Perhaps this is the result of a misleading prejudice deriving from too much initial emphasis on Heads and Tails (?).

We can also give a direct interpretation in terms of random quantities. The distributions arise if, instead of considering $Y_n/\sqrt{n}$, we consider $(Y_n + X)/\sqrt{n}$, where X is a random quantity independent of Y_n, and having the appropriate distribution (in the cases we mentioned, $f(x) = \frac{1}{2}$ $(|x| \leqslant 1)$, or $f(x) = \frac{1}{4}(2 - |x|)$ $(|x| \leqslant 2)$, respectively†). We observe immediately that, since no mass is shifted by more than $1/\sqrt{n}$ (or $2/\sqrt{n}$, respectively) in one direction or the other, F_n^I and F_n^{II} will, for each x, from some $n = N$ on, lie entirely between $F_n(x - \varepsilon)$ and $F_n(x + \varepsilon)$ (in fact, it suffices that $\varepsilon > 2/\sqrt{n}$; i.e. $n > N = 4/\varepsilon^2$). It follows that, so far as the passage to the limit is concerned, it will make no difference if we use these variants in place of the actual F_n (for notational convenience, we shall not make any distinctions in what follows; we shall simply write F_n). The change in the standard deviation also makes no difference, and can be obtained immediately, without calculation, from the previous representation: X has standard deviation $1/\sqrt{3}$ in the case of a uniform distribution (between ± 1), and $\sqrt{(2/3)}$ for a triangular distribution (between ± 2), and it therefore follows that the addition of X either changes the standard deviation to $\sqrt{(1 + 1/3n)}$ or to $\sqrt{(1 + 2/3n)}$ (i.e., asymptotically to $1 + 1/6n$ and $1 + 1/3n$).

Having given these basic details, we can proceed rather more rapidly, arguing in terms of the more convenient modified distribution.

By distributing the mass p_h uniformly over the interval $x_h \pm 1/\sqrt{n}$, we obtain a density $f_n(x_h) = p_h/(2/\sqrt{n}) = \frac{1}{2}p_h\sqrt{n} = \frac{1}{2}\sqrt{n}\binom{n}{h}/2^n$. By distributing it in a triangular fashion, the density at x_h remains the same, but, in every interval $[x_h, x_{h+1}]$, instead of preserving in the first and second half the values of the first and second end-points, respectively, it varies linearly (the graph = the jagged line joining the ordinates at the points $x = x_h$). In fact, the contribution of p_h decreases from x_h on, until it vanishes at x_{h+1} (and the contribution of p_{h+1} behaves in a symmetric fashion).

In the interval $x_h < x < x_{h+1}$, the derivative of the density, $f'_n(x)$, will therefore be constant:

$$f'_n(x) = (p_{h+1} - p_h)\,.\,\tfrac{1}{2}\sqrt{n}/(2/\sqrt{n}) = \tfrac{1}{4}n(p_{h+1} - p_h). \tag{21}$$

Recalling from Section 7.4.2 that $p_{h+1}/p_h = (n - h)/(h + 1)$, and from the expressions for $f_n(x_h)$ and x_h that $h = \frac{1}{2}(n + x_h\sqrt{n})$ (and similarly for x_{h+1}), we have the two alternative expressions

$$f'_n(x) = p_h \cdot \frac{n}{4}\left(\frac{n-h}{h+1} - 1\right) = p_h \cdot \tfrac{1}{4}n\,.\,\frac{n - 2h - 1}{h + 1}$$

$$= \tfrac{1}{2}\sqrt{n}\frac{-x_h\sqrt{n} - 1}{\frac{1}{2}(n + x_h\sqrt{n}) + 1} f_n(x_h) = -x_h \cdot f_n(x_h)\,.\,\frac{1 + 1/(x_h\sqrt{n})}{1 + (x_h/\sqrt{n}) + (2/n)},$$

† In the first case, X has a uniform distribution over $|x| \leqslant 1$; in the second case, $X = X_1 + X_2$, where X_1 and X_2 are stochastically independent, and each has this uniform distribution.

and similarly,

$$f'_n(x) = p_{h+1}\,.\,\tfrac{1}{4}n\left(1 - \frac{h+1}{n-h}\right) = -\,x_{h+1}\,.\,f_n(x_{h+1})\,.\,\frac{1 - 1/(x_{h+1}\sqrt{n})}{1 - (x_{h+1}/\sqrt{n}) + (2/n)}.$$

This proves that the logarithmic derivative $f'_n(x)/f_n(x)$ (which clearly has its own extreme values to the right of the left-hand end-point, and to the left of the right-hand end-point) always satisfies the equation

$$f'_n(x)/f_n(x) = \frac{\mathrm{d}}{\mathrm{d}x}\log f_n(x) = -x[1 + \varepsilon(x)] \tag{22}$$

(where, as n increases, $\varepsilon(x)$ tends uniformly to 0 in any finite interval which has a neighbourhood of the origin removed; e.g. we have $\varepsilon(x) < \varepsilon$, for some n, throughout the interval $2\varepsilon/\sqrt{n} < |x| < \sqrt{n}/2\varepsilon$; the apparent irregularity at the origin merely stems, however, from the fact that both x and $f'(x)$ go to zero, and the equation is automatically satisfied without there being any need to consider the ratio).

The limit distribution must, therefore, satisfy

$$f'(x)/f(x) = -x, \tag{23}$$

from which we obtain

$$\log f(x) = -\tfrac{1}{2}x^2 + \text{const.}, \qquad f(x) = K\,\mathrm{e}^{-\frac{1}{2}x^2} \qquad (K = 1/\sqrt{(2\pi)}). \tag{24}$$

The conclusion is therefore as follows: *the standardized binomial distribution* (the case of Heads and Tails, $p = \frac{1}{2}$) *tends, as $n \to \infty$, to the standardized normal distribution*. The same conclusion holds, however, in more general cases, and, because of its importance, is known as the *central limit theorem* of the calculus of probability. We see immediately that the conclusion holds in the binomial case for $p \neq \frac{1}{2}$ (except that we now require different coefficients in order to obtain the *standardized* form).

7.6.3. It is convenient at the beginning to dwell upon the rather special example of Heads and Tails, since this provides an intuitive and straightforward illustration of many concepts and techniques, which themselves have a much broader compass, but whose essential meaning could otherwise be obscured by the technical details of the general case.

The proof we have just given (based on a technique used by Karl Pearson for this and other examples) is probably the easiest (even more so if one omits the details of the inequalities and simply makes the heuristic observation that, for large enough n, $f'(x)/f(x)$ is practically equal to $-x$).

Remark. Geometrically, this means that the *subtangent* $-1/x$ of the graph of $y = f(x)$ is inversely proportional to the abscissa. The *tail* beyond x is, approximately, an exponential distribution, with density $f(\xi) = K\,e^{-x\xi}$† and prevision $1/x$; this is, in fact, as we can see asymptotically from (20), the prevision of the excess of X over x (provided it does exceed it). This means, essentially, that if an error X (with a standardized normal distribution) exceeds some large given value x, it is almost certain that it exceeds it by very little (about $1/x$): for example, if it exceeds 10σ (or 100σ), we can expect that it exceeds it by $\sigma/10$ (or $\sigma/100$).

Note that this is precisely what happens for the deviations of Heads and Tails (cf. the footnote to equation (4) of Section 7.4.2), provided we make appropriate allowances for the discreteness. If we know that Heads have occurred in more than 75% of the trials, the probabilities that it has occurred 1, 2, 3, 4, or more than 4 times beyond this limit are 0·67, 0·22, 0·074, 0·025, 0·012, respectively, no matter how many tosses n have been made. This means that for $n = 100$ it is almost certain that (with the probabilities given above) the number of successes is one of 76, 77, 78, 79, whereas, for $n = 1000$, the same holds for 751, 752, 753, 754, for $n = 1{,}000{,}000$, for 750,001, 750,002, 750,003, 750,004!

We shall present other (and more general) proofs of this theorem later, and it will be instructive to see it tackled from different standpoints. For the moment, however, we shall consider a useful corollary of it.

Using the fact that $f_n(x) \simeq f(x)$, and recalling the relation with p_h, we find that

$$\omega_h^{(n)} = p_h \simeq (2/\sqrt{n})f_n(x_h) \simeq (2/\sqrt{n})f(x_h) = \sqrt{(2/\pi n)}\exp\left[-\frac{1}{2n}(2h-n)^2\right]. \tag{25}$$

In particular, for $x = 0$, we obtain the maximum term, i.e. the central one ($h = \frac{1}{2}n$ if n = even, or either of $h = \frac{1}{2}(n \pm 1)$ if n = odd). We shall always denote this by a special symbol, u_n, and the formula we have arrived at gives the asymptotic expression $u_n \simeq \sqrt{(2/\pi n)}$; i.e., in figures, $u_n \simeq 0{\cdot}8/\sqrt{n}$ (this makes clear the meaning of the coefficient $\sqrt{(2/\pi)}$, which it is important to keep in mind). In fact, the probability u_n (the maximum probability among the $\omega_h^{(n)}$ of the Heads and Tails case) will crop up in many problems (a partial summary of which will be given in Chapter 8, 8.7.4). For the present, we shall just indicate a few of its properties.

In fact, we have

$$\begin{aligned} u_n = u_{2m} &= \mathbf{P}(Y_n = 0) = \omega_m^{(n)} && \text{for } n = 2m = \text{even},\\ u_n = u_{2m-1} &= \mathbf{P}(Y_n = 1) = \mathbf{P}(Y_n = -1) \\ &= \omega_m^{(n)} = \omega_{m-1}^{(n)} = u_{2m} && \text{for } n = 2m-1 = \text{odd}. \end{aligned} \tag{26}$$

† We have $\exp\{-\frac{1}{2}(x-\xi)^2\} = \exp(-\frac{1}{2}x^2)\exp(-x\xi)\exp(-\frac{1}{2}\xi^2)$, but only the first factor remains because the second is constant (with respect to ξ), and is incorporated into K, and the third is $\simeq 1$ (for small ξ).

The equality of the u_n for successive pairs of values (each odd one with the next even one) is obvious from the definition. In order that the gain after $2m$ tosses be zero, it is necessary that it was either $+1$ or -1 at the preceding toss, and that the final toss had the outcome required to bring it to 0; both possibilities have probability $u_{2m-1}\cdot\frac{1}{2}$, and their sum gives

$$u_{2m} = 2(\tfrac{1}{2}u_{2m-1}) = u_{2m-1}.$$

The same argument can be carried out for the binomial coefficients by applying Stiefel's formula. The central term, $\binom{2m}{m}$, for $n = 2m =$ even, is the sum of the two adjacent ones which are themselves equal,

$$\binom{2m-1}{m-1} = \binom{2m-1}{m},$$

and is therefore twice their value; in order to obtain the probability, however, we must divide by 2^{2m} rather than by 2^{2m-1}, and so $u_{2m} = u_{2m-1}$. We obtain, therefore,

$$(27) \qquad u_{2m-1} = u_{2m} = \binom{2m}{m}\Big/2^{2m} = \frac{(2m)!}{2^{2m}(m!)^2} \simeq \sqrt{(2/\pi n)} \simeq 0{\cdot}8/\sqrt{n}.$$

7.6.4. We see from this that the factor $\sqrt{(2/\pi)}$, hitherto regarded simply as the normalization factor for the standardized normal distribution, also has a link with the combinatorial calculus. This connection is given by Stirling's formula, which provides an asymptotic expression for the factorial, and which enables us to arrive at the central limit theorem for the binomial distribution by a different route (one which is more laborious, but is often used and is, in any case, useful to know).

Stirling's formula expresses $n!$ as follows:

$$(28) \qquad n! = n^n e^{-n}\sqrt{(2\pi n)}(1 + \varepsilon_n) \qquad (\text{where } \varepsilon_n \to 0;\ \text{more precisely, } 0 < \varepsilon_n < 1/11n\dagger).$$

Since the formula is used so often, we shall give a quick proof of it. We have

$$\log n! = \log 2 + \log 3 + \ldots + \log n$$

$$(29) \qquad \simeq \int_{\frac{3}{2}}^{n+\frac{1}{2}} \log x\,dx = [x\log x - x]_{\frac{3}{2}}^{n+\frac{1}{2}}$$

$$= (n + \tfrac{1}{2})\log n - n + \text{const.},$$

† Note that, if we neglect ε_n, Stirling's formula gives $n!$ with smaller and smaller *relative* error, but with greater and greater *absolute* error (i.e. the *ratio* tends to 1, but the *difference* between $n!$ and the approximation tends to $+\infty$). In practical terms, for $n \simeq 10^k$ we have $n!$ with about the first $k + 1$ digits correct; but $n!$ (for large k) has about 10^k digits, and the error has not many less. In any case, what matters in applications is the relative approximation, and this is adequate even for small values.

and we observe that the difference between the sum and the integral converges (substituting $\log n$ in place of

$$\int_{n-\frac{1}{2}}^{n+\frac{1}{2}} \log x \, \mathrm{d}x,$$

we see immediately that we have an error of order $1/n^2$). From this, it follows that $n! \simeq Kn^{n+\frac{1}{2}} \mathrm{e}^{-n}$ (a result known to De Moivre). As for the fact that $K = \sqrt{(2\pi)}$ (discovered by Stirling in 1730), we shall consider it as being established heuristically by virtue of the fact that, were we to leave it indeterminate, the limit of the $f_n(x)$ would be given by

$$f(x) = (1/K)\,\mathrm{e}^{-\frac{1}{2}x^2}$$

and we know that this multiplicative factor must be $1/\sqrt{(2\pi)}$.

Let us just evaluate u_n by this method (n even: $n = 2m$): we obtain

$$u_n = \binom{2m}{m}/2^{2m} = \frac{(2m)!}{2^{2m}(m!)^2} \simeq \frac{2^{2m}m^{2m}\mathrm{e}^{-2m}\sqrt{(2\pi 2m)}}{2^{2m}[m^m\,\mathrm{e}^{-m}\sqrt{(2\pi m)}]^2}$$

$$= \frac{1}{\sqrt{(\pi m)}} = \sqrt{(2/\pi n)} = 0{\cdot}8\sqrt{n}.$$

In order to evaluate

$$\omega_{m+k}^{(n)} = \frac{(2m)!}{2^{2m}(m-k)!(m+k)!} = \omega_m^{(n)} \frac{(m!)^2}{(m-k)!(m+k)!},$$

it is more convenient to make use of an alternative form of Stirling's formula, one which will turn out to be useful in a number of other cases. It is based on evaluating products of the form $(1 + a)(1 + 2a)\ldots(1 + ka)$, with k large, and $ka = c$ small; in our case, $[m!/(m-k)!]/[(m+k)!/m!]$ can be written as

$$[1\,.\,(1-a)(1-2a)\ldots(1-(k-1)a)]/[(1+a)(1+2a)\ldots(1+ka)]$$

by dividing both ratios by m^k, and setting $1/m = a$.

Taking the logarithm, we have

$$\log \prod_{h=1}^{k} (1 + ah) = \sum_{h=1}^{k} \log(1 + ah) \simeq \frac{1}{a}\int_1^{\lambda} \log x \, \mathrm{d}x$$

$$= \frac{1}{a}[(1+\lambda)\log(1+\lambda) - \lambda],$$

with $\lambda = (k + \frac{1}{2})a$,† and, expanding in a series, we have

$$\log \prod_{h=1}^{k} (1 + ah) = \frac{\lambda^2}{2a}\left(1 - \frac{\lambda}{3} + \frac{\lambda^2}{6} - \frac{\lambda^3}{10} + \ldots \pm \frac{\lambda^n}{\frac{1}{2}(n+1)(n+2)} \mp \ldots\right)$$
$$\simeq \tfrac{1}{2}(k + \tfrac{1}{2})^2 a.$$

It follows that

$$(30) \qquad (1 + a)(1 + 2a)\ldots(1 + ka) \simeq e^{\frac{1}{2}(k+\frac{1}{2})^2 a} \simeq e^{\frac{1}{2}ak^2}.$$

In our case, with $a = \pm 1/m$, the two products equal $e^{\pm\frac{1}{2}ak^2} = e^{\pm k^2/2m}$, and their ratio is

$$e^{-k^2/2m}/e^{k^2/2m} = e^{-k^2/m} = e^{-(h-m)^2/m} = e^{-(2h-n)^2/2n}$$

(since $k = h - m$ and $m = \frac{1}{2}n$). We thus obtain the result (which, of course, we already knew).

7.6.5. *Relation to the diffusion problem.* We give here a suggestive argument (due to Pólya), which is entirely heuristic, but is very useful as a basis for discussions and developments. The relation between random processes of the kind we have just exemplified with Heads and Tails and diffusion processes, which we shall meet later, will, in fact, provide a basis for interpreting the latter and even identifying them with the kinds of process already studied. The Wiener–Lévy process (cf. Chapter 8) can, with reference to our previous work, be thought of as a Heads and Tails process involving an enormous number of tosses with very small stakes, taking place at very small time intervals. This process has also been referred to (by P. Lévy) as the *Brownian motion* process, because it can be used (although only for certain aspects of the problem) to represent and study the phenomenon of the same name (which is, as is well-known, a diffusion process).

The Heads and Tails process can be thought of as a diffusion process in which a mass (a unit mass, initially—i.e. at $t = 0$—concentrated at the origin) moves, with respect to time t, through the lattice of Figure 7.2, splitting in half at each intersection (encountered at times t = integer). The mass (which represents the probability) would, according to this representation, divide up in a certain (i.e. deterministic) manner, and, formally, everything goes through (indeed, it will be even simpler than this).

† The simpler form $\lambda = ka$ is practically equivalent to the effect of an individual evaluation. In the case of products or ratios of a number of expressions of this kind, however, it can happen (and does in the example of Section 7.4.3) that it is the contributions deriving from the '$+\frac{1}{2}$' which are important, because the main contributions cancel out.

A more meaningful interpretation, however, and one more suited to our purpose, derives from consideration of a random process of the statistical type. Assume that, initially, a very large number of particles (N, say) are concentrated at the origin, and move at equal and constant rates to the right, through the lattice. At each time instant $t =$ integer, they meet an intersection, and each chooses its direction independently of the others. Equivalently, we could think of them as moving with constant speed on the y-axis, choosing directions at random at each time instant $t =$ integer (i.e. each time a point $y =$ integer is reached); alternatively, we could think of them at rest, but making a jump of ± 1 at each $t =$ integer.

Taking the total mass $= 1$, the mass crossing a given point can no longer be determined with certainty: where, in the deterministic case, it was ω, we can now only say that we have prevision ω, and that the number of particles has prevision $N\omega$, but could take any value h, lying between 0 and N, with probability $\binom{N}{h}\omega^h(1-\omega)^{N-h}$. If we want to give a rough idea of what happens, we could say (quoting the prevision $\pm$ the standard deviation) that the number of particles will be

$$N\omega \pm \sqrt{[N\omega(1-\omega)]}$$

($\simeq N\omega \pm \sqrt{(N\omega)}$ for small ω; the Poisson approximation).

This is what we are interested in: a normal distribution being attained as a result of a statistical diffusion process.

For the purposes of the mathematical treatment (whatever the interpretation), the mass crossing the point (vertex) (t, y), where t and y are integer, both even or both odd, is, as usual,

$$\mathbf{P}(Y_t = y) = \omega^{(t)}_{(t+y)/2} = \omega(t, y),$$

given by one half of that which has crossed $(t-1, y-1)$ or $(t-1, y+1)$:

$$\omega(t, y) = \tfrac{1}{2}[\omega(t-1, y-1) + \omega(t-1, y+1)].$$

The notation $\omega(t, y)$ has been introduced in order to allow us to think of the function as defined everywhere (no matter what the interpretation), even on those points where it has no meaning in the actual problem; in particular, for t and y integer, but $t + y$ odd, like $\omega(t-1, y)$. Subtracting this value from both sides of the previous equation, we obtain

(31) $$\Delta_t\omega = \tfrac{1}{2}\Delta_y^2\omega$$

and, in the limit,

(32) $$\frac{\partial\omega}{\partial t} = \frac{1}{2}\frac{\partial^2\omega}{\partial y^2},$$

provided that (taking the units of t and y to be very small) one considers it legitimate to pass from the discrete to the continuous.

Let us restrict ourselves here to simply pointing out that, in this way, one arrives at the correct solution. In fact, the general solution of the *heat equation*, (32), is given by

(33) $$\omega(t, y) = (K/\sqrt{t})\,e^{-\frac{1}{2}y^2/t},$$

a well-known result, which can easily be verified.

7.6.6. The form of the normal distribution is well-known, and is given in Figure 7.6 (where we show the density $y = f(x)$). We also provide a table of numerical values for both the density and the distribution function (the latter giving the probabilities of belonging to particular half-lines or intervals).

We shall confine ourselves to calling attention to a few points of particular importance.

The density function is symmetric about its maximum, which is at the origin, and decreases away from it, being convex (upwards) in the interval $[-1, +1]$, and concave outside this interval. As $x \to \pm\infty$, it approaches the x-axis, the approach being very rapid, as we already pointed out in Section 7.6.3 (the 'tails' are 'very thin'). In fact, the subtangent decreases, in our case, like $1/|x|$ (if $f(x)$ tends to 0 like a power, $|x|^{-n}$, with n arbitrarily large, the subtangent, in absolute value, increases indefinitely as $|x|$ increases; if the function decreases exponentially, the subtangent is constant).

The graph has two points of inflection at ± 1 (corresponding to the change from convexity to concavity that we already mentioned). The ordinate at these points is about 0·6 of the maximum value (the table gives 0·60652, but it is better to keep an approximate round figure in mind; this is enough to prevent one from making the all too usual distortions when sketching it—on the blackboard, for example). The subtangents are equal to ∓ 1; i.e. the slope is such that the tangents cross the x-axis at the points ± 2.

Since the tails are 'very thin', it is clear that the probabilities of the occurrence of extreme values beyond some given x are, in the case of the normal distribution, much smaller than is usual (in the case of densities decreasing like powers, or exponentially). They are, therefore, much smaller than the values provided by Tchebychev's inequality, which is valid under very general conditions.

We give below a few examples of the probabilities of $|X|$ exceeding $k\sigma$ (or, in the standardized case, $\sigma = 1$, of exceeding k), for $k = 1, 2, 3$ and $3\frac{1}{2}$:

Absolute value greater than:		σ	2σ	3σ	$3\frac{1}{2}\sigma$
Probability	normal distribution:	31·74%	4·55%	0·27%	0·05%
	Tchebychev inequality:	$\leqslant$100·00%	$\leqslant$25·00%	$\leqslant$11·11%	$\leqslant$8·16%

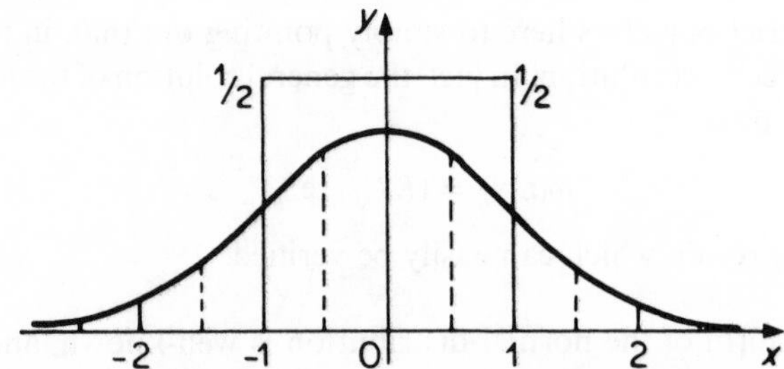

Figure 7.6 The standardized normal distribution ($m = 0$, $\sigma = 1$): the density function. The subdivisions ($0, \pm 1, \pm 2, \pm 3$) correspond to σ, 2σ, 3σ; at ± 1 we have points of inflection, between which the density is convex. The rectangle of height $\frac{1}{2}$ shows, for comparative purposes, the uniform distribution on the interval $[-1, +1]$ (which might well be called the 'body' of the distribution; see Chapter 10, 10.2.4). Note that the vertical scale is, in fact, four times the horizontal one, in order to avoid the graph appearing very flat (as it is, in fact), and hence not displaying the behaviour very clearly.

The table is not only useful for numerical applications but it should also be used in order to commit to memory a few of the significant points (e.g. a few ordinates, and, more importantly, the areas corresponding to abscissae 1, 2 and 3; i.e. to σ, 2σ and 3σ)

The reader is invited to refer to equation (20) in Section 7.5.4, and to the *Remarks* of Section 7.6.3, where we looked at asymptotic expressions for such probabilities ($\simeq K\,e^{-\frac{1}{2}x^2}/x$, $K = 1/\sqrt{(2\pi)} \simeq 0{\cdot}40$†), and at the order of magnitude for possible exceedances (prevision $\simeq 1/x$).

Table of values for the standardized normal (Gaussian) distribution

Abscissa	Ordinate (density)		Area ($\int f(x)\,dx$) in %		
x	$f(x) = \frac{1}{\sqrt{(2\pi)}} e^{-\frac{1}{2}x^2}$	$f(x)$ as % of central ordinate	from x to $+\infty$	in the individual intervals given	2 × (5)
(1)	(2)	(3)	(4)	(5)	(6)
0·0	0·398942	100·0	50·0		
0·1	0·396952	99·50	46·0172		
0·2	0·391043	98·02	42·0740		
0·3	0·381388	95·60	38·2089	19·15	38·30
0·4	0·368270	92·31	34·4978		
0·5	0·352065	88·25	30·8538		

† Writing $K(1 + \Theta/x^2)$, with $0 \leqslant \Theta \leqslant 1$, in place of K, we have an exact bound (and $\Theta \sim 1 - 3/x^2$).

Table (continued)

Abscissa	Ordinate (density)		Area ($\int f(x)\,dx$) in %		
x	$f(x) = \frac{1}{\sqrt{(2\pi)}} e^{-\frac{1}{2}x^2}$	$f(x)$ as % of central ordinate	from x to $+\infty$	in the individual intervals given	$2 \times (5)$
(1)	(2)	(3)	(4)	(5)	(6)
0·5	0·352065	88·25	30·8538		
0·6	0·333225	83·53	27·4253		
0·7	0·312254	78·27	24·1964	14·98	29·96
0·8	0·289692	72·61	21·1855		
0·9	0·266085	66·70	18·4060		
1·0	0·241971	60·652	15·8654		
1·1	0·217852	54·61	13·5666		
1·2	0·194186	48·68	11·5070	9·19	18·38
1·3	0·171369	42·96	9·6800		
1·4	0·149727	37·53	8·0756		
1·5	0·129518	32·47	6·6807		
1·6	0·110921	27·80	5·4799		
1·7	0·094049	23·57	4·4565	4·405	8·810
1·8	0·078950	19·79	3·5930		
1·9	0·065616	16·45	2·8717		
2·0	0·053991	13·53	2·2750		
2·5	0·017528	4·39	0·6210	2·140	4·280
3·0	0·004432	1·11	0·1350		
3·5	0·0008727	0·22	0·02326	0·135	0·270
∞	0·0	0	0·0		

Since the binomial distribution (and many others) approximates, under given conditions, as n increases, to the normal, the table of the latter can also be used in other contexts (with due care and attention‡). Such tables are often used in the case of empirical distributions (statistical distributions), under the confident assumption that the latter behave (at least approximately) like normal distributions. It is easily shown (cf. Section 7.6.9) that this confidence is often not, in fact, justified.

‡ If one were not careful, one might conclude that the probability of obtaining more than n Heads in n tosses (!) is very small, but not zero (about $2{\cdot}4 \times 10^{-23}$ for $n = 100$; about 10^{-2173} for $n = 10{,}000$).

7.6.7. In order to deal with certain other instructive and important features of the normal distribution, we shall have to refer to the multi-dimensional case (either just two dimensions, the plane, or some arbitrary number n; or even the asymptotic case, $n \to \infty$).

It will suffice to limit our discussion to the case of spherical symmetry, where the density has the form

$$f(x_1, x_2, \ldots, x_r) = K \exp(-\tfrac{1}{2}\rho^2), \qquad \rho^2 = x_1^2 + x_2^2 + \ldots + x_r^2.$$

This corresponds to assuming the X_h to be standardized ($m = 0$, $\sigma = 1$) and stochastically independent (for which, in the case of normality, it is sufficient that they be uncorrelated). In fact, we can always reduce the general situation to this special case provided we apply to S_r the affine transformation which turns the covariance ellipsoid into a 'sphere' (cf. Chapter 4, 4.17.6). In other words, we change from the X_h to a set of Y_k which are standardized and uncorrelated (and are linear combinations of the X_h). We shall have more to say about this later (Chapter 10, 10.2.4).

For the moment, let us evaluate the normalizing constant of the standardized normal distribution (which we have already stated to be $K = 1/\sqrt{(2\pi)}$). Integrating over the plane, we obtain

$$\iint e^{-\frac{1}{2}x^2} e^{-\frac{1}{2}y^2}\, dx\, dy = \int e^{-\frac{1}{2}\rho^2} \rho\, d\rho \int d\theta = 2\pi,$$

which is also equal to $[\int e^{-\frac{1}{2}x^2}\, dx]^2$. It follows that $K = 1/2\pi$ in the plane ($r = 2$), $K = (2\pi)^{-\frac{1}{2}}$ over the real line ($r = 1$), and, for general r, $K = (2\pi)^{-\frac{1}{2}r}$.†

There are another two important, interesting properties to note. They involve the examination of two conditions, closely linked with one another, each of which provides a meaningful characterization of the normal distribution. In both cases, it is sufficient to deal with the case of the plane.

The first of them is summarized in the following: the only distribution over the plane which has circular symmetry, and for which the abscissa X and the ordinate Y are stochastically independent (orthogonal), is that in which X and Y have normal distributions with equal variances (and zero prevision, assuming the symmetry to be about the origin). The second property concerns the *stability* of distributions (which we discussed in Chapter 6, 6.11.3). If we require, in addition, that the variance be finite, then stability is the *exclusive* property of the normal distribution.

† We should make it clear (because it is customary to do so—it is, in fact, obvious) that the integral of a positive function taken over the plane does not depend on how one arrives at the limit (by means of circles, $\rho < R$, or squares, $|x| \vee |y| < R$, or whatever); it is always the supremum of the values given on the bounded sets.

For the first property, if we denote by $f(\cdot, \cdot), f_1(\cdot), f_2(\cdot)$ the joint density for (X, Y), and the marginal densities for X and Y, respectively, the given conditions may be expressed as follows:

(*a*) *rotational symmetry*; $f(x, y) = \text{const.}$ for $x^2 + y^2 = \rho^2 = \text{const.}$, from which it follows immediately that $f(x, y) = f(\rho, 0) = f(0, \rho)$ for $\rho = \sqrt{(x^2 + y^2)}$;

(*b*) *independence*; $f(x, y) = f_1(x)f_2(y)$.

In our case, given the symmetry, we can simply write $f(\cdot)$ instead of $f_1(\cdot)$ and $f_2(\cdot)$, and hence obtain a single condition,

$$f(x, y) = f(x)f(y) = f(0)f(\rho).$$

In other words,

$$\frac{f(x)}{f(0)}\frac{f(y)}{f(0)} = \frac{f(\rho)}{f(0)},$$

and, if we put $f(x)/f(0) = \psi(x^2)$, this gives the functional equation

$$\psi(x^2)\psi(y^2) = \psi(\rho^2) = \psi(x^2 + y^2).$$

Taking logarithms, this gives the additive form

$$\log \psi(x^2 + y^2) = \log \psi(x^2) + \log \psi(y^2).$$

Under very weak conditions, which are usually satisfied, this implies linearity (e.g. it is sufficient that ψ is non-negative in the neighbourhood of some point; here, this holds over the whole positive real line). It follows that

$$\log \psi(x^2) = kx^2, \qquad \psi(x^2) = e^{kx^2}, \qquad f(x) = f(0)\, e^{kx^2}$$

(and, normalizing, that $k = -1/2\sigma^2$ and $f(0) = 1/\sqrt{(2\pi)\sigma}$); the required property is therefore established.

In certain cases, this property is in itself sufficient to make the assumption of a normal distribution plausible. A celebrated example is that of the distribution of the velocity of the particles in Maxwell's kinetic theory of gases. If one assumes: (a) *isotropy* (the same distribution for components in all directions), and (b) stochastic *independence* of the orthogonal components, then the distribution of each component is normal (with zero previsions and equal variances). In other words, the distribution of the velocity vector is normal and has spherical symmetry (with density $K\, e^{-\frac{1}{2}\rho^2/\sigma^2}$).

Given the assumptions, the above constitutes a mathematical proof. But however necessary they are as a starting point, the question of whether or not these (or other) assumptions should be taken for granted, or regarded as more or less plausible, is one which depends in part upon the actual physics, and in part upon the psychology of the author concerned.

The second property referred to above reduces to the first one. We must first of all restrict ourselves to the finite variance case (otherwise, we already know the statement to be false; cf. the stable, Cauchy distribution, mentioned at the end of Chapter 6, 6.11.3), and we might as well assume unit variance. We therefore let $f(x)$ denote the density of such a distribution (with $m = 0$, $\sigma = 1$), and X and Y be two stochastically independent random quantities having this distribution.

In order for there to be stability, $Z = aX + bY$ must, by definition, have the same distribution (up to a change of scale, since $\sigma^2 = a^2 + b^2$). If, by taking $a^2 + b^2 = 1$, we make $Z = X \cos \alpha + Y \sin \alpha$, we can avoid even the change of scale, and we can conclude that all projections of the planar distribution, $f(x, y) = f(x)f(y)$, in whatever direction, must be the same. In other words, the projections must possess circular symmetry, and it can be shown that a necessary condition for this (and clearly a sufficient one also) is that the density has circular symmetry (as considered for the first property).†

The conclusion is therefore the same: the property characterizes the normal distribution. The result contains within it an implicit justification (or, to be more accurate, a partial justification) of the 'central limit theorem'. In fact, if the distribution (standardized, with $\sigma = 1$) of the gain from a large number of trials at Heads and Tails follows, in practice, some given distribution, then the latter must be stable (and the same is true for any other case of stochastically independent gains). It is sufficient to note that if Y' and Y'' are the gains from large numbers of trials, n' and n'', respectively, then, *a fortiori*, $Y = Y' + Y''$ is the gain from $n = n' + n''$ trials. If the two independent summands belong to the limit distribution family, then so does their sum: this implies stability.

The justification is only partial because the above argument does not enable us to say whether, and in which cases (not even for that of Heads and Tails), there is convergence to a limit distribution. It does enable us to say, however, that *if a limit distribution exists* (with a finite standard deviation, and if the process is additive with independent summands—all obvious conditions) *it is necessarily the normal distribution.*

7.6.8. *An interpretation in terms of hyperspaces.* It is instructive to bear in mind, as an heuristic, but meaningful, interpretation, that which can be given in terms of hyperspaces. Compared with the previous example, it constitutes even less of a 'partial justification' of the appearance of the normal distribution under the conditions of the central limit theorem (sums of independent random quantities), but, on the other hand, it reveals how the result is often the same, even under very different conditions.

† This is intuitively entirely 'obvious'. The proof, which is rather messy if one proceeds directly, follows immediately from the properties of characteristic functions of two variables (Chapter 10, 10.1.2).

Let us begin by considering the uniform distribution inside the sphere (hypersphere) of unit radius in S_r, and the projection of this distribution onto the diameter, $-1 \leqslant x \leqslant +1$. The section at x has radius $\sqrt{(1-x^2)}$, 'area' equal to $[\sqrt{(1-x^2)}]^{r-1}$, and hence the density is given by

(34) $$f(x) = K(1-x^2)^{(r-1)/2}.$$

In particular, we have $K\sqrt{(1-x^2)}$ for $r=2$ (projection of the area of the circle); $K(1-x^2)$ for $r=3$ (projection of the volume of the sphere); etc.

As r increases, the distribution concentrates around the origin (as happened in the case of frequencies at Heads and Tails). In order to avoid this, and to see what, asymptotically, happens to the 'shape' of the distribution, it is necessary (again, as in the case of Heads and Tails) to expand it in the ratio $1:\sqrt{r}$ (i.e. by replacing x by $x/\sqrt{r}$). We then obtain

$$f(x) = K\left(1 - \frac{x^2}{r}\right)^{(r-1)/2} \to K\,\mathrm{e}^{-\frac{1}{2}x^2}.$$

In the limit, this gives the normal distribution, but without any of the assumptions of the central limit theorem. What is more surprising, however, is that the same conclusion holds under circumstances even less similar to the usual ones. For example, it also holds if one considers a hollow sphere, consisting of just a small layer between $1-\varepsilon$ and 1 ($\varepsilon > 0$ arbitrary). It is sufficient to note that the mass inside the smaller sphere contains $(1-\varepsilon)^r$ of the total mass. This tends to 0 as r increases, and hence its contribution to the determination of the shape of $f(x)$ becomes negligible.

Well: the central limit theorem, also, can be seen as a special case of, *so to speak*, this kind of tendency for distributions in higher dimensions to have normally distributed projections onto a certain straight line.

The case of Heads and Tails shows that one obtains this projection (asymptotically, for large n) by projecting (onto the diagonal) a distribution of equal masses $(\frac{1}{2})^n$ on the 2^n vertices of an n-dimensional hypercube. The same holds, however, for projections onto any other axis (provided it does not belong entirely to a face having only a small number of dimensions compared with n), and also if one thinks of the cube as a solid (with uniformly distributed mass inside it), or with a uniform distribution on the surface, and so on. To summarize; the interpretation in terms of hyperspaces holds in all cases where the central limit theorem holds (although it cannot be of any help in picking out these cases, except in those cases where there exists an heuristic argument by analogy with some known case).

More specifically useful is the conclusion that can be drawn in the opposite sense: namely, that of convergence to the normal distribution in many more cases than those, already numerous, which fall within the ambit of sums of independent random quantities, the case we are now considering (having started with the case of Heads and Tails). The solid cube does fall within such an interpretation (summands chosen independently and with a uniform distribution between ± 1), but that of a distribution on the surface (or on edges, or m-dimensional faces) does not, and this would be even less true for the case of the hypersphere (solid, or hollow).

A wide-ranging generalization of the central limit theorem was given by R. von Mises,† and shows that even the distributions of non-linear 'statistical functions' may be normal (under conditions which, in practice, are not very restrictive). Examples of non-linear statistical functions of the observed values $X_1, X_2, \ldots, X_n$ are the means (other than the arithmetic mean, or their deviations from it), the moments and the functions of the moments (e.g. μ_3^2/μ_2^3, or $(\mu_4/\mu_2^2) - 3$, where the μ_h are the moments about the mean, and the expressions are used as indices of asymmetry and 'kurtosis', respectively; cf. Chapter 6, 6.6.6), the concentration coefficient (of Gini; cf. Chapter 6, the end of 6.6.3), and so on. In general, they are the functions that can be interpreted as functionals of the $F_n(x) = (1/n)\sum(X_h \leqslant x)$ (jump $1/n$ for $x = X_1, X_2, \ldots, X_n$), i.e. of the statistical distribution, under conditions similar to differentiability (i.e. of 'local linearity'). In essence (the actual formulation is quite complicated, and involves long preliminary explanations before one can even set up the notation), one requires that the first derivative (in the sense of Volterra, for 'fonctions de ligne') satisfies the conditions for the validity of the 'central limit theorem' in the linear case, and that the second derivative satisfies a complementary condition.

This generalization, wide-ranging though it is, does not, however, include the cases that we considered in the hyperspace context. This emphasizes even further just how general is the 'tendency' for the normal distribution to pop up in any situation involving 'chaos'.

7.6.9. *Order out of chaos.* We shall postpone what we consider to be a *valid* proof of the central limit theorem until the next section (from a mathematical viewpoint it is a stronger result). Let us consider first the notion of 'order generated out of chaos', which has often been put forward in connection with the normal distribution (as well as in many other cases).

A general observation, which is appropriate at this juncture, concerns a phenomenon which often occurs in the calculus of probability; that of obtaining conclusions which are extremely precise and stable, in the sense that they hold unchanged even when we start from very different opinions or situations. This is the very opposite of what happens in other fields of mathematics and its applications, where errors pile up and have a cumulative effect, with the risk of the results becoming completely invalidated, no matter

† R. von Mises, *Selected Papers*, Vol. II, Providence (1964); cf. various papers, among which (pp. 388–394) the lectures given in Rome (Institute of Advanced Mathematics) provide one of the most up to date expositions (for an exposition of a more illustrative kind, see pp. 246–270).

how carefully the initial data were evaluated and the calculations carried out. This particular phenomenon compensates for the disadvantages inherent in the calculus of probability due to the subjective and often vague nature of the initial data. It is because of this peculiarity (which is, in a certain sense, something of a miracle, but which, after due consideration, can be seen, in a certain sense, as natural) that a number of conclusions appear acceptable to everyone, irrespective of inevitable differences in initial evaluations and opinions. This is a positive virtue, notwithstanding the drawbacks which stem from a too indiscriminate interpretation of it, leading one to accept as objective those things whose roots are, in fact, subjective, but have not been explicitly recognized as such.

In this connection, we shall now put forward an example which is basically trivial, but nonetheless instructive (because it is clear what is going on; the central limit theorem is less self-evident). We return to case (E) of Section 7.2.2, and we consider the probability of an odd number of successes out of n events, $E_1, E_2, \ldots, E_n$. If we assume them to be stochastically independent with probabilities $p_1, p_2, \ldots, p_n$, the probability in question is given by

$$q_n = \tfrac{1}{2} - \prod_{h=1}^{n} (1 - 2p_h)$$

(which can be verified by induction). As n increases, the difference between q_n and $\frac{1}{2}$ decreases (in absolute value). In other words, if one is interested in obtaining a probability close to $\frac{1}{2}$, it is always better to add in (stochastically independent) events, whatever the probabilities p_h might be, because the above-mentioned difference is multiplied by $2(\frac{1}{2} - p_h)$, which is $\leqslant 1$ in absolute value, and the smaller the difference is, the closer p_h is to $\frac{1}{2}$: if $p_h = \frac{1}{2}$, the difference becomes zero (as we remarked at the time). Suppose we now consider a cube or a parallelepiped which we wish to divide into two equal parts as accurately as possible. Making use of the above, instead of performing only one cut (parallel to a face) we could perform three cuts (parallel to the three pairs of faces) and then make up a half with the four pieces which satisfy one or all of the three conditions of being 'above', 'in front', or 'on the left' (and the other half with the four pieces satisfying two or none of these conditions). (What is the point of these digressions? They are an attempt to show that phenomena of this kind do not derive from the principles or assumptions of probability theory—in which case one might well have called them 'miraculous'. They may show up, in their own right, in any kind of applications whatsoever. The fact is simply that *the exploitation and the study* of methods based on disorder is more frequent and 'relevant' in probability theory than elsewhere.)

This should give some idea (as well as, in a sense, some of the reasons) of why it is, in complicated situations where some kind of 'disorder' prevails, that something having the appearance of 'order' often emerges.

A further fact, which serves to 'explain' why it is that this 'order generated out of chaos' often has the appearance of a normal distribution, is that out of all distributions having the same variance the normal has maximum entropy (i.e. the minimum amount of information).

Among the discrete distributions with preassigned possible values x_h, and prevision $\sum p_h x_h = m$, those which maximize $\sum p_h |\log p_h|$ (the sums $\sum p_h = 1$, $\sum p_h x_h$ and $\sum p_h x_h^2$ being fixed) are obtained by setting the $\partial/\partial p_h$ of $\sum p_h\{|\log p_h| + Q(x)\}$ equal to 0, where $Q(x)$ is a second degree polynomial. In other words, we set $-\log p_h + Q(x) = 0$, from which it follows that

$$p_h = \exp(-Q(x)) = K \exp\{-\tfrac{1}{2}(x - m)^2\}.$$

If we choose the x_h to be equidistant from each other, and then let this distance tend to zero, we obtain the normal distribution.

In Chapter 3, 3.8.5, we briefly mentioned the idea of information, but without going at all deeply into it. In the same way here, without going into the relevant scientific theory, we merely note that, when considering the distribution of velocities in the kinetic theory of gases, 'the same variance' corresponds to 'kinetic energy being constant' (and this may suggest new connections with Maxwell's conclusions; cf. Section 7.6.7, above).

We could continue in a like manner: there appear to be an endless variety of ways in which the tendency for the normal distribution to emerge occurs.†

It is easy to understand the wonder with which its appearance in so many examples of statistical distributions (e.g. in various characteristics of animal species, etc.) was regarded by those who first came across the fact, and it is also easy to understand the great, and somewhat exaggerated, confidence in its universal validity that followed.

A typical expression of this mood is found in the following passage of Francis Galton's (it appears in his book *Natural Inheritance*, published in 1889, in the chapter entitled, 'Order in apparent chaos', and the passage is reproduced by E. S. Pearson in one of his 'Studies in the history of probability and statistics', *Biometrika* (1965), pp. 3–18, which also contains a number of other interesting and stimulating quotations):

> 'I know of scarcely anything so apt to impress the imagination as the wonderful form of cosmic order expressed by the "Law of Frequency of Error". The Law would have been personified by the Greeks and deified, if they had known of it. It reigns with serenity and in complete self-effacement amidst the wildest confusion. The huger the mob, and the greater the apparent anarchy, the more perfect is its sway. It is the supreme law of Unreason. Whenever a large sample of chaotic elements are taken in hand and marshalled in the order of their magnitude, an unsuspected and most beautiful form of regularity proves to have been latent all along. The tops of the marshalled row form a flowing curve of invariable proportions; and each element, as it is sorted into place, finds, as it were, a pre-ordained niche, accurately adapted to

† A similar 'tendency' for the normal distribution to appear operates, although in a different manner, in problems of statistical inference, as a result of more and more information being acquired. We mention this now merely to make the above survey complete, and in no way to anticipate what will be said later (Chapter 11, 11.4.6–11.4.7, and Chapter 12, 12.6.5).

fit it. If the measurement of any two specified Grades in the row are known, those that will be found at every other Grade, except towards the extreme ends, can be predicted in the way already explained, and with much precision.'

Are statements of this kind acceptable? It seems to me the answer can be both yes and no. It depends more on the nuances of interpretation than on any general principle of whether such statements are correct or not.

The idea that all natural characteristics have to be normally distributed is one that can no longer be sustained: it is a question that must be settled empirically.† What we are concerned with in the present context, however (and not only in relation to the passage above, but also to the numerous other statements, of a more or less similar nature, that one comes across practically everywhere), are the attitudes adopted in response to the 'paradox' of a 'law' governing the 'accidental', which surely obeys no rules.

Perhaps the following couple of sentences will suffice as a summary of the circumstances capable of differentiating and revealing the attitudes which I, personally, would characterize as 'distorted' or 'correct', respectively:

(*a*) there exist chance phenomena which are really under control, in that they follow the 'rules of chance phenomena', and there are others which are even more chancy, accidental in a more extreme sense, irregular and unforeseeable, occurring 'at random', without even obeying the 'laws of chance phenomena';

(*b*) chance phenomena—the completely accidental, those which are to a large extent irregular or unforeseeable, those occurring 'at random'—are those which presumably 'obey the laws of chance phenomena'; these laws express no more and no less than that which can be expected in the absence of any factor which allows one to a large extent to foresee something falling outside the ambit formed by the overwhelming majority of the vast number of possible situations of chaos.

Even when expressed in this way, the two alternatives are very vague (and it would be difficult to avoid this—I certainly did not succeed). They may be sufficient, however, to remove some of the ambiguity from Galton's position, because they show up what the essential ambiguity is that has to be overcome.

Having said this, it remains for me to make clear that I consider (*a*), the *first* interpretation, to be '*distorted*', and (*b*), the *second* interpretation, to be the *correct one*.

The reasons for this are those that have been presented over and over

† One must not adopt the exaggerated view that all, or almost all, statistical distributions are normal (a habit which is still widespread, although not so much as it was in the past). Around 1900, Poincaré made the acute observation that 'everyone believes in it: experimentalists believing that it is a mathematical theorem, mathematicians believing that it is an empirical fact'.

again in the context of concrete problems. There is no need to repeat them here, nor is there any point in adding further general comments or explanations; these, I am afraid, would inevitably remain at a rather vague level.

7.7 PROOF OF THE CENTRAL LIMIT THEOREM

7.7.1. We now give the proof of the central limit theorem. This is very short if we make use of the method of *characteristic functions*—although this has the disadvantage of operating with purely analytic entities, having nothing to do with one's intuitive view of the problem. It has the advantage, however, that the very simple proof which can be given for the case of Heads and Tails (confirming something that we have already established in a variety of alternative ways) will turn out to be easily adapted, with very little effort, to provide a proof for very much more general cases.

For the gain at a single trial of Heads and Tails ($X_i = \pm 1$ with $p_i = \frac{1}{2}$) the characteristic function is given by

$$\tfrac{1}{2}(\mathrm{e}^{iu} + \mathrm{e}^{-iu}) = \cos u.$$

For the sum Y_n of n such trials (stochastically independent summands), we have $(\cos u)^n$. In the standardized form, $Y_n/\sqrt{n}$, this becomes $[\cos(u/\sqrt{n})]^n$, with logarithm equal to $n \log \cos(u/\sqrt{n})$.

Since $\log \cos x = -\frac{1}{2}x^2[1 + \varepsilon(x)]$ (where $\varepsilon(x) \to 0$ as $x \to 0$), we have $n \log \cos(u/\sqrt{n}) = -\frac{1}{2}n(u/\sqrt{n})^2[1 + \varepsilon(u/\sqrt{n})] = -\frac{1}{2}u^2[1 + \varepsilon(u/\sqrt{n})] \to -\frac{1}{2}n^2$, and, passing from the logarithm back to the characteristic function, we obtain

$$[\cos(u/\sqrt{n})]^n \to \mathrm{e}^{-\frac{1}{2}u^2} \quad \text{as } n \to \infty. \tag{35}$$

This is precisely the characteristic function of the standard normal distribution, and so the theorem is proved.

But this does not merely hold for the case of Heads and Tails. The essential property that has been used in the proof is not the fact that the characteristic function of the individual gain is given by $\phi(u) = \cos u$, but only that its qualitative behaviour in the neighbourhood of the origin is

$$\log \phi(u) = -\tfrac{1}{2}u^2[1 + \varepsilon(u)].$$

This requires only that the variance be finite (the value 1 is merely due to the convention adopted previously).

Therefore: *the central limit theorem holds for sums of independent, identically distributed random quantities provided the variance is finite.*†

It is clear, however, that the conclusion does not require the distributions to be identical, nor the variances to be equal: given the qualitative nature of the circumstances which ensure the required asymptotic behaviour, purely qualitative conditions should suffice.

It is perhaps best to take one step at a time, in order to concentrate attention on the two different aspects separately. Let us begin with the assumption that the distributions do not vary, but that the variances may differ from trial to trial (to be accurate, we should say that the type of distribution does not vary; for the sake of simplicity, we shall continue to assume the prevision to be zero).

In other words, we continue to consider the X_i to be independently and identically distributed, standardized random quantities ($\mathbf{P}(X_i) = 0$, $\mathbf{P}(X^2) = 1$), but with the summands X_i replaced by $\sigma_i X_i$ (with $\sigma_i > 0$, and varying with i). Explicitly, we consider the sums

$$Y_n = \sigma_1 X_1 + \sigma_2 X_2 + \ldots + \sigma_n X_n.$$

We again let $\phi(u)$ denote the characteristic function of the X_i, and $\varepsilon(u)$ the correction term defined by $\log \phi(u) = -\tfrac{1}{2}u^2(1 + \varepsilon(u))$. The characteristic function of $\sigma_i X_i$ is then given by $\phi(\sigma_i u)$, with

$$\log \phi(\sigma_i u) = -\tfrac{1}{2}u^2[\sigma_i^2 + \sigma_i^2 \varepsilon(\sigma_i u)].$$

By taking the product of the ϕ, and the sum of the logarithms, we obtain, for the sum Y_n,

$$\text{(36)} \quad \begin{aligned} \log \prod_{i=1}^{n} \phi(\sigma_i u) &= \sum_{i=1}^{n} \log \phi(\sigma_i u) = -\tfrac{1}{2}u^2 \left[\sum_{i=1}^{n} \sigma_i^2 + \sum_{i=1}^{n} \sigma_i^2 \varepsilon(\sigma_i u) \right] \\ &= -\tfrac{1}{2}s_n^2 u^2 \left[1 + \sum_{i=1}^{n} \frac{\sigma_i^2 \varepsilon(\sigma_i u)}{s_n^2} \right], \end{aligned}$$

where s_n^2 denotes the variance of Y_n; i.e.

$$s_n^2 = \mathbf{P}(Y_n^2) = \sum_{i=1}^{n} \sigma_i^2.$$

† If the variance is infinite, the central limit theorem can only hold in what one might call an anomalous sense; i.e. by not dividing Y_n by $\sqrt{n}$, as would be the case for the normal distribution itself, but rather, if at all, through some other kind of standardization procedure, $(Y_n - A_n)/B_n$, with A and B appropriate functions of n. This holds (see Lévy, *Addition*, p. 113) for those distributions for which the mass outside $\pm x$, if assumed concentrated at these points, has a moment of inertia about the origin which is negligible compared to that of the masses within $\pm x$ (i.e. the ratio tends to zero as $x \to \infty$). These distributions, plus those with finite variances, constitute the 'domain of attraction' of the normal distribution.

There exist other stable distributions (with infinite variances), each having its own domain of attraction (cf. Chapter 8, Section 8.4).

For the standardized Y_n, i.e. Y_n/s_n, we have (substituting u/s_n for u)

$$(37) \qquad -\tfrac{1}{2}u^2\left[1 + \sum_{i=1}^{n}\frac{\sigma_i^2}{s_n^2}\varepsilon\left(\frac{\sigma_i u}{s_n}\right)\right],$$

and, hence, the validity of the central limit theorem depends on the fact that the 'correction term', given by the sum, tends to 0 as $n \to \infty$.

The sum in question is a weighted mean (with weights σ_i^2) of the $\varepsilon(\sigma_i u/s_n)$. Each term tends to 0 as n increases, provided $s_n \to \infty$, because then we will have $(\sigma_i u/s_n) \to 0$. This means that the series formed by summing the variances σ_i^2 must diverge (and this becomes the first condition). This is not sufficient, however. For example, if we took each σ_i very much greater than the previous ones we could make the ratios σ_n/s_n arbitrarily close to 1 and tending to 1, and the correction term would be $\varepsilon(u)$; this would not be improved by dividing u by s_n. The same problem arises if all the ratios, or an infinite number of them, are greater than some given positive number. To ensure that the correction term tends to 0, it is therefore necessary to have $\sigma_n/s_n \to 0$; this also turns out to be sufficient† (and will be the second condition).

To summarize: *the central limit theorem holds for sums of independent random quantities whose distributions, apart from the variances,‡ are the same, provided the total variance diverges* ($s_n \to \infty$) *and the ratios* $\sigma_n/s_n \to 0$. In other words, the theorem holds if, roughly speaking, the contribution of each term becomes negligible compared with that of the total of the preceding terms.

7.7.2. In particular, this holds for bets on Heads and Tails (or at dice, or other games of chance; trials with $p \neq \frac{1}{2}$) when we allow the stakes, S_i, to vary from trial to trial. The individual random gains are $S_i(E_i - p)$, the variance is $\sigma_i = S_i\sqrt{(p\tilde{p})}$, and the standardized random quantity is

$$X_i = (E_i - p)/\sqrt{(p\tilde{p})}$$

(for $p = \frac{1}{2}$, $\sigma_i = \frac{1}{2}S_i$ and $X_i = 2(E_i - \frac{1}{2}) = 2E_i - 1$, as is always used in the case of Heads and Tails).

In order to fix ideas, we can develop considerations of more general validity in the context of this case; this should clarify the result we have obtained. Recall that the σ_i are the same as the S_i, apart from a change in the unit of measurement.

† This is intuitively obvious, but it is perhaps best to give the proof, because it is a little less immediate than it might appear at first sight. Fixing $\varepsilon > 0$, we have $\sigma_n/s_n < \varepsilon$ for all n greater than some given N; each σ_i will therefore satisfy $\sigma_i < \varepsilon s_i < \varepsilon s_n$ for $n > i > N$, and $\sigma_i < s_i \leqslant s_N$ for $i \leqslant N$. Given that $s_n \to \infty$, for all n greater than some given M we have $s_n > s_N/\varepsilon$, i.e. $s_N/s_n < \varepsilon$, and hence we have, for $i < N$, also $\sigma_i/s_n < s_i/s_n < s_N/s_n < \varepsilon$.

‡ See the statement of the theorem for the full meaning of this phrase.

If the sum of the σ_i^2 were convergent, it would be like having a sum with a finite number of terms (one could stop when the 'remainder' becomes negligible when modifying the distribution obtained). Not only would the argument used to prove that such a distribution is normal not then be valid any longer, but a different argument would even allow one to exclude it being so (except in the trivial case in which all the summands are normal).† The condition $s_n \to \infty$ is therefore necessary.

So far as the condition $\sigma_n/s_n \to 0$ is concerned, notice that it is satisfied, in particular, if the σ_n are bounded above (in the above example, this would be the case if the stakes could not exceed some given value), and that this is the only case in which the conclusion holds independently of the order of summation. Were this not the case, one could, in fact, alter the original order, $\sigma_1, \sigma_2, \ldots, \sigma_n, \ldots$, in such a way as to every now and again (and hence infinitely often) make the ratio σ_n^2/s_n^2 greater than $\frac{1}{2}$ (say). One possible procedure would be the following: after, say, 100 terms, if the next one (σ_{101}) is too small to give $\sigma_{101}^2/s_{101}^2 > \frac{1}{2}$, insert between the 100th and the 101st the first of the succeeding σ which is $> s_{101} \cdot \sqrt{2}$. Proceed for 100 more terms, and then repeat the operation; and so on.‡

The conclusion is, therefore, the following: if we have a countable number of summands with no preassigned order, then only the more restrictive condition of *bounded variance* (all the $\sigma_i \leqslant K$) ensures the validity of the central limit theorem (the integers serve as subscripts, but these are merely used by convention to distinguish the summands). On the other hand, if the order has some significance—for example, chronological—then things are different, and the previous conclusion ($s_n \to \infty$, $\sigma_n/s_n \to 0$) is, in fact, valid and less restrictive.§

If, in particular, we wish to consider the case in which all the stakes (and, therefore, the variances) are increasing, the condition means that the σ_n must increase more slowly than any geometric progression ('eventually'; i.e. at least from some given point on).

The reason for such bounds is also obviously intuitive. In fact, if a very large bet arises, it, by itself, will influence the shape of the distribution in such a way as to destroy the approach to the normal which might have resulted from all the preceding bets.

† By virtue of Cramèr's theorem (Chapter 6, Section 6.12).

‡ There is no magic in the figure 100; it was chosen in this example because it seemed best to have a number neither too large, nor too small. The rule must guarantee that all the terms of the original sequence appear in the rearranged sequence (part of the original sequence might be permanently excluded if at each place one term were chosen on the basis of the exigencies of magnitude, or whatever).

§ It seems to be important, both from a conceptual and practical point of view, to distinguish the two cases. In general, however (and, indeed, always, so far as I know), it seems that one only thinks in terms of the case of ordered sequences. It is always necessary to ask oneself whether the symbols actually have a genuine meaning.

It remains to consider now what happens if we let not only the σ_i vary, but also the (standardized) distributions of the X_i. All the expressions that we wrote down in the previous case remain unaltered, except that, in place of $\phi(\sigma_i u)$ and $\varepsilon(\sigma_i u)$, we must now write $\phi_i(\sigma_i u)$ and $\varepsilon_i(\sigma_i u)$, allowing for the fact that the distribution (and hence the ϕ and ε) may vary with i.

All that we must do, then, is to examine the 'correction term' in the final expression. The single ε is now replaced by the ε_i, and, in order to be able to draw the same conclusion, it will be sufficient to require that the $\varepsilon_i(u)$ all tend to zero in the same way as $u \to 0$. In other words, it is sufficient that there exists a positive $\varepsilon(u)$, tending to 0 as $u \to 0$, which provides an upper bound for the $\varepsilon_i(u)$; $|\varepsilon_i(u)| \leqslant \varepsilon(u)$.

As far as the meaning of this condition is concerned, it requires that (for the standardized summands, X_i) the masses far away from the origin tend to zero in a sufficiently rapid, uniform manner. More precisely, it requires that $\mathbf{P}(|X_i| \geqslant x)$ be less than some $G(x)$, the same for all the X_i, which is decreasing and tending to zero rapidly enough for $\int x^2 |\,\mathrm{d}G(x)| < \infty$ (see Lévy, *Addition*, p. 106).

A sufficient condition is that of Liapounov, which is important from a historical point of view in that it provided the basis of the first rigorous proof of the central limit theorem under fairly unrestrictive conditions (1901). The condition requires that, for at least one exponent $2 + \delta > 2$, the moment exists for all the X_i

$$\mathbf{P}(|X_i|^{2+\delta}) = a_i < \infty \tag{38}$$

and that

$$(a_1 + a_2 + \ldots + a_n)/s_n^{2+\delta} \to 0 \qquad \text{as } n \to \infty.$$

7.7.3. All that remains is to ask whether the three conditions

$$(s_n \to \infty, \sigma_n/s_n \to 0, |\varepsilon_i(u)| < \varepsilon(u))$$

that we know to be sufficient for the validity of the central limit theorem are also necessary. The answer, a somewhat unusual one, perhaps, but one whose sense will become clearer later, is that they are not necessary, but almost necessary.

The necessary and sufficient conditions constitute the so-called Lindeberg–Feller theorem. This improves upon the version which we gave above, and which is known as the Lindeberg–Lévy theorem. The range of questions involved is very extensive and has many aspects, the theory having been developed, more or less independently, and in various ways, by a number of authors, especially in the period 1920–1940. In a certain sense, Lindeberg was the one who began the enterprise, and Lévy and Feller produced the greatest number of contributions (along with Cramèr, Khintchin and many others). Our treatment has looked at just a few of the most important, but straightforward, aspects of the theory. The presentation is original, however, in that we have made an effort to unify everything (arguments, choice of notation and terminology, emphasis on what is fundamental, and what is peripheral), and because of the inclusion of examples and comments, perhaps novel, but, in any case, probably useful, at least for clarification.

This digression, in addition to providing some historical background, serves to give warning of the impossibility of giving a brief, complete clarification of the phrase, 'not necessary, but almost necessary', which constituted our temporary answer. Not only would we need to include even those things which we intended to omit, but we would also need to give the reasons why we intended omitting them. As an alternative, we shall give the gist of the matter, together with a few examples: the gist is that the conditions only need be weakened a little—'tinkered with' rather than substantially altered. We have already seen the trivial case (summands all normal) which holds without requiring $s_n \to \infty$; the necessary and sufficient conditions are analogous to the necessary ones, but refer to the sums, Y_n, rather than to the summands (allowance is therefore made for intuitive cases of compensation among the effects of different summands, or, for an individual summand, of compensation between a large value of σ_i and a very small $\varepsilon_i(u)$, i.e. an X_i with a distribution which is almost exactly normal).

An extension of a different kind is provided by the following, which seems, for a variety of reasons, worth mentioning: *a sum of X_h with infinite variances can also tend to a normal distribution* (although within pretty narrow confines, and with rather peculiar forms of normalization). The condition (for X_h with the same distribution F and prevision 0) is that $U(a) = \int_{-a}^{a} x^2\,\mathrm{d}F$ 'varies slowly' as $a \to \infty$ (i.e. for every $k > 0$ we must have

$$U(ka)/U(a) \to 1, \qquad \text{as } a \to \infty,$$

although, by hypothesis, $U(a) \to \infty$). This implies, however, that, *for every* $\alpha < 2$, the moments of order α are finite (this involves the same integral as above, but with $|x|^\alpha$ replacing x^2), and that one does not have convergence for the distributions of the $Y_n/\sqrt{n}$ (but instead for some other sequence of constants, to be determined for each case separately). These are the two remarks we made above; note that the second takes up and clarifies the remarks of Chapter 6, 6.7.1, and the first footnote of that section: an example of this is provided by $f(x) = 2|x|^{-3}\log|x|(|x| \geqslant 1)$, where the normalization is given by $Y_n/(\sqrt{n}\log n)$ (cf. Feller, Vol. II, in several places).

7.7.4. *A complement to the 'law of large numbers'.* This complement (and we present here the important theorem of Khintchin) is included at this point simply for reasons of exposition. In fact, the method of proof is roughly the same as that given above.

We know that for the arithmetic mean, Y_n/n, of the first n random quantities, X_i, with $\mathbf{P}(X_i) = 0$, we have $Y_n/n \stackrel{\cdot}{\to} 0$, and hence $Y_n/n \stackrel{\sim}{\to} 0$ (the quadratic and weak laws of large numbers, respectively), provided that the variances σ_i^2 are bounded and have a divergent sum. Khintchin's result states that $Y_n/n \stackrel{\sim}{\to} 0$ also holds if the variances are not finite, provided the X_i all have the same distribution. (Other cases also go through, under appropriate restrictions.)

If $\mathbf{P}(X_i) = 0$, we have $\log \phi(u) = u\varepsilon(u)$, with $\varepsilon(u) \to 0$ as $u \to 0$. For Y_n/n, the logarithm of the characteristic function is therefore

$$n \,.\, u/n \,.\, \varepsilon(u/n) = u\varepsilon(u/n) \to 0 \qquad \text{as } n \to \infty,$$

and hence the characteristic function tends to $e^0 = 1$, and the distribution to $F(x) = (x > 0)$ (all the mass concentrated at the origin): in other words, the limit of Y_n/n (in the weak sense) is 0; $Y_n/n \stackrel{\sim}{\to} 0$.

If the distributions of the X_i are not all equal, the $\phi_i(u)$, and therefore the $\varepsilon_i(u)$, will be different. The logarithm of the characteristic function of Y_n/n will then be equal to $(u/n)\sum \varepsilon_i(u/n) = u \times$ the (simple) arithmetic mean of the $\varepsilon_i(u/n)$. Here, too, it suffices to assume that the $\varepsilon_i(u)$ tend to 0 in the same way; i.e. for all i we must have $|\varepsilon_i(u)| < \varepsilon(u)$, with $\varepsilon(u)$ positive and tending to zero as $u \to 0$. The condition concerning the distributions of the X_i is similar to the previous one (except that it entails the first moment rather than the second): $\mathbf{P}(|X_i| \geqslant x)$ all bounded by one and the same $G(x)$, decreasing and tending to 0 rapidly enough to ensure that $\int x|\,\mathrm{d}G(x)| < \infty$. Clearly, this is a much less restrictive condition than the previous one: the present condition concerns the influence of the far away masses on the evaluation of the *prevision* (whereas the variance can be infinite); the previous condition was concerned with the influence on the evaluation of the *variance* (which had to exist, or, perhaps better, to be finite).

In order to understand the Khintchin theorem, it is necessary to recollect that we here assume for $\mathbf{P}(X)$ the value given by $\hat{\mathbf{P}}(X)$ (by virtue of the convention we adopted in Chapter 6, 6.5.6).

The theorem states that the mean Y_n/n of $X_1, X_2, \ldots, X_n, \ldots$ (independent, with the same distribution F), converges in a *strong* way to a constant a if and only if the mean value $F(\square) = \hat{\mathbf{P}}(X)$ of F exists and equals a (weak convergence can hold even without this condition).†

This property shows $\hat{\mathbf{P}}$ to have an interesting probabilistic significance and hence to appear as something other than merely a useful convention.

† More general results, with simple proofs, are given in Feller, Vol. II (1966), pp. 231–234.

CHAPTER 8

Random Processes with Independent Increments

8.1 INTRODUCTION

8.1.1. In Chapter 7, we saw that viewing Heads and Tails as a random process enabled us to present certain problems (laws of large numbers, the central limit theorem) in a more expressive form—as well as giving us insights into their solution. It was, essentially, a question of obtaining a deeper understanding of the problems by looking at them from an appropriate dynamic viewpoint.

This same dynamic viewpoint lends itself, in a natural way, to the study of a number of other problems. Not only does it serve as an aid to one's intuition, but also, and more importantly, it reveals connections between topics and problems that otherwise appear unconnected (a common circumstance, which results in solutions being discovered twice over, and hence not appearing in their true perspective); in so doing, it provides us with a unified overall view.

We have already seen, in the case of Heads and Tails, how the representation as a process enabled us to derive, in an elegant manner, results which could then easily be extended to more general cases. We now proceed by following up this idea in two different, but related, directions:

making precise the kind of process to be considered as a first development of the case of Heads and Tails;

considering (first for the case of Heads and Tails, and then in the wider ambit mentioned above) problems of a more complicated nature than those studied so far.

8.1.2. The random processes that we shall consider first are those *with independent increments*,† and we shall pay special attention to the *homogeneous* processes. This will be the case both for processes *in discrete time* (to

† Here, and in what follows, *independent* always means *stochastically independent*.

which we have restricted ourselves so far) and for those in *continuous time*.

In *discrete time* (where t assumes integer values), the processes will be of the same form as those already considered: $Y(t) = X_1 + X_2 + \ldots + X_t$, the sum of *independent* random quantities (increments) X_i. In terms of the Y, one can describe the process by saying that the increments in $Y(t)$ over disjoint time intervals are independent: i.e. $Y(t_2) - Y(t_1)$ and $Y(t_4) - Y(t_3)$ are independent for $t_1 < t_2 \leqslant t_3 < t_4$. The increments are, in fact, either the Xs themselves, or sums of distinct Xs, according as we are dealing with unit time, or with a longer time interval.

Such a process is called *homogeneous* if all the X_i have precisely the same distribution, F. More generally, all increments $Y(t + \tau) - Y(t)$ relating to intervals with the same length, τ, have exactly the same distribution (given by the convolution $F^{\tau *}$).

In *continuous time* (a case which we have so far only mentioned in passing) the above conditions, expressed in terms of the increments of $Y(t)$, remain unchanged; except that now, of course, they must hold for arbitrary times t (instead of only for integer values, as in the discrete time case). This makes the conditions *much more restrictive*. The consequence of this is that whereas in the discrete case all distributions were possible for the Xs (and all distributions decomposable into a t-fold convolution were admissible for the $Y(t)$), in continuous time we can only have, for the $Y(t)$ and their increments $Y(t + \tau) - Y(t)$, those distributions whose decomposition can be taken a great deal further: in other words, the *infinitely divisible* distributions (which we mentioned in Chapter 6, 6.11.3).†

In such a process, the function $Y(t)$ can be thought of as decomposed into two components, $Y(t) = Y_J(t) + Y_C(t)$, the first of which varies *by jumps*, and the second *continuously*. The arguments we shall use, based on this idea, incomplete though they are at the moment, are essentially correct so far as the conclusions are concerned, although critical comments are required at one point (Section 8.3.1) in connection with the interpretation of these conclusions and the initial concepts.

The conclusion will be that the distribution of $Y(t)$ (or of the increment $Y(t + \tau) - Y(t)$, respectively) can be completely and meaningfully expressed by considering the two components Y_J and Y_C separately:

in order to characterize the distribution of $Y_J(t)$ it is necessary and sufficient to give *the prevision*, over the interval $[0, t]([t, t + \tau]$, respectively), *of the number of jumps of various sizes*;

† These statements are not quite correct as they stand. Firstly, X_i could be a *certain* number; secondly, the jumps might occur at *known* instants. These are trivial special cases, however, which could be considered separately (cf. Section 8.1.4).

in order to characterize the distribution of $Y_C(t)$ it is necessary and sufficient to give *the prevision and the variance of the continuous component* (since, as we shall see, its distribution is necessarily *normal*);

taken together, these characterize $Y(t)$.

All the previsions are additive functions of the intervals, and, except for that of the increment of the continuous component, Y_C, they are essentially non-negative, and therefore non-decreasing.

In the *homogeneous* case, they depend only on the length of the interval, and are therefore proportional to it.

8.1.3. We shall illustrate straightaway, by presenting some of the most important cases, the structure which derives from what we have described as the most general form of random process with independent increments. Although this will only be a summary, it should help to make clear the scope of our investigation, and also give an idea of the kinds of problems we shall encounter.

Let us first restrict ourselves to the *homogeneous* case (it will be seen subsequently—Section 8.1.4—that the extension to the general case is reasonably straightforward, and only involves minor additional considerations).

It is convenient to indicate and collect together at this point the notation that will be used in what follows. This will be given for the homogeneous case, and hence a single distribution suffices for the increments $Y(t_0 + t) - Y(t_0)$. This will depend on t but not on t_0, and will also be the distribution of $Y(t)$ if the initial condition $Y(0) = 0$ is assumed (as will usually be convenient). The distribution function, density (if any), and characteristic function of this distribution will be denoted by $F^t(y)$, $f^t(y)$ and $\phi^t(u)$, respectively. The t is, in fact, an exponent of $\phi(u)$ (as is obvious from the homogeneity and the independence of the increments), and is used as a superscript for F and f, both for uniformity, and to leave room for possible subscripts. Its use is also partially justified by the fact that F^t and f^t are, actually t-fold convolutions, $(F^1)^{t*}$ and $(f^1)^{t*}$, of F^1 and f^1 with themselves, provided that the concept (where it makes sense, as is the case here for all t, because of infinite divisibility) is suitably extended to the case of non-integer exponent t. The distribution and density (if any) of the *jumps* will be denoted by $F(x)$ and $f(x)$, respectively (with no superscripts), and the characteristic function by $\chi(u)$.

Let us examine now the various cases.

The simplest example, and the one which forms the basis for the construction of all the random processes of the type under consideration, is that of the *Poisson process*. This is a jump process, all jumps being of the same size, x; for the time being, we shall take $x = 1$, so that $Y(t) = N(t) =$ the *number of jumps* in $[0, t]$. We shall denote by μ the *prevision* of the number of jumps occurring per unit time (i.e. μt is the prevision of the number of jumps in a time interval of length t). In an infinitesimal time, $\mathrm{d}t$, the prevision of the number of jumps is $\mu\,\mathrm{d}t$, and, up to an infinitesimal quantity of greater order, this is also the probability that *one* jump occurs within the small time interval (the probability of more than one jump occurring is, in comparison, negligible). We call μ the *intensity* of the process.

The distribution of $N(t)$, the number of jumps occurring before time t, is Poisson, with prevision $a = \mu t$ (cf. Chapter 6, 6.11.2, (39), for the explicit form of the probabilities and the characteristic function).

We recall that the variance in this case is also equal to μt. It is better, however, to note explicitly that the prevision is $x\mu t$, and the variance is $x^2\mu t$, where x is the magnitude of the jump. In this way, one avoids the ambiguities which arise from ignoring dimensional questions (i.e. taking $x =$ pure number) and, subsequently, from assuming the special value $x = 1$ (for which $x = x^2$).

A superposition of several Poisson processes, having jumps of different sizes, x_k, and different intensities, $\mu_k (k = 1, 2, \ldots, m)$, is also a jump process, homogeneous with independent increments, $Y(t) = \sum x_k N_k(t)$. It also has an alternative interpretation as a process of intensity $\mu = \mu_1 + \mu_2 + \ldots + \mu_m$, where each jump has a random size X (independently of the others), X taking the value x_k with probability μ_k/μ.

Instead of considering the sum of a finite number of terms, we could also consider an infinite series, or even an integral (in general, a Stieltjes integral). Provided the total intensity remains finite, the above interpretations continue to apply (except that X now has an arbitrary distribution, rather than the discrete one given above). This case is referred to as a *compound Poisson process*, and provides the most general homogeneous process with independent increments and *discrete jumps* (i.e. finite in number in any bounded interval).

Using the same procedure, one can also obtain the *generalized Poisson processes*: i.e. those *with everywhere dense jumps* (discontinuity points), with $\mu = \infty$. It is necessary, of course, to check that the process does not diverge: for this, we require that the intensity μ_ε of jumps X with $|X| > \varepsilon (\varepsilon > 0)$ remains finite, diverging only as $\varepsilon \to 0$, and then not too rapidly.

The Poisson processes, together with the compound and generalized forms, exhaust all the possibilities for the jump component $Y_J(t)$.

It remains to consider the continuous component $Y_C(t)$. In every case, $Y(t)$ could be considered as the sum of N increments (N arbitrary), corresponding to the N small intervals of length t/N into which the interval $[0, t]$ could be decomposed. If one makes precise the idea of separating off the 'large' increments (large in absolute value), which correspond to the jumps, it can be shown that the others (the 'small' increments) satisfy the conditions of the central limit theorem, and hence the distribution is necessarily normal (as we mentioned above).

As we have already stated, its prevision varies in a linear fashion, $\mathbf{P}[Y_C(t)] = mt$, and the same is true of the variance, $\mathbf{P}[\{Y_C(t) - mt\}^2] = \sigma^2 t$ (where m and σ^2 denote the prevision and variance corresponding to $t = 1$).† In some cases, it may also turn out to be convenient to separate the certain linear function, mt, and the fair component (with zero prevision), $Y_C(t) - mt$ (whose variance is still $\sigma^2 t$).

The set-up we have just described is called the *Wiener–Lévy process* (cf. also Chapter 7, 7.6.5).

8.1.4. So far as the increment over some given interval is concerned, the conclusions arrived at in the homogeneous case carry over without modification to the general case. This is clear, because the conclusions depend on the prevision (of the number of certain jumps, etc.) over such an interval as a whole, and not on the way the prevision is distributed over the subintervals. In the inhomogeneous case, the only new feature is that, within an interval, each prevision can be increasing in an arbitrary way, not necessarily linearly.

The one thing that is required is that one exclude (or, better, consider separately, if they exist) the points of discontinuity for such a prevision. In fact, at such points one would have a non-zero probability of discontinuity (a 'fixed discontinuity'; see the remark in the footnote to Section 8.1.2). This is equivalent to saying that at these points $Y(t)$ receives a random (instantaneous) increment, incompatible with the nature of the 'process in continuous time', because (being instantaneous) it is not decomposable, and therefore does not have to obey the 'infinite divisibility' requirement. In what follows, we shall always tacitly assume that such fixed discontinuities have been excluded.

Observe that, if all the previsions are proportional to one another, the process can be said to be homogeneous with respect to a different time scale (one which is proportional to them). In general, however, the previsions will vary in different ways, and then we have no recourse to anything of this kind.

8.1.5. Let us now consider more specifically some of the problems that one encounters. These are of interest for a number of reasons: by virtue of their probabilistic significance and their range of application; because of the various mathematical aspects involved; and, above all, because of the unified and intuitively meaningful presentation that can be given of a vast collection of seemingly distinct problems. The problems that we shall mention are only a tiny sample from this collection, and the treatment that we shall give will only touch upon some of the more essential topics, presenting them in their simplest forms.

† These formulae hold, of course, for every $Y(t)$ which is homogeneous with independent increments (with finite m and σ^2). We mention them here, for the particular case of Y_C, only because of their importance in specifying the distribution in this case.

First of all, we have to translate into actual mathematical terms (through the distribution, by means of the characteristic function) the characterization of the general process with independent increments, and hence of the most general infinitely divisible distribution (which we have already given above, in terms of the intensities of the jumps and the normal component).

Particular attention must be paid, however, to the limiting arguments which lead to the generalized Poisson process, since the latter gives rise to rather different problems. For example, in our preliminary remarks we did not mention that sometimes convergence can only be achieved by 'compensating' the jumps by means of a certain linear function, and that, in this case, the intuitive idea of behaviour similar to that of the discrete case (apart from minor details) must undergo a radical change.

Even the behaviour of the continuous component (the Wiener–Lévy process), despite what one might, at first sight, be led to expect from the regular and familiar shape of the normal distribution, turns out to be extremely 'pathological'. The study of the behaviour of the function (or, more precisely, the behaviour which it 'almost certainly' enjoys) is, however, a more advanced problem. We shall begin by saying something about the distribution.

How does the distribution of $Y(t)$ behave as t increases? We already know that it tends to normality in the case of finite variances, but there also exist processes (of the generalized Poisson type) with infinite variances. The answer in this case in that there exist other types of stable distribution, all corresponding to generalized Poisson processes (more precisely, as we shall see in Sections 8.4.1–8.4.4, there are a doubly infinite collection of them, reducible, essentially, to a single infinite collection). The processes which are not stable either tend to a stable form, or do not tend to anything at all.

The key to the whole question lies in the behaviour at the other extreme, as $t \to 0$. This is directly connected with the intensity of the jumps of different sizes: one has stability if the intensity of the jumps $>x$ or $<-x$ decreases proportionally to $x^{-\alpha}(x > 0, 0 < \alpha < 2)$; one has a tendency to a stable form if the process, in some sense, approximately satisfies this condition. Referring back to a remark made previously, we note that the 'sum of the jumps' does not require 'compensation' if $\alpha < 1$, but does if $\alpha > 1$ ($\alpha = 1$ constitutes a subcase of its own). It follows that the stable distributions generated by just the positive (negative) jumps extend only over the positive (negative) values if $\alpha < 1$, whereas they extend over all positive and negative values if $\alpha \geqslant 1$.

An important special case is that of 'lattice' processes, in which $Y(t)$ can only assume integer values (or those of an arithmetic progression; there is no essential difference). They must, of course, be compound Poisson processes, but this does not mean that they cannot tend to a stable (continuous distribution as t increases. Indeed, if the variances are finite they necessarily tend to the normal distribution (as we already saw, in the first instance, in the case of Heads and Tails).

8.1.6. The study of the way in which the distribution of $Y(t)$ varies with t does not necessarily entail the study of the behaviour of the actual process $Y(t)$ (i.e. of the function $Y(t)$). The essential thing is to examine the characteristics of the behaviour of the latter which are of interest to us, behaviour which can only be studied by the simultaneous consideration and comparison of values of $Y(t)$ at different times t (possibly a large or infinite number of them).

So far as the phrase 'infinite number' is concerned, we should make it clear at once that it means 'an arbitrarily large, but finite, number' (unless, in the case under consideration, one makes some additional assumption, such as the validity of countable additivity, or gives some further explanation). However, in order to avoid making heavy weather of the presentation with subtle, critical arguments, we shall often resort to intuitive explanations of this kind (as well as to the corresponding 'practical' justifications).

Problems that have already been encountered in the discrete cases—like that of the asymptotic behaviour of $Y(t)/t$ (as $t \to \infty$), to which the strong law of large numbers gives the solution—are restated, and shown to have, generally speaking, similar solutions in the continuous case. One also meets, in the latter case, problems which are, in a certain sense, reciprocal; involving what happens as $t \to 0$ ('local' behaviour—like 'continuity', etc.—at the origin, and hence, in the homogeneous case, at any arbitrary point in time). In the Wiener–Lévy process, the two problems correspond exactly to one another by reciprocity.

If one confines attention to considerations based on the Tchebychev inequality, the conclusions hold for every homogeneous process with independent increments for which the variance is finite. One such process, viewing the jumps and possible horizontal segments 'in the large' (i.e. with respect to large time intervals and intervals of the ordinate), in such a way as to make them imperceptible, is the Wiener–Lévy process. Indeed, Lévy calls it a Brownian motion process, which corresponds to the perception of an observer who is not able to single out the numerous, tiny collisions that, in any imperceptibly small time interval, suddenly change the motion of the particle under observation.

More generally, even for arbitrary random processes (provided they have finite variance), answers can be obtained to a number of problems, even though, in general, they may only be qualitative and based on second-order characteristics (as in the case of a single random quantity, or of several). Here, the random quantities to be considered are the values $Y(t)$, and the characteristics to be used are the previsions, variances and covariances (or, equivalently, $\mathbf{P}[Y(t)]$, $\mathbf{P}[Y^2(t)]$, $\mathbf{P}[Y(t_1)Y(t_2)]$). In our case, this is trivial:

evaluated at $t = 0$, the prevision and variance of $Y(t)$ are, as we know, mt and $\sigma^2 t$, and the covariance of $Y(t_1)$ and $Y(t_2)$ is $\sigma^2 t_1$ (if $t_1 \leqslant t_2$); the correlation coefficient is, therefore, $r(t_1, t_2) = \sigma^2 t_1/\sigma\sqrt{t_1} \cdot \sigma\sqrt{t_2} = \sqrt{(t_1/t_2)}$ (it is sufficient to observe that $Y(t_2) = Y(t_1) + [Y(t_2) - Y(t_1)]$, and that the two summands are independent).

There are other problems, however, which require one to take all the characteristics into account, and to have recourse to new methods of approach. (This also applies to the problems considered previously, if one wishes to obtain conclusions which are more precise, in a quantitative sense, and more specifically related to the particular process under consideration.) A concept which can usefully be applied to a number of problems is that of a *barrier* (a line in the (t, y)-plane on which $y = Y(t)$ is represented). One observes when the barrier is first reached, or when it is subsequently reached, or, sometimes, one assumes that the barrier modifies the process (it may be absorbing—i.e. the process stops—or reflecting, and so on).

The classical problem of this kind, and one that finds immediate application, is that of the 'gambler's ruin' (corresponding to the point when his gain reaches the level $-c$, where c is his initial capital). There are a number of obvious variants: one could consider the capital as being variable (an arbitrary barrier rather than the horizontal line $y = -c$), or one could think of two gamblers, both having a bounded initial capital (or variable capital), and so on.

In addition to this and other interesting and practical applications, problems of this kind also find application in studying various aspects of the behaviour of the function $Y(t)$. In particular, and this question has been studied more than any other, they are useful for specifying the asymptotic behaviour, indicating which functions tend to zero too rapidly, or not rapidly enough, to provide a (practically certain) bound for $Y(t)/t$ from some point $t = T$ on (and similarly as $t \to 0$).

Reflecting barriers, and others which modify the process, take us beyond the scope of this present chapter. In any case, they will enter, in a certain sense, into the considerations we shall make later, and will provide an instructive and useful technique. In particular, they will enable us to make use of the elegant and powerful arguments of Desiré André (and others) based on symmetries.

8.2 THE GENERAL CASE: THE CASE OF ASYMPTOTIC NORMALITY

8.2.1. Let us first give a precise, analytic statement of what we have hitherto presented in a descriptive form concerning the structure and properties of the general homogeneous process with independent increments.

When the intensity of the jumps, μ, and the variance, σ^2, are finite, the process is either normal (if $\mu = 0$), or asymptotically normal (as we have already seen from the central limit theorem). In other words, we either have the Wiener–Lévy process, or something which approximates to it asymptotically. We are not saying that the restrictions made are necessary for such behaviour: the restriction on μ, i.e. $\mu < \infty$, has no direct relevance, and the restriction $\sigma^2 < \infty$ could be weakened somewhat (cf. Chapter 7, 7.7.3). However, for our immediate purpose it is more appropriate to concentrate attention on the simplest case, avoiding tiresome complications which do not really contribute anything to our understanding.

Our first task, once we have set up the general analytical framework, is to provide insight into the way in which, and in what sense, processes of this kind (i.e. those which are asymptotically normal) can be considered as an approximation, on some suitable scale, to the Wiener–Lévy process, and conversely. In this way, any conclusion established for a special case, for example, that of Heads and Tails, turns out to be necessarily valid (in the appropriate asymptotic version) in the general case. Note that this enables us, among other things, to establish properties of the Wiener–Lévy process by means of elementary combinatorial techniques, which are, in themselves, applicable only to the case of Heads and Tails. Conversely, it enables us to derive properties of the latter, and of similar examples, that cannot be obtained directly (for example, asymptotic properties) by using approaches—often much simpler—which deal with the limit case of the Wiener–Lévy process (itself based on the normal distribution). This is just one example, out of many, where the possibility exists of advantageously switching from a discrete schematization to a continuous one, or vice-versa, as the case may be.

8.2.2. The *Wiener–Lévy process* can be derived as the limit case of the *Heads and Tails* process (in discrete time).

Suppose, in fact, that we change the scale, performing tosses at shorter time intervals, and with smaller stakes, so that the variance per unit time remains the same. For this to be so, if the stake is reduced by a factor of N ($a = 1/N$) the number of tosses per unit time must be increased to N^2 (with time intervals $\tau = 1/N^2$). To see this, note that the variance for each small time interval τ is $a^2 = (1/N)^2$, and in order for it to be equal to 1 per unit time, the number of small intervals by which it has to be multiplied must be N^2.

By taking N sufficiently large, one can arrange things in such a way that the increment per unit time has a distribution arbitrarily close to that of the standardized normal. Taking N even larger, one can arrange for the same properties to hold, also, for the small intervals. In other words, one can arrange for the distribution of $Y(t)/\sqrt{t}$ to be arbitrarily close, to any preassigned degree, to the standardized normal, for every t exceeding some arbitrarily chosen value. This could also be expressed by saying that the Heads and Tails process can be made to resemble (with a suitable change of scale) a process in discrete time with normally distributed jumps, and (with a more pronounced change of scale) even to resemble the Wiener–Lévy process (provided, of course, that one regards as meaningless any claim that the scheme is valid, or, anyway, observable, for arbitrarily small time periods).

In terms of the characteristic function, these considerations reduce, in the first case, to the straightforward and obvious observation that if we substitute e^{-u^2} for $\cos u$ then $[\cos(u/\sqrt{n})]^n \to e^{-u^2}$ becomes the identity $[e^{-(u/\sqrt{n})^2}]^n = e^{-u^2}$; whereas, in the second case, they simply repeat the procedure used in Chapter 7, 7.7.1. In the Heads and Tails process, $Y(t)$ has characteristic function $\phi^t(u) = (\cos u)^t$ (t = integer); under the above-mentioned change of scale, this becomes $[\cos(u/N)]^{tN^2}$ ($t = \text{integer}/N^2$), and in the limit (as $N \to \infty$) it becomes e^{-tu^2} (t arbitrary).

8.2.3. The Wiener–Lévy process can be obtained in an analogous manner from the *Poisson version of the Heads and Tails process* (a compound Poisson process, with the intensity of the jumps given by $\mu = 1$, and with jumps ± 1, each with probability $\frac{1}{2}$). The difference is that instead of there certainly being a toss after each unit time interval the tosses occur at random, with a *prevision* of one per unit time (the probability being given by dt in each infinitesimal time interval of length dt). Alternatively (as we mentioned already in Section 8.1.3), we can say that $Y(t) = Y_1(t) - Y_2(t)$, where Y_1 and Y_2 are the number of positive and negative gains, respectively, both occurring at random and independently, each with intensity $\frac{1}{2}$.

The distribution of $Y(t)$ in such a process is the Poisson mixture of the distributions of Heads and Tails. In terms of the characteristic function, $\phi^t(u)$ is the Poisson mixture (with 'weights' given by the probabilities $e^{-t}t^n/n!$ of there being n jumps in $[0, t]$) of the $(\cos u)^n$ (the characteristic functions of the sums of n jumps; i.e. of $Y(t)$, assuming that there are n jumps up until time t):

$$\text{(1)} \qquad \phi^t(u) = \sum e^{-t}(t\cos u)^n/n! = e^{t(\cos u - 1)}.$$

In this case, too, the same change of scale (jumps reduced by $1/N$, and the intensity increased by N^2) leads to the Wiener–Lévy process. In fact, as $N \to \infty$, the characteristic function $\exp\{tN^2[\cos(u/N) - 1]\}$ tends to e^{-tu^2}.

We therefore obtain the conclusion mentioned above, and it is worth pausing to consider what it actually means. It establishes that the distribution of the gain in a game of Heads and Tails is, after a sufficiently long period of time has elapsed, practically the same, no matter whether tosses were performed in a regular fashion (one after each unit time period), or were randomly distributed (with a Poisson distribution, yielding, *in prevision*, one toss per each unit time period).

8.2.4. The three examples given above (the Heads and Tails process, in both the regular, discrete case and its Poisson variant, and the normally distributed jump process in discrete time) provide the simplest approaches to approximate representations of the Wiener–Lévy process and should be borne in mind in this connection. If we wished, we could also include a fourth such example; the Poisson variant of the normal jump process:

$$\phi^t(u) = \sum \mathrm{e}^{-t}(t\,\mathrm{e}^{-u^2})^n/n! = \exp[t(\mathrm{e}^{-u^2} - 1)]. \tag{2}$$

Strictly speaking, however, if one ignores the psychological case for presenting these introductory examples, the above discussion is entirely superfluous. We have merely anticipated, in a few special cases, ideas which can be examined with equal facility in the general case.

Let us now turn, therefore, to a systematic study of the general case. We begin with the Poisson process, and then proceed to a study of the compound Poisson processes.

8.2.5. The (simple) *Poisson process* deals with the number of occurrences, $N(t)$, of some given phenomenon within a time period $[0, t]$. In other words, it counts the *jumps*, each considered as being of unit size (like a meter that clicks once each time it records a phenomenon—such as the beginning of a telephone conversation, a particle hitting a screen, a visitor entering a museum, or a traveller entering an underground station, etc.).

The conditions given in Section 8.1.3 simply mean that we must be dealing with a homogeneous process with independent increments, and with jumps all of unit size. It is instructive to go back to these conditions, and, in the context of the present problem of deriving the probabilities of the Poisson distribution, to provide two alternative derivations in addition to that given in Chapter 6, 6.11.2.

First method. Let a ($a = \mu t$) be the prevision of the number of jumps occurring in a given interval (of length t, where μ denotes the intensity). If we subdivide the interval into n equal parts (n large, so that a/n is small compared with 1; i.e. $a/n < \varepsilon$, for some preassigned $\varepsilon > 0$), then a/n is the prevision of the number of jumps occurring in each small interval. We also have $(a/n) = q_n m_n$, where q_n is the probability of *at least one* jump occurring in a small time interval of length t/n, and m_n is the prevision of the number of jumps occurring in intervals containing at least one jump. It follows that we must have $m_n \to 1$ (as $n \to \infty$). If this were not so, it would mean that each discontinuity point had a positive probability of having further jumps in any arbitrarily small neighbourhood of itself; in other words, practically speaking, of being a multiple jump (contrary to the hypothesis that all jumps are of unit size).†

The probability that h out of the n small intervals contain discontinuities, and $n - h$ do not, is given by $\binom{n}{h}q_n^h(1 - q_n)^{n-h}$. As $n \to \infty$, the probability that there are small intervals containing more than one jump becomes negligible (so that h gives the actual number of jumps). On the other hand, we also have $q_n \simeq a/n$, and therefore

$$p_h(t) = \lim \binom{n}{h}\left(\frac{a}{n}\right)^h\left(1 - \frac{a}{n}\right)^{n-h}.$$

In this way, we reduce to the formulation and procedure that we have already seen (in Chapter 6, 6.11.2) for so-called 'rare events' (which the occurrences of jumps in very small intervals certainly are).

Second method. We can establish immediately that $p_0(t)$, the probability of no jumps in a time interval of length t, must be of the form e^{-kt}. To see this, we note that, because of the independence assumption,

$$p_0(t' + t'') = p_0(t')p_0(t'')$$

and this relation characterizes the exponential function. The probability that the *waiting time*, T_1, until the occurrence of the first jump does not exceed t is given by $F(t) = 1 - p_0(t) = 1 - \mathrm{e}^{-kt}$ (which is equivalent to saying that it is not the case that no jump occurred between 0 and t). From the distribution function $F(t)$, we can derive the density function $f(t) = k\,\mathrm{e}^{-kt}$, and we then know that the characteristic function is given by

$$\phi(u) = 1/(1 - ku).$$

We recall (although, in fact, it follows directly from the above) that the gamma distribution is obtained by convolution:

$$[\phi(u)]^h = (1 - ku)^{-h}, \qquad f^{h*} = Kx^{h-1}\,\mathrm{e}^{-kt}(x \geqslant 0),$$

$$F^{h*} = 1 - \mathrm{e}^{-kt}\left[1 + \frac{kt}{1!} + \frac{(kt)^2}{2!} + \ldots + \frac{(kt)^h}{h!}\right].$$

This, therefore, gives the distribution of S_h, the waiting time for the occurrence of the hth jump, which is the sum of the first h waiting times (independent, and exponentially distributed):

$$S_h = T_1 + T_2 + \ldots + T_h.$$

† It would certainly be more direct to impose an additional condition requiring that the probability $p^*(t) = 1 - p_0(t) - p_1(t)$ of there being *two or more* jumps in an interval of length t be an infinitesimal of second order or above. This is, in fact, the approach adopted in many treatments, but it carries the risk of being interpreted as an additional restriction, without which there could be different processes, each compatible with the initial assumptions.

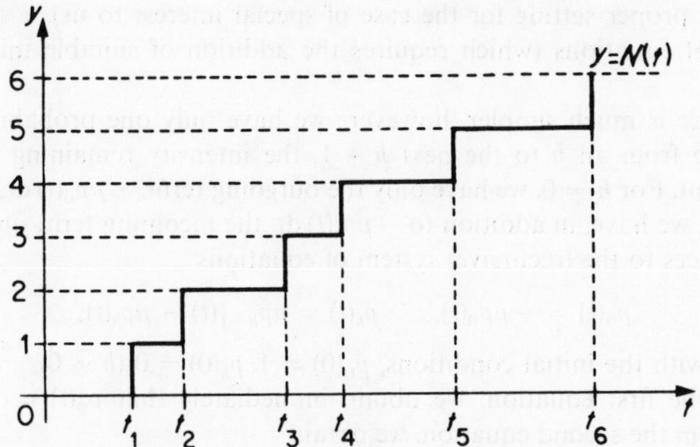

Figure 8.1 The simple Poisson process

This method of approach is, in a sense, the converse of the first one. The connection is provided by noting that $N(t) < h$ is equivalent to $S_h > t$ ('less than h jumps in $[0, t]$' = 'the hth jump takes place after time t'), and that $N(t) = h$ is equivalent to

$$\{N(t) < h + 1\} - \{N(t) < h\} = \{S_{h+1} > t\} - \{S_h > t\}.$$

From this we see that

$$\text{(3)} \qquad p_h(t) = \mathbf{P}\{N(t) = h\} = \mathbf{P}\{S_{h+1} > t\} - \mathbf{P}\{S_h > t\} = \mathrm{e}^{-kt}(kt)^h/h!,$$

again yielding the Poisson distribution. Given that its prevision is kt, it turns out that k, introduced as an arbitrary constant, is actually μ: imagine the latter in place of k, therefore, in the preceding formulae.

Third method. This is perhaps the most intuitive approach, and the most useful in that it can be applied to any scheme involving passages through different 'states' with intensities μ_{ij}, constant or variable, where $\mu_{ij}\,\mathrm{d}t$ = the probability that from an initial state 'i' at time t one passes to state 'j' within an infinitesimal time $\mathrm{d}t$.

Let us denote by $p_h(t)$ the functions (assumed to be unknown) which express the probabilities of being in state h at time t (in the Poisson process, state h at time t corresponds to $N(t) = h$, $h = 0, 1, 2, \ldots$). In the general case, the probability of a passage from i to j in an infinitesimal time $\mathrm{d}t$ is given by $p_i(t)\mu_{ij}\,\mathrm{d}t$ (the probability of two passages, from i to some h, and then from h to j, within time $\mathrm{d}t$, is negligible, since it is an infinitesimal of the second order). The change in $p_h(t)$ is given by $\mathrm{d}p_h = \dot{p}_h\,\mathrm{d}t$ and consists of the positive contribution of all the incoming terms (from all the other i to h), and the negative contribution of the outgoing terms (from h to all the other j). One has, therefore, in the general case (which has been mentioned merely to provide a proper setting for the case of special interest to us), a system of differential equations (which requires the addition of suitable initial conditions).

Our case is much simpler, however: we have only one probability, that of passing from an h to the next $h + 1$, the intensity remaining constant throughout. For $h = 0$, we have only the outgoing term, $-\mu p_0(t)\,\mathrm{d}t$, whereas for $h > 0$, we have, in addition to $-\mu p_h(t)\,\mathrm{d}t$, the incoming term $\mu p_{h-1}(t)\,\mathrm{d}t$. This reduces to the (recursive) system of equations

$$(4) \qquad p_0'(t) = -\mu p_0(t), \qquad p_h'(t) = \mu p_{h-1}(t) - \mu p_h(t),$$

together with the initial conditions, $p_0(0) = 1$, $p_h(0) = 0$ $(h \neq 0)$.

From the first equation, we obtain immediately that $p_0(t) = \mathrm{e}^{-\mu t}$, and hence, from the second equation, we obtain

$$p_1(t) = \mu t\,\mathrm{e}^{-\mu t},$$

and so on. If (realizing from the first terms that it is convenient to extract the factor $\mathrm{e}^{-\mu t}$) we set $p_h(t) = \mathrm{e}^{-\mu t} g_h(t)$ (with $g_0(t) = 1$, and $g_h(0) = 0$ for $h \neq 0$), we can virtually eliminate any need for calculations: the recursive relation for the $g_h(t)$ reduces to the extremely simple form $g_h'(t) = \mu g_{h-1}(t)$, so that $g_h(t) = (\mu t)^h/h!$.

8.2.6. As an alternative to the method used in Chapter 6, 6.11.2, the characteristic function of the Poisson distribution can be obtained by a direct calculation:

$$(5) \qquad \phi^t(u) = \sum_h \mathrm{e}^{-\mu t}(\mu t)^h\,\mathrm{e}^{\mathrm{i}uh}/h! = \exp\{\mu t(\mathrm{e}^{\mathrm{i}u} - 1)\}.$$

So far as the random process is concerned, it is very instructive and meaningful to observe that, as $t \to 0$, we have, asymptotically,

$$\phi^t(u) = 1 + \mu t(\mathrm{e}^{\mathrm{i}u} - 1) = (1 - \mu t) + \mu t\,\mathrm{e}^{\mathrm{i}u}$$

(probability $1 - \mu t$ of 0, and μt of 1). This is the 'infinitesimal transformation' from which the process derives. The simplest way of seeing this is, perhaps, to observe that

$$\phi^t(u) = \lim_{n\to\infty}\left[1 + \frac{1}{n}\mu t(\mathrm{e}^{\mathrm{i}u} - 1)\right]^n.$$

The Poisson process also tends to a normal form (as is obvious, given that it has finite variance); in other words, asymptotically it approximates the Wiener–Lévy process. However, the prevision is no longer zero, but equals μt (as does the variance). In order to obtain zero prevision, i.e. in order to have a finite process, it is necessary to subtract off a linear term and to consider a new process consisting of $N(t) - \mu t$: instead of the number of jumps, one considers the difference between this number and its prevision. The behaviour of

$$Y(t) = N(t) - \mu t$$

gives rise to the saw-tooth appearance of Figure 8.4:† all the jumps are equal to $+1$, and the segments in between have slope $-\mu$. With the introduction of the correction term $-\mu t$, the characteristic function is multiplied by $e^{-i\mu t u}$, and we obtain

$$\exp\{\mu t(e^{iu} - 1 - iu)\}. \tag{6}$$

This was obvious: when we take the logarithm of the characteristic function, the linear term in u must vanish, and we have

$$e^{iu} - 1 - iu = -\tfrac{1}{2}u^2[1 + \varepsilon(u)] \qquad (\varepsilon(u) \to 0 \quad \text{as } u \to 0).$$

Replacing u by $u/\sqrt{(\mu t)}$, in order to obtain the standardized distribution, we obtain

$$\exp\{\mu t[-\tfrac{1}{2}(u/\sqrt{(\mu t)})^2][1 + \varepsilon(u/\sqrt{(\mu t)})]\} \to e^{-\frac{1}{2}u^2}.$$

This provides, if required, a fifth approach to approximating the Wiener–Lévy process. Simple though it is, we thought it worth spending some time on this example, because the idea of adjusting (in the mean) the jumps by a certain linear term, i.e. of considering jumps with respect to an inclined line rather than a horizontal one, turns out, in a number of cases, to be necessary in order to ensure the convergence of certain procedures (and we shall see examples of this in Sections 8.2.7 and 8.2.9).

8.2.7. The compound Poisson process can be developed along the same lines as were followed in the case of the (simple) Poisson process. The analysis of the most general cases can then be attempted, recognizing that these are, in fact, the generalized Poisson processes.

In the case of a compound Poisson process—with intensity μ, and each jump X having distribution function $F(x)$ and characteristic function $\chi(u) = \mathbf{P}(e^{iuX})$—the characteristic function $\phi^t(u)$ is obtained in exactly the same way as for the simple Poisson process: i.e. by substituting $\chi(u)$ in place of e^{iu} (the latter being the $\chi(u)$ of the simple case, where $X = 1$ with certainty; i.e. $F(x)$ consists of a single mass concentrated at the point $x = 1$).

This is immediate: one can either note that the 'infinitesimal transformation' is now $1 + \mu t[\chi(u) - 1] = (1 - \mu t) + \mu t\chi(u)$ (probability $1 - \mu t$ of 0, and probability μt distributed according to the distribution of a jump), from which it follows that

$$\phi^t(u) = \lim_{n\to\infty}\left\{1 + \frac{1}{n}\mu t[\chi(u) - 1]\right\}^n = \exp\{\mu t[\chi(u) - 1)\}, \tag{7}$$

† Figure 8.4 has been placed later in the text (see Section 8.4.3), in order to emphasize its connection with Figure 8.5.

or one can simply observe that, conditional on the number of jumps $N(t)$ being equal to h, the characteristic function is given by $\chi^h(u)$, and hence that $\phi^t(u)$ is a mixture of the latter, with weights equal to the probabilities of the individual h. In other words,

$$\text{(8)} \qquad \phi^t(u) = \sum_{h=0}^{\infty} [e^{-\mu t}(\mu t)^h/h!]\chi^h(u) = e^{-\mu t} \sum_{h=0}^{\infty} [\mu t \chi(u)]^h/h!,$$

which is the series expansion of the form given in (7). Here too, of course, we are merely rewriting (5) with $\chi(u)$ substituted in place of e^{iu}.

If one wishes to give an expression in terms of the distribution of the jumps, $F(x)$, one can write the characteristic function in the form

$$\text{(9)} \quad \phi^t(u) = \exp\left\{\mu t\left[\left[\int e^{iux}\, dF(x) - 1\right]\right]\right\} = \exp\{t\textstyle\int (e^{iux} - 1)\mu\, dF(x)\},$$

or, alternatively,

$$\text{(10)} \qquad \phi^t(u) = \exp\left\{t \int (e^{iux} - 1)\, dM(x)\right\},$$

where $M(x) = \mu F(x)$. As another alternative (in order to deal more satisfactorily with the arbitrariness of the additive constant), we could take

$$\text{(11)} \quad \begin{aligned} M(x) &= \mu F(x) && \text{for } x < 0, \\ M(x) &= \mu[F(x) - 1] = -\mu[1 - F(x)] && \text{for } x > 0, \end{aligned}$$

so that, to make the situation clear in words,

$M(x) =$ The intensity of jumps having the same sign as x, and greater then x in absolute value, taken with the opposite sign to that of x.

With this definition, $M(x'') - M(x')$ is always the intensity of jumps whose magnitude lies between x' and x'', provided they have the same sign and $x'' > x'$. For $x = 0$, $M(x)$ has a jump $M(+0) - M(-0) = -\mu$, because $M(-0)$ is precisely the intensity of negative jumps, and $M(+0)$ is (with a minus sign) the intensity of positive jumps. The intensity of jumps between some $x' < 0$ and $x'' > 0$ is given by $M(x'') - M(x') + \mu$ (but, usually, one needs to consider separately jumps of opposite sign).

Remark. One can always assume (and we shall always do so, unless we state to the contrary) that there does not exist a probability concentrated at $x = 0$ (i.e. that one can speak of $F(0)$ without having to distinguish between '$+0$' and '-0', as we have tacitly done when stating that $M(+0) - M(-0) = -\mu[1 - F(0)] - \mu F(0) = -\mu$). In actual fact, so far as any effect on the process is concerned, a 'jump of magnitude $x = 0$' and 'no jump' are the same thing. Mathematically speaking, an increment of F (and hence of M) at $x = 0$ gives a contribution of zero to the integral in (9), since the integrand vanishes at that point. Sometimes, however, one may make the convention of including in $N(t)$ occurrences of a phenomenon 'able to give rise to a jump', even if the jump does not take place, or, so to speak, is zero. An example of this arises in the field of motor car insurance: if the process $Y(t)$ of interest is the total compensation per accident occurring before time t, it is quite natural (as well as more convenient and meaningful) to count up all accidents, or, to be technical, all claims arising from accidents, without picking out, and excluding, the occasional case for which the compensation was zero. The same principle applies if the process of interest concerns the number of dead, or injured, or those suffering damage to property, and so on.

Formally, in this case one would merely replace μ (the intensity of the jumps) by $\mu + \mu_0$ (where μ_0 is the intensity of the phenomenon with 'zero jumps'), including in $M(x)$ a jump μ_0 at $x = 0$, and consequently altering $F(x)$ and the characteristic function $\chi(u)$, which would be replaced by a mixture of $\chi(u)$ and 1, with weights μ and μ_0. This would be irrelevant, as it must be, since the product $\mu[\chi(u) - 1]$ remains unchanged, and this is all that really matters.

Recall (from Chapter 6, 6.11.6, (69)) that, in order to obtain expressions in normal form ($\mu_0 = 0$), it is necessary and sufficient that we have $(1/2a)\int_{-a}^{a}\chi(u)\,du \to 0$ (as $a \to \infty$). Were the limit to equal $c \neq 0$ (necessarily >0), it would suffice to remove $\chi(u)$ and replace it by $[\chi(u) - c]/(1 - c)$.

In the case of a compound process, consisting of a finite number of simple processes (like that considered in Section 8.1.3, the notation of which we continue to use), we have the following:

the $dM(x)$ are the masses (intensities) μ_k concentrated at the values x_k;

$M(x)$ is the sum of the μ_k corresponding to the x_k lying between x and $+\infty$ if $x > 0$, and to those lying between $-\infty$ and x if $x < 0$ (in this case, with the sign changed);

$F(x)$ is the same sum, but running always from $-\infty$ to x, and normalized (i.e. divided by $\mu = \mu_1 + \mu_2 + \ldots + \mu_n$);

the characteristic function for the jumps is given by $\chi(u) = \sum_k e^{iux_k}\mu_k/\mu$, that of the process by

$$\phi^t(u) = \exp\left\{t\sum_k \mu_k(e^{iux_k} - 1)\right\}, \tag{12}$$

which, as is obvious, can also be obtained as the product of the characteristic functions of the superimposed simple processes; i.e. of the $\exp\{t\mu_k(e^{iux_k} - 1)\}$.

Equation (10) has the same interpretation in the case of an arbitrary compound Poisson process: it reveals it to be a mixture of simple processes, but no longer necessarily a finite mixture.

Finally, we observe that the prevision $\mathbf{P}[Y(t)]$, and the variance $\boldsymbol{\sigma}^2[Y(t)]$, both exist and are determined by the distribution (both in the strict sense, and in terms of $\hat{\mathbf{P}}$; cf. Chapter 6, 6.5.7), so long as the same holds true for the jumps; i.e. if $\mathbf{P}(X)$ and $\boldsymbol{\sigma}^2(X)$, respectively, exist (in the same sense). In this case, we have

$$\mathbf{P}[Y(t)] = \mu t\mathbf{P}(X), \qquad \boldsymbol{\sigma}^2[Y(t)] = \mu t\boldsymbol{\sigma}^2(X)$$

(where $\mathbf{P}$ can be replaced by $\hat{\mathbf{P}}$, provided we do so on both sides).

If the prevision makes sense, it also makes sense to consider the process minus the prevision; in other words, modified by subtracting the certain linear function $\mu t\mathbf{P}(X)$, so that we obtain a process with zero prevision (i.e. a finite process). In other words, one considers $Y(t) - \mu t\mathbf{P}(X)$, which is the amount by which $Y(t)$ exceeds the prevision, as in the simple case of (6). The characteristic function is also similar to that of the latter case, and has the form

$$\phi^t(u) = \exp\left\{\mu t \int (e^{iux} - 1 - iux)\,dF(x)\right\} = \exp\left\{t\int (e^{iux} - 1 - iux)\,dM(x)\right\}. \tag{13}$$

8.2.8. We are now in a position to characterize the most general form of homogeneous process with independent increments; that is to say, the most general infinitely divisible distribution (cf. Chapter 6, 6.11.3 and 6.12). In fact, in either formulation it is a question of characterizing the characteristic functions $\phi(u)$ for which $[\phi(u)]^t$ turns out to be a characteristic function for any arbitrary $t > 0$.†

We have already encountered an enormous class of infinitely divisible characteristic functions; those of the form $\phi(u) = \exp\{a[\chi(u) - 1]\}$, where $\chi(u)$ is a characteristic function (the $\phi^t(u)$ of the compound Poisson processes). A function which is a limit of characteristic functions of this kind (in the sense of uniform convergence in any bounded interval) is again an infinitely divisible characteristic function. To see this, note that the limit of characteristic functions is again a characteristic function, and if some sequence of $\phi_n(u)$, such that $\phi_n \to \phi$, are infinitely divisible, then ϕ_n^t is a characteristic function, $\phi_n^t \to \phi^t$, and hence ϕ^t is a characteristic function (for every $t > 0$); in other words, ϕ is infinitely divisible. Conversely, it can be shown that an infinitely divisible characteristic function is necessarily either of the 'compound Poisson' form, or is a limit case; i.e. *the set of infinitely divisible characteristic functions coincides with the closure of the set of characteristic functions of the form* $\phi(u) = \exp\{a[\chi(u) - 1]\}$, *where* $\chi(u)$ *is a characteristic function.*

In order to prove this, it is sufficient to observe that if $[\phi(u)]^t$ is a characteristic function for every t, then $\phi_n(u) = \exp\{n[[\phi(u)]^{1/n} - 1]\}$ is also a characteristic function of the compound Poisson type, and tends to $\phi(u)$ (in fact, we are dealing with the well-known elementary limit $n(x^{1/n} - 1) \to \log x$).

† For any $t < 0$, this is impossible (except in the degenerate case $\phi(u) = e^{iua}$, in which $|\phi(u)| = 1$; in other cases, for some u we have $|\phi(u)| < 1$, and then, for $t < 0$, we would have $|\phi(u)|^t > 1$). Moreover, it is sufficient to verify the condition for the sequence $t = 1/n$ (or any other sequence tending to zero), rather than for all t. In fact, it holds for all multiples, and hence for an everywhere dense set of values; by the continuity property of Chapter 6, 6.10.3, it therefore holds for all $t > 0$.

The process $\phi^t(u)$ is thus approximated by means of the processes $\phi_n^t(u)$ having, perhaps only apparently (cf. the *Remark* in Section 8.2.7), intensities $\mu_n = n$, and jump distributions $\chi_n(u) = [\phi(u)]^{1/n}$. More precisely, this 'apparently' holds in the cases we have already considered (the compound Poisson processes), and *actually* holds in the new limit cases which we are trying to characterize. In fact, in the compound Poisson cases, with finite intensity μ, the probability

$$p_t(0) = \mathbf{P}[Y(t) = 0] = \lim(1/2a)\int_{-a}^{a} \phi^t(u)\,\mathrm{d}u \qquad (\text{as } a \to \infty)$$

(the mass concentrated at 0 in the distribution having characteristic function $\phi^t(u)$) is $\geqslant \mathrm{e}^{-\mu t}$ (which is the probability of no jump occurring before time t).† In this case, all the $\chi_n(u) = \phi^{1/n}(u)$ contain a constant term at least equal to $\mathrm{e}^{-\mu/n}$ (corresponding to the mass at 0), and the actual intensity, instead of being $\mu_n = n$, is at most $n(1 - \mathrm{e}^{-\mu/n}) \sim \mu$ (and it is easily verified that it actually tends to μ, as we might have guessed).

The new cases arise, therefore, when $Y(t) = 0$ has zero probability for every $t > 0$, no matter how small; i.e. when there is zero probability of $Y(t)$ remaining unchanged during any finite time interval, however small. We must have either a continuous variation, or a variation whose jumps are everywhere dense; i.e. with infinite intensity. In the approximation we have considered, the μ_n will all actually be equal to n.

These remarks and the treatment to follow are rather informal. We shall subsequently often have occasion to dwell somewhat more closely on certain critical aspects of the problem, but for the more rigorous mathematical developments we shall refer the reader to other works (for example, Feller, Vol. II, Chapter XVII, Section 2).

8.2.9. As a first step in getting to grips with the general case, let us begin by extending to this case the considerations concerning the distribution of the intensity of the jumps, $M(x)$, as defined in Section 8.2.7 for the compound Poisson process. A definition which (making the previous considerations more precise) would be equivalent, and which is also directly applicable to the general case, is the following: $M(x)$ (taken with a plus or minus sign opposite to that of x) is the prevision of the number of increments having the same sign as x, and greater than x in absolute value, that occur in a unit

† We have $p_t(0) > \mathrm{e}^{-\mu t}$ if and only if there are jump values having non-zero probabilities (concentrated masses in the distribution whose characteristic function is $\chi(u)$) and some sum of them is 0. For example, in the case of Heads and Tails, values ± 1, we have $1 + (-1) = 0$ (i.e. we can return to 0 after two jumps). If, in this same example, the gains had been fixed at $+2$ and -3, then it would be possible to return to 0 after 5 jumps $(2 + 2 + 2 - 3 - 3 = 0)$; etc. In general, $p_t(0) = \sum_{h=2}^{\infty} \mathbf{P}[N(t) = h]\mathbf{P}(X_1 + X_2 + \ldots + X_h = 0)$, where $\mathbf{P}[N(t) = h) = \mathrm{e}^{-\mu t}(\mu t)^h/h!$, and $\mathbf{P}(X_1 + X_2 + \ldots + X_h = 0)$ is the mass concentrated at 0 in the distribution whose characteristic function is $[\chi(u)]^h$.

time interval (subdivided into a large number of very small time intervals). More simply, and more concretely, we shall restrict ourselves to considering the subdivision into n small subintervals, each of length $1/n$, subsequently passing to the limit as $n \to \infty$.

The increment of $Y(t)$ in any one of these subintervals,

$$Y(t + 1/n) - Y(t),$$

has distribution function $F^{1/n}(y)$ (cf. Section 8.1.3). The probability that it is greater than some positive x is $1 - F^{1/n}(x)$, and the prevision of the number of increments greater than x is $n[1 - F^{1/n}(x)]$, or, if one prefers, $[1 - F^t(x)]/t$. Similarly, for increments 'exceeding' some negative x (i.e. negative, and greater in absolute value), the probability and prevision are given by $F^t(x)$ and $F^t(x)/t$ $(t = 1/n)$. We define $M(x)$ as the limit (as $t = 1/n \to 0$) of $-[1 - F^t(x)]/t$ for positive x, and of $F^t(x)/t$ for negative x. Assuming (as is, in fact, the case) that these limits exist, we can say, to a first approximation, that, for $t \sim 0$, we have $F^t(x) = 1 + tM(x)$ (for $x > 0$) and $F^t(x) = tM(x)$ (for $x < 0$). In other words (in a unified form),

$$F^t(x) = F^0(x) + tM(x),$$

where $F^0(x)$ (the limit case for $t = 0$) represents the distribution concentrated at the origin ($F^0(x) = 0$ for $x < 0$, and $= 1$ for $x > 0$).

This agrees intuitively with the idea that $M(x)$ is the intensity of the jumps 'exceeding' x, and, in particular, in the compound Poisson case, with $M(x) = \mu[F(x) - F^0(x)]$. In the general case, the meaning is the same, except that $M(-0)$ and $M(+0)$ can become infinitely large (either $M(-0) = +\infty$, or $M(+0) = -\infty$, or both), as shown in Figure 8.2.

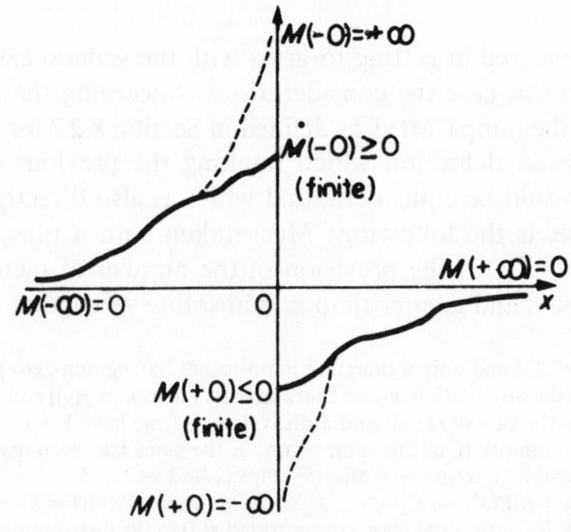

Figure 8.2 Distribution of the intensity of the jumps

The passage to the limit, which enables one to obtain the generalized Poisson processes, thus reduces to the construction of the $\phi^t(u)$ on the basis of formulae (10) and (13) of Section 8.2.7, allowing the function $M(x)$ to become infinite as $x \to \pm 0$, along with appropriate restrictions to ensure that the function converges, and that the process it represents makes sense (but we limit ourselves here to simply indicating how this can be done, and that it is, in fact, possible).

8.2.10. A new form, intermediate between the two previous forms in so far as it provides compensation only for the small jumps, proves more suitable as a basis for a unified account. This is defined by (14) below, and has to be constructed (within largely arbitrary limits) in such a way that it turns out to be equivalent to (13) in the neighbourhood of $x = \pm 0$, and to (10) in the neighbourhood of $x = \pm\infty$. We consider

(14) $$\phi^t(u) = \exp\left\{t \int [e^{iux} - 1 - iux \,.\, \tau(x)]\, dM(x)\right\},$$

where $\tau(x)$ is an arbitrary bounded function, tending to 1 as $x \to 0$, and to 0 as $x \to \pm\infty$.

Possible choices are:

$$\tau(x) = (|x| < 1) \quad \text{(i.e. 1 in } [-1, 1]\text{, and 0 elsewhere; P. Lévy)},$$

or

$$\tau(x) = 1/(1 + x^2) \quad \text{(Khintchin)},$$

or

$$\tau(x) = \sin x/x \quad \text{(Feller).}\dagger$$

A necessary and sufficient condition for the expression (14) (with any $\tau(x)$ whatsoever) to make sense as the characteristic function of a random process—and, in this way, to provide all the infinitely divisible distributions, apart from the normal, which derives from it as a limit case—is that the contribution to the variance from the 'small jumps' be finite. In other words, we must have $\int x^2\, dM(x) < \infty$ (the integral being taken, for example, over $[-1, 1]$; the actual interval does not matter, provided it is finite, and contains the origin).‡ We note that, in the more regular case in which the intensity admits a density $M'(x)$, and in which it makes sense to speak of an 'order of

† Some authors prefer, in place of $dM(x)$ as the differential element in the integral, to adopt variants like $dK(x) = [x^2/(1 + x^2)]\, dM(x)$ (Khintchin), or $dH(x) = x^2\, dM(x)$ (Feller). These have distinct formal advantages, but do not seem to me to compensate for the loss of direct meaning (cf. Feller, Vol. II, p. 536 *et passim*, and P. Lévy (1965), p. 141).

‡ Cf. P. Lévy and Feller, *loc. cit.*

infinity' as it tends to 0 (from the left or right), the necessary and sufficient condition is that this order of infinity (from both sides) be <2: if

$$M'(x) \sim 1/x^\alpha \; (\alpha < 2),$$

things go through; if $M'(x) \sim 1/x^2$, they no longer do.

Having said this, it is now easy to make precise the circumstances under which, and the reasons why, the expression in (14) can be replaced by one or other of the two simpler forms given previously. These can be considered as special cases of (14), in which $\tau(x)$ (instead of satisfying the imposed conditions) is set equal to 0 for (10) (the term iux is omitted), and equal to 1 for (13) (the term iux is always present).

The term iux is innocuous (it merely produces an addition to $Y(t)$ of a certain linear function ct) so long as it is applied to jumps which are neither too large nor too small (e.g. for $\varepsilon < |x| < 1/\varepsilon$, with $\varepsilon > 0$ arbitrarily small). When applied in a neighbourhood of $x = 0$, it is either innocuous or *useful*; when applied in a neighbourhood of ∞ (e.g. for $1/|x| < \varepsilon$), it is either innocuous or *harmful*. It can be useful, and indeed *necessary*, when the *small jumps*, if not 'compensated', do not lead to convergence; this happens if $\int |x| \, dM(x)$ diverges in $[-1, 1]$ (or, equivalently, in any neighbourhood of 0). It can be harmful, because the 'compensation' of the '*big jumps*' destroys convergence if they give too large a contribution, and this happens precisely when the previous integral diverges over $|x| \geqslant 1$. Observe, also, that this may very well happen even in a compound Poisson process (μ finite); it is enough that the distribution of the jumps should fail to have a 'mean value' (as, for example, with the Cauchy distribution). This does not affect the process, but any attempt to pass to the limit for the 'compensated' jumps would destroy the convergence rather than assist it.

In conclusion: the condition $\int x^2 \, dM(x) < \infty$ over $[-1, 1]$ is necessary and sufficient, and expression (14) always holds if the condition is satisfied. Both of the simple forms (10) and (13) can be applied if $\int |x| \, dM(x) < \infty$ over $[-\infty, +\infty]$ (and we observe that this condition implies the general condition, and is, therefore, itself sufficient). If, instead, the integral diverges, it is necessary to distinguish whether this is due to contributions in the neighbourhood of the origin, or in the neighbourhood of $\pm\infty$; in the first case (10) is ruled out, and in the second (13) is ruled out (and both are if there is trouble both at the origin and at infinity).

8.3 THE WIENER–LÉVY PROCESS

8.3.1. We now turn to an examination of the continuous component of a homogeneous random process with independent increments, which we described briefly in Section 8.1.3: this is the Wiener–Lévy process. The points already made, together with some further observations, will suffice here to provide a preliminary understanding of the process, and will be all that is required for the discussion to follow.

Let us first make clear what is meant by calling the process 'continuous'. It amounts to saying that, for any preassigned $\varepsilon > 0$, if we consider the increments of $Y(t)$ in $0 \leqslant t \leqslant 1$, divided up into N equal intervals, the probability that even one of the increments is, in absolute value, greater than ε, i.e. the possibility that $|Y(t + 1/N) - Y(t)| > \varepsilon$ for some $t = h/N < 1$, tends to 0 as N increases. In short, we are dealing with 'what escapes the sieve laid down for the selection of the jumps'. There is no harm in thinking (on a superficial level) of this as being equivalent to the continuity of the function $Y(t)$. From a conceptual point of view, however, this would be a distortion of the situation, as can be seen from the few critical comments we have already made, and from the others we shall be presenting later, albeit rather concisely, in a more systematic form (cf., for example, Section 8.9.1; especially the final paragraph).

We shall usually deal with the *standardized* Wiener–Lévy process, having zero prevision, $m(t) = 0$, unit variance per unit time period, $\sigma^2(t) = t$, and initial condition $Y(0) = 0$.

In this case, the density $f^t(y)$ and the characteristic function $\phi^t(u)$ (both at time t) are given by

$$f^t(y) = Kt^{-\frac{1}{2}}\,\mathrm{e}^{-\frac{1}{2}y^2/t}\ (K = 1/\sqrt{(2\pi)}), \tag{15}$$

$$\phi^t(u) = \mathrm{e}^{-\frac{1}{2}tu^2}. \tag{16}$$

The general case ($Y(0) = y_0, m(t) = mt, \sigma^2(t) = t\sigma^2$) can be reduced to the standardized form by noting that it can be written as $a + mt + \sigma Y(t)$, where $Y(t)$ is in standardized form. If we wish to consider it explicitly in the general form, we have

$$f^t(y) = Kt^{-\frac{1}{2}}\,\mathrm{e}^{-\frac{1}{2}[(y-y_0-mt)/\sigma]^2/t} \qquad (K = 1/\sqrt{(2\pi)}\sigma),$$

$$\phi^t(u) = \mathrm{e}^{iuy_0}\,.\,\mathrm{e}^{t(ium-\frac{1}{2}u^2\sigma^2)}$$

(where, for greater clarity, the terms depending on the initial value, y_0, and those depending on the process, i.e. on t, are written separately).

8.3.2. The same approach holds good when we wish to examine the random process and its behaviour, rather than just isolated values assumed by it. In fact, the joint distribution of the values of $Y(t)$ at an arbitrary number of instants $t_1, t_2, \ldots, t_n$, is also a normal distribution, with density† given by

$$f(y_1, y_2, \ldots, y_n) = K\,\mathrm{e}^{-\frac{1}{2}Q(y_1, y_2, \ldots, y_n)},$$

† We assume that $m = 0$, $\sigma = 1$ (the standardized case); only trivial modifications are required for the general case.

where Q is a positive definite quadratic form determined by the covariances $\mathbf{P}[Y(t_i)Y(t_j)] = \sigma_i\sigma_j r_{ij}$ (if $i = j$, $r_{ii} = 1$, and covariance = variance = σ_i^2). We have already seen (in Section 8.1.6) that (if $t_i \leqslant t_j$) the covariance is t_i^2, and therefore $r_{ij} = \sqrt{(t_i/t_j)}$; this gives all the information required for any application.

It is simpler and more practical, however, to observe that all the $Y(t_i)$ can be expressed in terms of increments, $\Delta_i = Y(t_i) - Y(t_{i-1})$, which follow consecutively and are independent: $Y(t_i) = \Delta_1 + \Delta_2 + \ldots + \Delta_i$. But Δ_i has a centred normal distribution, with variance $(t_i - t_{i-1})$, and Q, as a function of the variables $(y_i - y_{i-1})$, is a sum of squares:

$$Q(y_1, y_2, \ldots, y_n) = \sum_i [(y_i - y_{i-1})^2/(t_i - t_{i-1})]. \tag{17}$$

We shall now make use of this to draw certain conclusions which we shall require in what follows.

What we have been considering so far, to be absolutely precise, is the Wiener–Lévy process on the half-line $t \geqslant 0$, given that $Y(0) = 0$; the case of $t \geqslant t_0$, given that $Y(t_0) = y_0$, is identical (and the same is true for $t \leqslant 0$, $t \leqslant t_0$, respectively).†

In order to consider the case in which several values are given (at $t = t_1, t_2, \ldots$), it is sufficient to consider the problem inside one of the (finite) intervals; on the unbounded intervals things proceed as above. Let us therefore consider the process over the interval (t_1, t_2), given the values $Y(t_1) = y_1$ and $Y(t_2) = y_2$ at the end-points. In order to characterize it completely, it will suffice, in this case also, to determine the prevision (no longer necessarily zero!) and the variance of $Y(t)$ for each $t_1 \leqslant t \leqslant t_2$, and the covariance (or correlation coefficient) between $Y(t')$ and $Y(t'')$ for each pair of instants $t_1 \leqslant t' \leqslant t'' \leqslant t_2$.

We now decompose $Y(t)$ into the sum of a certain linear component (a straight line through the two given points) and the deviation from it:

$$Y(t) = y_1 + [(t - t_1)/(t_2 - t_1)](y_2 - y_1) + Y_0(t),$$

where $Y_0(t)$ corresponds to the same problem with $y_1 = y_2 = 0$. We now consider the two components as if the end-points were not yet fixed, so that the increments $\Delta_1 = Y - Y_1$ and $\Delta_2 = Y_2 - Y$ are independent, with standard deviations $\sigma_1 = \sqrt{(t - t_1)}$ and $\sigma_2 = \sqrt{(t_2 - t)}$. The linear component

† Provided that the process is assumed to make sense, even in the past, and that, in fact, no knowledge of the past leads us to adopt different previsions (these assumptions are not, in general, very realistic).

is then the random quantity

$$[1/(t_2 - t_1)][(t_2 - t)Y_1 + (t - t_1)Y_2],$$

and, by subtraction, $Y_0(t)$ is given by

$$Y_0(t) = [1/(t_2 - t_1)][(t_2 - t)\Delta_1 - (t - t_1)\Delta_2].$$

It is easily seen that $Y_0(t)$ has zero prevision (as was obvious) and standard deviation given by

$$\boldsymbol{\sigma}(Y_0(t)) = \sqrt{[(t - t_1)(t_2 - t)/(t_2 - t_1)]}\,; \tag{18}$$

moreover, it is uncorrelated with the linear component

$$Y_1 + [(t - t_1)/(t_2 - t_1)](\Delta_1 + \Delta_2)$$

(and hence, by normality, they are independent). In fact, we have

$$\begin{aligned}\text{covariance} &= K[(t - t_1)(t_2 - t)\mathbf{P}(\Delta_1^2) - (t - t_1)^2\mathbf{P}(\Delta_2^2)]\\ &= K[(t_2 - t)\sigma_1^2 - (t - t_1)\sigma_2^2] = 0.\end{aligned}$$

It can be seen (as a check, and in order to realize the difference between this and $\sqrt{(t - t_1)}$, which would apply in the absence of the condition on the second end-point) that the standard deviation of the linear component (considering the value of the first end-point, Y_1, as fixed) is given by

$$[(t - t_1)/(t_2 - t_1)]\sqrt{[\mathbf{P}(\Delta_1 + \Delta_2)^2]} = \sqrt{(t - t_1)}\,.\sqrt{[(t - t_1)/(t_2 - t_1)]}.$$

Summing the squares of the standard deviations of the two summands, we obtain the square of $\sigma_1 = \sqrt{(t - t_1)}$ (as, indeed, we should). It is useful to note that, as is shown in Figure 8.3, the standard deviation (18) of $Y(t)$, given the values at t_1 and t_2, is represented by the (semi) circle resting on the segment (t_1, t_2) (provided the appropriate scale is used; i.e. taking the segment $t_2 - t_1$ on the t-axis equal to the unit of measure on the y-axis: on the other hand, this is really irrelevant, except as an aid to graphical representation and description). If we consider the parabolae which represent, in a similar fashion, $\sigma(t)$ given only Y_1 (i.e. $y = \sqrt{(t - t_1)}$), or given only Y_2 (i.e. $y = \sqrt{(t_2 - t)}$), we see that the product of these two functions is represented by the circle, which, therefore, touches the parabolae (at the end-points), because when one of the two factors vanishes, the other has value 1.

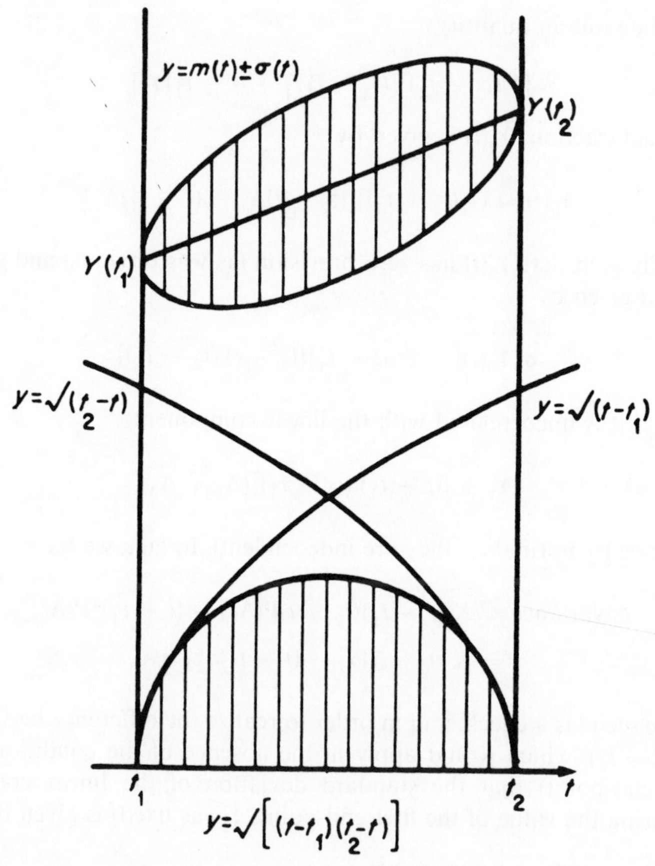

Figure 8·3 Interpolation between two known points in the Wiener–Lévy process (straight line and semi-ellipses: graphs of *prevision* and *prevision* ± *standard deviation*). The lower diagram represents the behaviour of the standard deviation given the point of origin, the final point, or both

We can determine, in a similar manner, the covariance between

$$Y(t') \text{ and } Y(t''), \qquad t_1 \leqslant t' \leqslant t'' \leqslant t_2.$$

We denote the successive, independent increments by

$$\Delta_1 = Y(t') - Y(t_1), \qquad \Delta_2 = Y(t'') - Y(t'), \qquad \Delta_3 = Y(t_2) - Y(t'');$$

$Y_0(t)$ will be the same as before, but writing $y_2 - y_1 = \Delta_1 + \Delta_2 + \Delta_3$, $T = t_2 - t_1$ and assuming, simply for notational convenience, that $t_1 = 0$

and $y_1 = 0$, we have

$$Y_0(t') = \Delta_1 \qquad - t'(\Delta_1 + \Delta_2 + \Delta_3) = (T - t')\Delta_1 \qquad - t'\Delta_2 - t'\Delta_3,$$

$$Y_0(t'') = \Delta_1 + \Delta_2 - t''(\Delta_1 + \Delta_2 + \Delta_3) = (T - t'')\Delta_1 + (T - t'')\Delta_2 - t''\Delta_3,$$

and hence

$$Y_0(t')Y_0(t'') = (T - t')(T - t'')\Delta_1^2 - t'(T - t'')\Delta_2^2 + t't''\Delta_3^2$$

$$+ \text{ cross-product terms (in } \Delta_i\Delta_j,\ i \neq j, \text{ with zero prevision).}$$

Taking prevision, and bearing in mind that the previsions of the Δ_i^2 are, respectively, $\sigma_1^2 = t'$, $\sigma_2^2 = t'' - t'$, $\sigma_3^2 = T - t''$, we have

$$\begin{aligned}\text{Covar}(t', t'') &= (T - t')(T - t'')t' - t'(T - t'')(t'' - t') + t't''(T - t'') \\ &= t'(T - t'')[(T - t') - (t'' - t') + t''] = t'(T - t'')T.\end{aligned}$$

Dividing by $\sigma' = \sqrt{[t'(T - t')/T]}$ and $\sigma'' = \sqrt{[t''(T - t'')/T]}$, we obtain, finally, the correlation coefficient

$$\text{(19)} \qquad r(t', t'') = \sqrt{\left[\frac{t'(T - t'')}{(T - t')t''}\right]} = \sqrt{\left[\frac{(t' - t_1)(t_2 - t'')}{(t_2 - t')(t'' - t_1)}\right]},$$

and, in this way, we return to our original notation.

8.3.3. It will certainly be no surprise to find that if we put $t_2 = \infty$ we reduce to the results obtained in the case of the single condition at t_1. This is, however, simply a special case of a remarkable fact first brought to light by P. Lévy, and which is often useful for inverting this conclusion by reducing general cases (with two fixed values) to the special limit case (with only one fixed value). The fact referred to is the *projective invariance* of problems concerning processes of this kind,† and derives from the expression under the square root in $r(t', t'')$ being the cross-ratio of the four instants involved. It therefore remains invariant under any homographic substitution for the time t, provided the substitution does not make the *finite* interval $[t_1, t_2)$ correspond to the complement of a finite interval (but instead, to either a finite interval, or to a half-line; in other words, the inequalities $t_1 \leqslant t' \leqslant t'' \leqslant t_2$ must either be all preserved, or all inverted). Consequently, the stochastic nature of the function $Y(t)$ remains invariant if we ignore multiplication by a certain arbitrary function; invariance holds for any random function of

† The meaning of this is most easily understood by introducing the projective coordinate

$$\tau = \tau(t) = (t - t')(t'' - t_2)/(t'' - t)(t_2 - t');$$

i.e. (as is obvious) taking t', t_2 and t'' to 0, 1 and ∞: $\tau' = \tau(t') = 0$, $\tau_2 = \tau(t_2) = 1$, $\tau'' = \tau(t'') = \infty$. It follows that $r = \sqrt{\tau_1}$, where $\tau_1 = \tau(t_1)$ is the abscissa of the point t_1 after the projective transformation has been performed.

the form $Z(t) = g(t)Y(t)$, or, in particular, for the standardized process (always having $\sigma = 1$) that can be obtained by taking

$$g(t) = 1/\boldsymbol{\sigma}[Y(t)] = 1/\sqrt{t}. \tag{20}$$

This device will prove useful for, among other things, reducing the study of the asymptotic behaviour of $Y(t)$ in the neighbourhood of the origin (for $0 < t < \varepsilon$, $\varepsilon \to 0$) to that at infinity (for $t > T$, $T \to \infty$); cf. Section 8.9.7.

The basic properties of the Wiener–Lévy process will be derived later, in contexts where they correspond to actual problems of interest.

8.4. STABLE DISTRIBUTIONS AND OTHER IMPORTANT EXAMPLES

8.4.1. We have already encountered (in Chapter 6, 6.11.3) two families of stable distributions: the normal (which, in Chapter 7, 7.6.7, we saw to be the only stable distribution having finite variance) and the Cauchy (which has infinite variance).

We are now in a position to determine all stable distributions. It is clear—and we shall see this shortly—that they must be infinitely divisible. Our study can therefore be restricted to a consideration of distributions having this latter property, and, since we know what the explicit form of their characteristic functions must be, this will not be a difficult task.

Knowledge of these new stable distributions will also prove useful for clarifying the various necessary conditions and other circumstances occurring in the study of the asymptotic behaviour of random processes.

We begin by observing that the convolution of two compound or generalized Poisson distributions, represented by the distributions of the intensities of the jumps, $M_1(x)$ and $M_2(x)$, is obtained by summing them: $M(x) = M_1(x) + M_2(x)$. In fact, $M(x)$ determines the logarithm of the characteristic function in a linear fashion, and the sum in this case corresponds to the product of the characteristic functions; i.e. to the convolution.

This makes clear, conversely, what the condition for an infinitely divisible distribution to be a factor in the decomposition of some other distribution (also infinitely divisible) must be. The distribution defined by $M_1(x)$ 'divides' the distribution defined by $M(x)$ if and only if the difference

$$M_2(x) = M(x) - M_1(x)$$

is also a distribution function of intensities. This implies that it must never decrease, so that every interval receives positive mass (or, at worst, zero mass), and this implies, simply and intuitively, that in any interval (of the positive or negative semi-axis) $M_1(x)$ must have an increment not exceeding that of $M(x)$. In particular, the concentrated masses and the density (if they exist) must, at every individual point, not exceed those of $M(x)$. If one wishes to include in the statement the case in which a normal component exists (and then we have the most general infinitely divisible distribution) it is sufficient to state that here, too, this component must be the smaller (as a measure, one can take the variance).†

In order to prove that stability implies infinite divisibility, it is sufficient to observe that, in the case of stability, the sum of n independent random quantities which are identically distributed has again the same distribution (up to a change of scale). It is itself, therefore, a convolution product of an arbitrary number, n, of identical factors, and is therefore infinitely divisible. If, for each factor, the distribution of the intensities of the jumps is $M(x)$, then for the convolution of n factors it is $nM(x)$.

For stability it is necessary and sufficient that the distribution defined by $nM(x)$ belongs to the same family as that defined by $M(x)$. This requires that it differs only by a (positive) scale factor $\lambda(n)$: i.e. $nM(x) = M(\lambda(n)x)$. It follows immediately that

$$knM(x) = kM(\lambda(n)x) = M(\lambda(k)\lambda(n)x) = M(\lambda(kn)x);$$

in other words, $\lambda(k)\lambda(n) = \lambda(kn)$ for k and n integer. The same relation also holds for rationals if we set $\lambda(1/n) = 1/\lambda(n)$, and hence $\lambda(m/n) = \lambda(m)/\lambda(n)$. By continuity, we then have $\lambda(v)$ for all positive reals v. The functional equation $\lambda(v_1)\lambda(v_2) = \lambda(v_1 v_2)$ characterizes powers, so we have an explicit expression for λ:

$$(21) \qquad \lambda(v) = v^{-1/\alpha}; \qquad \text{in particular, } \lambda(n) = n^{-1/\alpha}.$$

We have written the exponent in the form $-1/\alpha$, because it is the reciprocal, $-\alpha$, which appears as the exponent in the expression for $M(x)$, and which will be of more direct use in what follows. It is for this reason that α is known as the 'characteristic exponent' of the distribution for which we have

$$(22) \quad nM(x) = M(n^{-1/\alpha}x) \ \ \text{(and, in general, } vM(x) = M(v^{-1/\alpha}x), 0 < v < +\infty).$$

In fact, we can immediately obtain an explicit expression for $M(x)$. When $x = 1$, the above expression reduces to $vM(1) = M(v^{-1/\alpha})$, and when $x = v^{-1/\alpha}$, we have $M(x) = -Kx^{-\alpha}$, where $-K = M(1)$, a constant: this holds for all positive x (running from 0 to $+\infty$ as v varies in the opposite direction from $+\infty$ to 0). Setting $x = -1$ (instead of $+1$), we can obtain the same result for negative x, except that we must now write $|x|^{-\alpha}$ in place of $x^{-\alpha}$.

† Note that what we have said concerns divisibility within the class of infinitely divisible distributions. However, there may exist indivisible factors of infinitely divisible distributions (and, of course, conversely), as we mentioned already in Chapter 6, Section 6.12.

Allowing for the fact that the constant K could assume different values on the positive and negative semi-axes, we have, finally,

$$M(x) = -K^+|x|^{-\alpha}(x > 0) + K^-|x|^{-\alpha}(x < 0) = K^{\pm}|x|^{-\alpha}, \tag{23}$$

where K^+ and K^- are positive, and are therefore written preceded by the appropriate sign (this ensures that $M(x)$ is increasing, in line with what we said in Section 8.2.9). It is obviously unnecessary to write $|x|$ when x is positive, but it serves to stress the identical nature of the expressions over the two semi-axes: $K^{\pm}$ is merely a compact way of writing either $-K^+$ or $+K^-$ for $x \gtrless 0$ (it could be written in the form $K^{\pm} = K^-(x<0) - K^+(x>0)$).

It remains to examine which values are admissible for the characteristic exponent, We see immediately that these are the α for which $0 < \alpha \leqslant 2$, and it turns out that there are good reasons for considering these as four separate subcases;

$$0 < \alpha < 1, \qquad \alpha = 1, \qquad 1 < \alpha < 2, \qquad \alpha = 2.$$

The value $\alpha = 1$ is a somewhat special case, and corresponds to the Cauchy distribution (we have already met this in Chapter 6, 6.11.3; the correspondence is established by examining the characteristic function given in equation (59) of the section mentioned).

8.4.2. For $\alpha = 2$, we cannot proceed in the above manner—the expression for the characteristic function diverges—but we can consider it as a limit case (or we could include it by using the kind of procedures mentioned in the first footnote to Section 8.2.10). The limit case turns out to be the by now familiar normal distribution.

In fact, this corresponds to the characteristic exponent

$$\alpha = 2 \qquad (\text{or, } 1/\alpha = \tfrac{1}{2})$$

because the scale factor (in this case, the standard deviation) for the sum Y_n of n identically distributed summands is multiplied by $\sqrt{n}$, i.e. $n^{\frac{1}{2}}$ (and hence, for the mean Y_n/n, we multiply by $n^{-\frac{1}{2}}$).

More generally, even in the case of distributions from the same family but with different scale factors, the well-known formula for the standard deviation (for sums of independent quantities) holds for the normal distribution,

$$\sigma = (\sigma_1^2 + \sigma_2^2 + \ldots + \sigma_n^2)^{\frac{1}{2}}, \tag{24}$$

and also for all stable distributions if adapted to the appropriate characteristic exponent α. Explicitly, if $X_1, X_2, \ldots, X_n$ have distributions all belonging to the same family (stable, with characteristic exponent α), and $a_1, a_2, \ldots, a_n$ are the respective scale factors, then the random quantity defined by the sum

$aX = a_1X_1 + a_2X_2 + \ldots + a_nX_n$ again has a distribution belonging to this family, with scale factor given by

$$(25) \qquad a = (a_1^\alpha + a_2^\alpha + \ldots + a_n^\alpha)^{1/\alpha} \qquad (\text{i.e. } a^\alpha = \sum a_i^\alpha).$$

In particular, if all the a_i are equal to 1,

$$(26) \qquad a = n^{1/\alpha}.$$

This is an immediate consequence of the expression for $M(x)$ given in (23); setting $v = v_1 + v_2$ in $vM(x) = M(v^{-1/\alpha}x)$, we obtain

$$M(v^{-1/\alpha}x) = M(v_1^{-1/\alpha}x) + M(v_2^{-1/\alpha}x).$$

We began with the case $\alpha = 2$, not only because of its importance and familiarity, but also because it enables us to establish immediately that values $\alpha > 2$ are not admissible. This is not only because, *a fortiori*, the integral would diverge; there is another, elementary, or, at least, familiar, reason (which we shall just mention briefly). If the variance is finite, the formula for standard deviations holds when $\alpha = 2$; if it is infinite, we must have $\alpha \leqslant 2$, because $\alpha = 2$ holds for any bounded portion of the distribution (considering, for example, the *truncated* X_h; $-K \vee X_h \wedge K$).

In connection with the idea of *compensation* (e.g. for errors of measurement), there is, by virtue of the (magic?) properties of the arithmetic mean, a point of some conceptual and practical importance which is worth making. (Of course, we are dealing with a mathematical property which we would have had to mention anyway, in order to deal with an important aspect of the behaviour of the means Y_n/n of n summands, each of which follows a stable distribution with some exponent α).

The form given in (26) asserts that, compared with the individual summands, Y_n has scale factor $n^{1/\alpha}$; it follows therefore that, for the arithmetic mean Y_n/n, the scale factor is $n^{(1/\alpha)-1}$.

Taking, for example,	$\alpha =$	2	3/2	4/3	1	$\frac{3}{4}$	$\frac{1}{2}$	$\frac{1}{4}$
the factor is	$=$	$n^{-\frac{1}{2}}$	$n^{-\frac{1}{3}}$	$n^{-\frac{1}{4}}$	1	$n^{\frac{1}{3}}$	n	n^3
giving, for example, for	$n = 2$	0·707	0·793	0·841	1	1·189	2	8
	$n = 10$	0·316	0·464	0·681	1	1·779	10	1000

The usual amount of 'compensation' (i.e. an increase in precision in the ratio $1:\sqrt{n}$ for the mean of n values) is only attained for $\alpha = 2$; the increase diminishes as α moves from 2 to 1, and for $\alpha = 1$ (the Cauchy distribution) the precision is unaltered, implying no advantage (or disadvantage) in using a mean based on several values rather than just using a single value; for $\alpha < 1$, the situation is reversed, and worsens very rapidly as we approach zero (a limit value which must be excluded since it would give ∞!). The values given above should be sufficient to provide a concrete numerical feeling for the situation.

One should not conclude, however, that in these latter cases there is no advantage in having more information (this would be a narrow, short-sighted mistake, resulting from the assumption that the only way to utilize repeated observations is by forming their mean). There is always an advantage in having more observations (there is more information!), but, to make the most of it, it is necessary to pose the problem properly, in a form corresponding to the actual circumstances. This kind of problem of mathematical statistics is dealt with by the Bayesian formulation given in Chapters 11 and 12.

8.4.3. For $\alpha < 2$, we have, in fact, a generalized Poisson distribution, with $M(x)$ corresponding to the distribution function of the intensities of the jumps. Moreover, since $M(x) = K^+ . |x|^{-\alpha}$, the density exists and is given by

(27) $$M'(x) = \alpha|K^\pm||x|^{-(\alpha+1)}.$$

It is simpler and clearer (apart from an exceptional case which arises for $\alpha = 1$) to consider separately the distribution generated by the positive jumps (the other is symmetric). We then have, taking $K^+ = -1/\alpha$ in order to obtain the simplest form of the density,

(27) $$M(x) = -(1/\alpha)x^{-\alpha}, \qquad M'(x) = x^{-(\alpha+1)}, \qquad \mathrm{d}M(x) = \mathrm{d}x/x^{\alpha+1}.$$

It is here that we meet the circumstance which forces us to distinguish between the two cases of α less than or greater than 1 (the case $\alpha = 1$ will appear later on). The fact is that in the first case we have convergence without requiring the correction term iux in (13), whereas in the second case this term is required. The reason for this is (roughly speaking) that $\mathrm{e}^{iux} - 1 \sim iux$ is an infinitesimal of the first order in x for $x \sim 0$; if multiplied by $\mathrm{d}M(x) = \mathrm{d}x/x^{\alpha+1}$, it gives $\mathrm{d}x/x^{\alpha}$, and the integral converges or diverges as $x \to 0$, according as $\alpha < 1$ or $\alpha > 1$. This is not merely a question of analysis, however; there is a point of substance involved here. For $\alpha < 1$, the generalized Poisson random process produced by the positive jumps only, with distribution of intensities $M(x) = Kx^{-\alpha}$ (which is therefore always increasing), makes sense. For $\alpha > 1$, however, only the *compensated sum* of the jumps makes sense. With reference to Figures 8.4 and 8.5,† we can give some idea of this behaviour by saying that (as more and more of the numerous very small jumps are added) the sum of the jumps (per unit time) becomes infinite, but the sloping straight line from which we start also becomes infinitely inclined downwards. The process, under these conditions, cannot have monotonic behaviour (in any time interval; no matter how short).

---

† Figure 8.4 shows a simple Poisson process with its prevision subtracted off (cf. Section 8.2.6). Figure 8.5 shows what happens in compound processes obtained by superposing, successively, on top of the previous one, simple processes having, in each case, a smaller jump. If one imagines, following on from the three shown, a fourth step, a fifth step, . . . , and so on, with the slope of the straight line of the prevision increasing indefinitely, one gets some idea of the generalized Poisson processes with only positive jumps.

Note that, in order not to make the diagram too complicated, it has been drawn as if all the additional processes vanish at the instants of the preceding jumps (this is very unlikely, but does not alter the accuracy of the visual impression; we merely wish to warn those who realize that this device has been adopted not to imagine that it reflects some actual property of the processes in question).

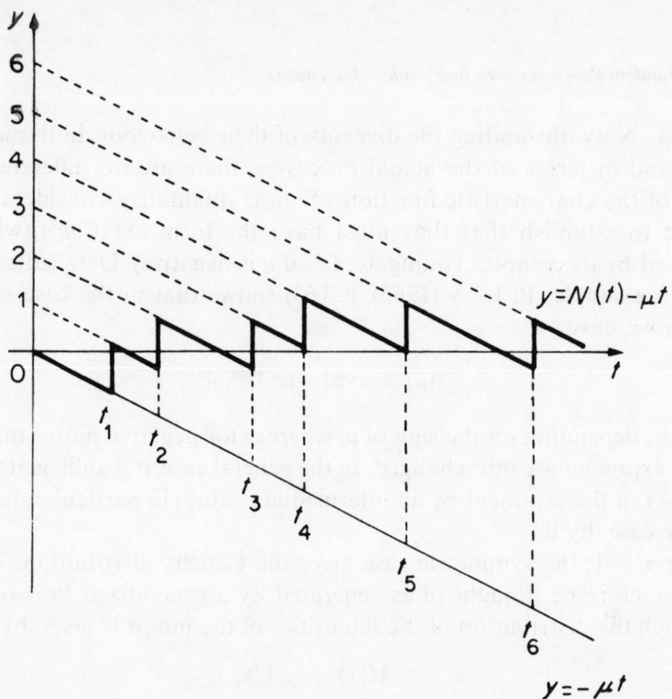

Figure 8.4 The simple, compensated Poisson process (prevision = 0)

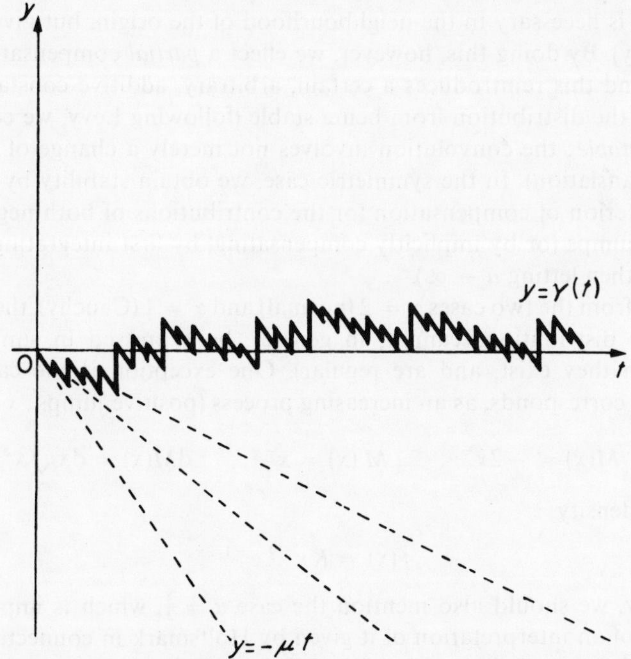

Figure 8.5 The compound Poisson process with successive sums from simple, compensated Poisson processes

8.4.4. Notwithstanding the diversity of their behaviour, both mathematically and in terms of the actual processes, there are no differences in the form of the characteristic functions. Simple qualitative considerations will suffice to establish that they must have the form $\exp(C|u|^\alpha)$ (where C is replaced by its complex conjugate, C^*, if u is negative). Detailed calculation (cf., for example, P. Lévy (1965), p. 163) shows that in the case of positive jumps we have

$$\phi(u) = \exp\{-\mathrm{e}^{\pm\frac{1}{2}i\pi\alpha}|u|^\alpha\}, \tag{28}$$

with $\pm$, depending on the sign of u, whereas for negative jumps the $\mp$ signs in the exponent are interchanged. In the general case, it is sufficient to replace α or $-\alpha$ in the exponent by an intermediate value; in particular, in the symmetric case, by 0.

For $\alpha = 1$, the symmetric case gives the Cauchy distribution. The latter can therefore be thought of as generated by a generalized Poisson process in which the distribution of the intensities of the jumps is given by

$$M(x) = \pm 1/x,$$

with density $M'(x) = 1/x^2$. In this case, however, we can only ensure convergence by having recourse to the form given in (14) (this is because the term iux is necessary in the neighbourhood of the origin, but gives trouble at infinity). By doing this, however, we effect a *partial* compensation in the jumps, and this reintroduces a certain, arbitrary, additive constant, which prevents the distribution from being stable (following Lévy, we could term it *quasi-stable*; the convolution involves not merely a change of scale, but also a translation). In the symmetric case, we obtain stability by using the same criterion of compensation for the contributions of both negative and positive jumps (or by implicitly compensating, by first integrating between $\pm a$ and then letting $a \to \infty$).

Apart from the two cases $\alpha = 2$ (normal) and $\alpha = 1$ (Cauchy), the densities of stable distributions cannot, in general, be expressed in simple forms (although they exist, and are regular). One exception is the case $\alpha = \frac{1}{2}$, to which corresponds, as an increasing process (positive jumps; $x > 0$),

$$M(x) = -2x^{-\frac{1}{2}}, \qquad M'(x) = x^{-\frac{3}{2}}, \qquad \mathrm{d}M(x) = \mathrm{d}x/\sqrt{x^3},$$

and the density

$$f(x) = Kx^{-\frac{3}{2}}\mathrm{e}^{-1/2x}. \tag{29}$$

Finally, we should also mention the case $\alpha = \frac{3}{2}$, which is important on account of an interpretation of it given by Holtsmark in connection with a problem in astronomy (and by virtue of the fact that it precedes knowledge of the problem on the part of mathematicians): the case $\alpha = \frac{4}{3}$ can be given an analogous interpretation (in a four-dimensional world). Cf. Feller, Vol. II, pp. 170, 215.

8.4.5. Other important examples of Poisson-type processes are the *gamma* processes (and those derived from them), and those of *Bessel* and *Pascal*.

The gamma distribution (see (55) and (56) in Chapter 6, 6.11.3) has density and characteristic function defined by

$$f^t(x) = Kx^{t-1}\,\mathrm{e}^{-x} \quad (x \geqslant 0), \qquad K = \frac{1}{\Gamma(t)}, \tag{30}$$

$$\phi^t(u) = 1/(1 - \mathrm{i}u)^t. \tag{31}$$

As $t \to 0$, $f^t(x)/t \to \mathrm{e}^{-x}/x$, and so the gamma process, in which $Y(t)$ has a gamma distribution with exponent t, derives from jumps whose intensities have the distribution

$$M(x) = \int_x^\infty (\mathrm{e}^{-x}/x)\,\mathrm{d}x, \qquad M'(x) = \mathrm{e}^{-x}/x. \tag{32}$$

In interpreting this, we note the connection with the Poisson process: $f^t(x)$, for $t = h$ integer, gives the distribution of the waiting-time, T_h, for the hth occurrence of the phenomenon. We can also obtain this by arguing in terms of the $p_h(t)$ (see Section 8.2.5); this makes it completely obvious, because for $h = 1$ we have the exponential distribution (for T_1, and, independently, for any waiting-time $T_h - T_{h-1}$), and for an arbitrary integer h we have the convolution corresponding to the sum T_h of the h individual waiting-times.

It is interesting to note that one possible interpretation of the gamma process is as the *inverse* of the (simple) Poisson process. To see this, we interchange the notation, writing this process as $t = T(y)$, the inverse of the other, for which we keep the standard notation, $y = Y(t)$. The inverse function $y = T^{-1}(t)$ (which, of course, does not give a process with independent increments), considered only at those points for which $y =$ integer (or taking $Y(t) =$ the integral part of $y = T^{-1}(t)$), gives precisely the simple Poisson process (with $\mu = 1$).

Remark. We obtain a perfect interpretation if we think, for example, of $y = T^{-1}(t)$ as representing the number of turns (or fractions thereof) made by a point moving in a series of jerks around the circumference of a circle. The standard Poisson situation then corresponds to that of someone who is only able to observe the point when it passes certain given marks.

We are in no way suggesting, however, that this mathematical possibility of considering an 'explanation' in terms of some 'hidden mechanism' provides any automatic justification for metaphysical flights of fancy leading on to assertions about its 'existence' (whereas it may, of course, be useful to explore the possibility if there are some concrete reasons for considering it plausible). There are a number of cases (but not, so far as I know, the present one) in which these kinds of metaphysical interpretation (or so they appear to me, anyway) are accepted, or, at any rate, seriously discussed.

More generally, changing the scale and intensity, we have

(33) $$f^t(x) = Kx^{\mu t - 1}\,\mathrm{e}^{-\lambda x} \quad (x > 0), \qquad K = \lambda^{\mu t}/\Gamma(\mu t),$$

(34) $$\phi^t(u) = (1 - \mathrm{i}u/\lambda)^{\mu t}.$$

Moreover, we can also reflect the distribution onto the negative axis; $f^t(x)$ is unchanged, apart from writing $|x|$ in place of x and changing the final term to $(x < 0)$, and the characteristic function is given by

(35) $$\phi^t(u) = (1 + \mathrm{i}u/\lambda)^{\mu t}.$$

By convolution, we can construct other processes corresponding to sums of gamma processes. The most important case is that obtained by symmetrization (cf. (57) and (59) in Chapter 6, 6.11.3), which, for $t = 1$, gives the double-exponential distribution. The general case is obtained from products of the form

$$\phi^t(u) = (1 - \mathrm{i}u/\lambda_1)^{\mu_1 t}(1 - \mathrm{i}u/\lambda_2)^{\mu_2 t} \ldots (1 - \mathrm{i}u/\lambda_n)^{\mu_n t}$$

(where the signs can be either $-$ or $+$; we could, alternatively, say that the λ_h can be positive or negative); the symmetric case arises when the products are taken in pairs with opposite signs: i.e.

$$\phi^t(u) = (1 + u^2/\lambda_1^2)^{\mu_1 t} \ldots (1 + u^2/\lambda_n^2)^{\mu_n t}.$$

8.4.6. The Bessel process acquires its name because the form of the density involves a Bessel function, $I_t(x)$ (for $x > 0$):

(36) $$f^t(x) = (\mathrm{e}^{-x}/x)tI_t(x), \qquad \text{where } I_t(x) = \sum_{k=0}^{\infty} \frac{1}{k!\Gamma(k + t + 1)}\left(\frac{x}{2}\right)^{2k+t};$$

the characteristic function is given by

(37) $$\phi^t(u) = \{1 - \mathrm{i}u - \sqrt{[(1 - \mathrm{i}u)^2 - 1]}\}^t.$$

The process deserves mentioning because it has the same interpretation as we put forward for the simple Poisson process, but referred instead to the Poisson variant of the Heads and Tails process (each occurrence of the phenomenon consists of a toss giving a gain of ± 1). Using the same notation, $t = T(y)$ and $y = T^{-1}(t)$ for the inverse function, we could say that the points at which $y =$ an integer correspond to the instants at which the gain $Y(t)$ first reaches level y (or that the 'integer part of $y = T^{-1}(t)$' is the maximum of $Y(t)$ in $[0, t]$, where $Y(t)$ is a Poisson Heads and Tails process).

8.4.7. The Pascal process is that for which $Y(t)$ has, for $t = 1$, a geometric distribution; for $t =$ integer, a Pascal distribution; and, for general t, a negative binomial distribution (cf. Chapter 7, 7.4.4). Our previous discussion reveals that we are dealing with a compound Poisson process having a *logarithmic* distribution of jumps (see Chapter 6, 6.11.2) and having the positive integers as the set of possible values (with intensity $\mu = 1$).

The interpretation is similar to that of the two previous cases, except that here we do not have a density, but, instead, we have concentrated masses (at the integer values). For $t =$ integer, $Y(t)$ is the number of failures (in a Bernoulli process with probability p, where $\tilde{p}$ is the factor in the geometric distribution corresponding to $t = 1$) occurring before the tth success. In any interval of unit length (from t to $t + 1$), the increment has the geometric distribution; here, it can be thought of as generated by logarithmically distributed jumps, one per interval 'on average'. A non-integer t could be interpreted as 'the number of successes already obtained, plus the *elapsed fraction* of the next one'; $Y(t)$ would then be the number of failures that had actually occurred so far.

Of course, the remark made in the context of the gamma process (Section 8.4.5) applies equally to the other two processes (Sections 8.4.6 and 8.4.7).

8.5 BEHAVIOUR AND ASYMPTOTIC BEHAVIOUR

8.5.1. We now turn to a probabilistic study of various aspects of the behaviour of the function $Y(t)$: as we pointed out in Section 8.1.6, these are, in fact, the most important questions. They might be concerned with the behaviour of $Y(t)$ in the neighbourhood of some given instant t (local properties), or in an interval $[t_1, t_2]$, or in the neighbourhood of $t = \infty$ (asymptotic properties). We might ask, for instance, whether $Y(t)$ vanishes somewhere in a given interval (and, if so, how many times), or if it remains bounded above by some given value M_1, or below by M_2 (or, more generally, by functions $M_1(t)$ and $M_2(t)$ rather than constants), and so on. Asymptotically, we might ask whether these or other circumstances will occur from some point on, or locally, or only in the neighbourhood of some particular instant.

Phrases such as we have used here will have to be interpreted with due care, especially if one is not admitting the assumption of countable additivity. Particular attention must then be paid to expressing things always in terms of a finite number of instants (which could be taken arbitrarily large) and not in terms of an infinite number.

We have already come across a typical example of just this kind of question in Chapter 7, 7.5.3. This was the strong law of large numbers, which, in the case of a discrete process, consisted in the study of the asymptotic validity of inequalities of the form $C_1 \leqslant Y(t)/t \leqslant C_2$, or, equivalently,

$$C_1 t \leqslant Y(t) \leqslant C_2 t.$$

We shall see now how this problem, and generalizations of it, together with a number of other problems of a similar kind, can be better formulated, studied and appreciated by setting them within the more general context of the study of random processes. In the most straightforward and intensively studied cases, the processes which will turn out to be useful are precisely those which we have been considering; specifically, the homogeneous processes with independent increments, and, from time to time the Wiener–Lévy process (in situations where it is valid as an asymptotic approximation).

It is both interesting and instructive to observe how the conjunction of the two different modes of presentation and approach serves to highlight conceptual links which are otherwise difficult to uncover, and to encourage the use of the most appropriate methods and techniques for each individual problem. In particular, later in this chapter (in which we shall only deal with the simplest cases) we shall see how the studies of the Heads and Tails process (developed directly using a combinatorial approach) and of the Wiener–Lévy process (which can be considered in a variety of ways) complement each other, enabling us in each case to use the more convenient form, or even to use both together.

8.5.2. We now consider two sets of questions, each set related to the other, and each collecting together, in a unified form, problems of different kinds which admit a variety of interpretations and applications, both theoretical and practical.

The first set brings together problems which reduce to the consideration of whether or not the path $y = Y(t)$ leaves some given region. In other words, whether or not it crosses some given barrier (the region may consist of a strip, $C_1 \leqslant y \leqslant C_2$, or $C_1(t) \leqslant y \leqslant C_2(t)$, or one of the bounds may not be present; i.e. $C_1 = -\infty$ or $C_2 = +\infty$). For this kind of problem, what we are usually interested in is knowing whether or not the process leaves the region, and, if so, when this occurs for the first time, and if the process then comes to a halt. In the latter case, we have a so-called absorbing barrier. In general, however, it is useful to argue as if the process were to continue.

An analysis of this kind will serve to make the strong law of large numbers more precise, by examining the rate of convergence that can be expected (we shall, of course, have to make clear what we mean by this!). From the point of view of applications, there will be a number of possible interpretations, among which we note: the gambler's ruin problem (which could be thought of in the context of an insurance company); the termination of a sequential decision-making process because sufficient information has been acquired; the end of a random walk—for example, the motion of a particle—due to arrival at an absorbing barrier, and so on.

The second set of problems are all concerned with recurrence, and involve the repetition of some given phenomenon (such as return to the origin, passing a check point, etc.). We shall see that in this case the process divides rather naturally into segments, and that it may be of interest to study various characteristics of these segments, which, in turn, often shed new light on the original recurrence problems.

In both cases, we shall first consider the Heads and Tails process and the Wiener–Lévy process, afterwards extending the study to the asymptotically normal cases. We shall also have occasion to consider other processes (the Bernoulli process with $p \neq \frac{1}{2}$, stable processes with $\alpha < 2$ and the Poisson process), mainly in order to have the opportunity of presenting further points of possible interest (explanations of unexpected behaviour, drawing attention to unexpected properties, and so on).

8.5.3. The three cases which we shall consider first involve the comparison of $|Y(t)|$ with $y = C, y = Ct, y = C\sqrt{t}$ $(c > 0)$. Even though we shall dwell, in each case, on possible interpretations in an applied context, developing each topic as required, it will be useful to bear in mind that these are in the way of preliminaries, enabling us subsequently to determine the foreseeable order of increase of $Y(t)$.

The first case, that of comparison with a constant, is the one of greatest practical interest, and corresponds to the gambler's ruin problem under the assumption of fixed capital (i.e. with no increases or decreases, other than those caused by the game). We shall now consider the complex of problems which arise in this case, beginning with the simplest.

We first observe that the probability of $Y(t)$ lying between $\pm C$ tends to zero at least as fast as $K/\sqrt{t}$ (if $Y(t)$ has finite variance σ^2 per unit time, it tends to zero as $K/\sqrt{t}$, with $K = 2C/[\sqrt{(2\pi)}\sigma]$; if the process has infinite variance, then, whatever K may be, it tends faster than $K/\sqrt{t}$). The same holds if we are dealing with just a single t. If $\sigma = 1$, we have, in a numerical form, $K = 0{\cdot}8C$ (cf. Chapter 7, 7.6.3), and, for ease of exposition, we shall, as a rule, refer to this case (the bound is $0{\cdot}8C/\sqrt{t}$, and is approximately attained if $\sigma = 1$; if $\sigma = \infty$, we have strict inequality). The same holds for every interval $[a - C, a + C]$ of length $2C$.

To see this, note that, for $t = n$, in the case of Heads and Tails the maximum probability is given by $u_n \simeq 0{\cdot}8/\sqrt{n}$ and there are $2C$ integers lying between $\pm C$ (give or take ± 1†). In the normal case (Wiener–Lévy), the maximum density is $1/[\sqrt{(2\pi)}\sigma(t)]$, and $\sigma(t) = \sqrt{t}$. In the general case, with $\sigma < \infty$ (and, without loss of generality, we assume $\sigma = 1$), we have, asymptotically, the same process.

† In order for this difference not to matter, it is, of course, necessary that C be much greater than 1. In general, in the case of lattice distributions, or distributions of similar kind, it is necessary that C be large enough for the concentrated masses to be regarded as a 'density'.

If $\sigma = \infty$, the bound corresponding to an arbitrary finite σ is *a fortiori* satisfied from a certain point onwards. Suppose $a > 0$ and arbitrary, but such that

$$\mathbf{P}(|X_h| < a) = p > 0,$$

further, let us distinguish the increments $X_h = Y(h) - Y(h-1)$ according as they are $<a$ or $>a$ in absolute value; $Y(n) = \sum X_h$ then contains about np summands which are $<a$ (*truncated* distribution, with $\sigma < a$ finite), and the sum is already practically normal with density $< 1/[\sqrt{(2\pi)}\sigma\sqrt{t}]$. Adding the sum of the other terms, we have, *a fortiori*, the same circumstance holding (theorem of the increase in dispersion; Chapter 6, 6.9.8).

What matters more than the quantitative result, is the qualitative conclusion: *however large C is chosen to be, the probability that* $|Y(t)|$ (*or* $|Y_n|$) *is greater than C differs from 1 by less than any given* $\varepsilon > 0$, *provided that t* (*respectively, n*) *is taken sufficiently large* (more precisely, from

$$t = n = (2/\pi)C^2/\varepsilon^2$$

onwards).

A fortiori, it follows that the probability of $|Y(\tau)| > C$ (or $Y_h > C$) for at least one $\tau \leqslant t$, or $h \leqslant n = t$, tends to 1 (more rapidly). In terms of the gambler's ruin problem, this implies that in a game composed of identical and independent trials between two gamblers each having finite initial capital, provided the game goes on long enough, the probability of it ending with the ruin of one of them tends to 1. Equivalently, the probability that the game does not come to an end is zero.

8.5.4. A warning against superstitious interpretations of the 'laws of large numbers'. *It is not only true that the absolute deviations, i.e. the gains and losses* (unlike the relative gains, the average gains per toss), *do not tend to offset one another, and, in fact, tend to increase indefinitely in quadratic prevision, but also that it is 'practically certain that, for large, they will be large'* (and it is only in the light of the above considerations that we are now able to see this).

One should be careful, however, not to exaggerate the significance of this statement, turning it, too, into something misleading or superstitious. It holds for each individual instant t (or number of tosses n), but not for a number of them simultaneously. In fact, it does not exclude the possibility (and, indeed, we shall see that this is practically certain) of the process *returning to equilibrium* (and hence of there being segments in which $|Y(t)| < C$) every now and again (and although this happens more and more rarely, it is a never-ending process).

8.6 RUIN PROBLEMS; THE PROBABILITY OF RUIN; THE PREVISION OF THE DURATION OF THE GAME

8.6.1. We shall use p_h and q_n to denote the probabilities of ruin at the hth trial, or before the nth, respectively

$$(q_n = p_1 + p_2 + \ldots + p_n, \qquad p_h = q_h - q_{h-1}).$$

In the case of two gamblers, G_1 and G_2, we shall use p_h' and p_h'', q_n' and q_n'', for the probabilities of ruin of G_1 and G_2, respectively

$$(p_h = p_h' + p_h'', \qquad q_n = q_n' + q_n''),$$

and c' and c'' for their respective initial fortunes.

By q' and q'' (or q_∞' and q_∞''), we shall denote the probabilities of ruin within an infinite time, to be interpreted as limits as $n \to \infty$: under the assumptions of Section 8.5.3, we have $q_n' + q_n'' \to 1$, and therefore $q' + q'' = 1$.

The probability of ruin, in a fair game, is an immediate consequence of the fairness condition: the previsions of the gains of the two gamblers must balance; i.e. $q'c' = q''c''$, from which we deduce that $q' = K/c'$, $q'' = K/c''$ $(K = c'c''/(c' + c''))$. In other words, the probabilities of ruin are inversely proportional to the initial fortunes. More explicitly,

$$q' = c''/(c' + c''), \qquad q'' = c'/(c' + c''). \tag{38}$$

Comments. In these respects, we might also apply the term 'fair game' to a non-homogeneous process with non-independent increments, provided that the prevision conditional on any past behaviour is zero (such processes are called *martingales*). One could think, for example, of the game of Heads and Tails with the stakes depending in some way on the preceding outcomes. With these assumptions, any mode of participation in the game is always fair: it does not matter if one interrupts the play in order to alter the stakes, or even if one decides to stop playing on account of a momentary impulse, or when something happens—such as someone's ruin.

The relation $q'c' = q''c''$ is an exact one if ruin is taken to mean the exact loss of the initial capital with no unpaid residue; in the latter case, it would be necessary to take this into account separately. If, for example, the jumps which are disadvantageous to G_1 and G_2 cannot exceed Δ' and Δ'', respectively, then c' must be replaced by some $c' + \theta'\Delta'$ $(0 \leqslant \theta' \leqslant 1)$, with a similar substitution for c''; the error is negligible if the probable residues are small compared with the initial capital.

The conclusion holds exactly for the Wiener–Lévy process because the continuity of $Y(t)$ ensures that it cannot exceed c' and c'' by jumping past them. The same holds true for the Heads and Tails process—including the Poisson variant—provided c' and c'' are integers (because it is then not possible, with steps of ± 1, to jump over them).

It is clear, particularly if we use the alternative form

$$q' = 1 - 1/[1 + (c''/c')],$$

that the probability of G_1's ruin tends to 1 if the opponent has a fortune which is always greater than his. If he plays against an opponent with infinite capital, the probability of ruin is therefore $q' = 1$ (and this is also true if one plays against the general public—who cannot be ruined). This is the *theorem of gambler's ruin* (for fair games).

The case of unfair games reduces to the previous case if one employs a device which goes back to De Moivre: in place of the process $Y(t)$, we consider $Z(t) = \exp[\lambda Y(t)]$. If λ is chosen in such a way as to make the prevision of $Z(t)$ constant $(=1$, say), then the process $Z(t)$ is fair, and ruin (starting from $Z(0) = 1$, which corresponds to $Y(0) = 0$) occurs when one goes down by $\bar{c}' = 1 - \exp(-\lambda c')$, or up by

$$\bar{c}'' = \exp(\lambda c'') - 1.$$

The probabilities of ruin are therefore inversely proportional to $\bar{c}'$ and $\bar{c}''$. It only remains to say how λ is determined. We observe that $\exp[\lambda Y(t)] = \phi^t(-i\lambda)$, and that, if it exists (cf. Chapter 6, 6.10.4), ϕ is real and concave on the imaginary axis, taking the value 1 (apart from at the origin) only at the point $u = -i\lambda$, with λ positive if the game is unfavourable ($\mathbf{P}[Y(t)] < 0$).

Example. We consider the case of Heads and Tails with an unfair coin ($p \neq \frac{1}{2}$), and with gains ± 1. We have $\exp[\lambda Y(1)] = p\mathrm{e}^{\lambda} + \tilde{p}\mathrm{e}^{-\lambda} = 1$, in other words (putting $x = \mathrm{e}^{\lambda}$), $px^2 - x + (1-p) = 0$ for $x = 1$ and $x = \tilde{p}/p$; $x = \mathrm{e}^{\lambda} = 1$ would yield $\lambda = 0$ (which is meaningless), and so we take $\mathrm{e}^{\lambda} = \tilde{p}/p$, $\mathrm{e}^{-\lambda c'} = (\tilde{p}/p)^{-c'}$, $\mathrm{e}^{\lambda c''} = (\tilde{p}/p)^{c''}$, from which we obtain

$$q' = \frac{(\tilde{p}/p)^{c''} - 1}{(\tilde{p}/p)^{c''} - (\tilde{p}/p)^{-c'}}, \qquad q'' = \frac{1 - (\tilde{p}/p)^{-c'}}{(\tilde{p}/p)^{c''} - (\tilde{p}/p)^{-c'}}. \tag{39}$$

If one plays against an infinitely rich opponent, the passage to the limit as $c'' \to \infty$ gives two different results, according as the game is favourable, $(\tilde{p}/p) < 1$, or unfavourable $(\tilde{p}/p) > 1$; in the latter case, $q' = 1$ (as was obvious *a fortiori*; ruin is practically certain in the fair case); if, instead, the game is favourable, the probability of ruin is $q' = (\tilde{p}/p)^{c'}$, and $1 - q' = 1 - (\tilde{p}/p)^{c'}$ is the probability that the game continues indefinitely.

8.6.2. The prevision $\mathbf{P}(T)$ of the duration T of the game until ruin occurs can also be determined in an elementary fashion for the game of Heads and Tails (even for the unfair case), and then carried over to the Wiener–Lévy process.

Instead of merely determining $\mathbf{P}(T)$ (starting from $Y(0) = 0$), it is convenient to determine the prevision of the future duration for any possible initial value y $(-c' \leqslant y \leqslant c'')$ using a recursive argument; we denote this general prevision by $\mathbf{P}_y(T)$. Obviously, we must have $\mathbf{P}_y(T) = 0$ at the end-points $(y = -c', y = c'')$ because there ruin has already occurred. For y in the interval between the end-points, we have, instead, the relation

$$\mathbf{P}_y(T) = 1 + \tfrac{1}{2}[\mathbf{P}_{y-1}(T) + \mathbf{P}_{y+1}(T)]$$

(because we can always make a first toss, and then the prevision of the remaining duration can be thought of as starting from either $y + 1$ or $y - 1$, each with probability $\frac{1}{2}$). We then obtain a parabolic form of behaviour (the second difference is constant!) with zeroes at the end-points; explicitly, we obtain

$$\mathbf{P}_y(T) = -(y + c')(y - c''), \qquad \text{and hence } \mathbf{P}(T) = \mathbf{P}_0(T) = c'c''. \tag{40}$$

As c'' increases, $\mathbf{P}(T)$ tends to ∞ no matter what $c' > 0$ is; it follows that $\mathbf{P}(T) = \infty$ for a game against an infinitely rich opponent, so that although ruin is practically certain ($q' = 1$), the expected time before it occurs is infinite.

Even in the case where c' and c'' are finite, the expected duration of the game, although finite, is much longer than one might at first imagine. For example, in the symmetric case, $c' = c'' = c$, the expected duration of the game is c^2 tosses: 100 tosses if each gambler starts with 10 lire; 40,000 tosses if each starts with 200 lire; 25 million tosses if each starts with 5000 lire. In the most extremely asymmetric case, $c' = 1$, $c'' = c$, the expected duration is c tosses; 1000 tosses if initially the fortunes are 1 lira versus 1000 lire; 1 million if initially we have 1 lira versus 1 million. One should note, however, that in this asymmetric case the gambler whose initial fortune is 1 lira always has the same (high) probabilities of coming to grief almost immediately, whatever the initial capital of his opponent (be it finite or infinite), provided that it is sufficient to preclude the opponent's ruin within a few tosses. Specifically, the probability is 75% that the gambler with 1 lira is ruined within 10 tosses, 92% that he is ruined within 100 (in general, $1 - u_n \simeq 1 - 0{\cdot}8/\sqrt{n}$); in these cases, fortunes of 10 or 100 lire, respectively, will ensure that the opponent cannot be ruined within this initial sequence of tosses. On the other hand, there is a chance, albeit very small, that the opponent will be the one who is ruined (this is about $1/c$; one thousandth if $c = 1000$, for example). In order for this to happen, it is necessary that the gambler who begins with 1 lira reaches a situation of parity (500 versus 500) without being ruined; after this, the expected duration of the game will be $500^2 = 250{,}000$ tosses, there then being equal probabilities of ruin for the two parties. There is, therefore, a probability of two in a thousand of reaching parity, but, in such a case, the subsequent duration of the game is almost certainly very long. As always, one should remember that prevision is not prediction.

For the Wiener–Lévy process, thinking of it as a limit case of Heads and Tails, one sees immediately that exactly the same conclusion holds. It is sufficient to observe that the change of scale ($1/N$ for the stakes, $1/N^2$ for the intervals between tosses) leaves the duration unchanged: the initial capitals are Nc' and Nc'', the duration $N^2c'c''$, with $1/N^2$ as unit. More generally, we can say that, roughly speaking, the conclusion holds for all processes with finite variances ($\sigma = 1$ per unit time; otherwise, $\mathbf{P}(T) = c'c''/\sigma^2$) provided c' and c'' are large enough to make the ruin very unlikely after a few large jumps.

In the case of games which are not fair, one can apply the same argument, but the result is different. In the case of Heads and Tails with an unfair coin ($p \neq \frac{1}{2}$), the relation

$$\mathbf{P}_y(T) = 1 + (1-p)\mathbf{P}_{y-1}(T) + p\mathbf{P}_{y+1}(T)$$

reduces to the characteristic equation $py^2 - y + (1-p) = 0$, with roots 1 and $(1-p)/p$, which gives $A + B(\tilde{p}/p)^y$ as the solution of the homogeneous equation. It is easily seen that $y(1-2p)$ (or $y/(\tilde{p}-p)$) is a particular solution of the complete equation, and, taking into account the fact that $\mathbf{P}_y(T) = 0$ for $y = -c'$ and $y = c''$, we have

$$\text{(41)} \qquad \mathbf{P}_y(T) = \frac{1}{1-2p}\left[(y+c') - (c'+c'')\frac{1-(\tilde{p}/p)^{y+c'}}{1-(\tilde{p}/p)^{c'+c''}}\right].$$

For the extension to the Wiener–Lévy process (and, more or less as we have said, to cases which approximate to it), it is sufficient to observe that, in the case we have studied, $m = 2p - 1$, $\sigma^2 = 1 - m^2$, from which we obtain $p = \frac{1}{2} + \frac{1}{2}m/\sqrt{(m^2 + \sigma^2)}$. Given the m and σ of a Wiener–Lévy process (or a general process), it suffices to evaluate p in this way.

If one plays against an infinitely rich opponent ($c'' = \infty$), we have $\mathbf{P}(T) = \infty$, provided the game is advantageous ($p > \frac{1}{2}$), and given that, with non-zero probability, it can last indefinitely. If it is disadvantageous ($p < \frac{1}{2}$), only the first term remains:

$$\mathbf{P}(T) = c'/(1 - 2p). \tag{42}$$

8.6.3. *The probabilities of ruin within given time periods* (i.e. within time t, or within $n = t$ tosses) provide the most detailed answer to the problem. Let us consider, for the time being, the case of one barrier ($c' = c$, $c'' = \infty$), and let us begin with Heads and Tails. We shall attempt to determine the probability, q_n, of ruin within n tosses; i.e. the probability that $Y_h = -c$ for at least one $h \leqslant n$.

The solution is obtained by making use of the celebrated, elegant *argument of Desiré André*. In the case of the game of Heads and Tails, we adopt the following procedure for counting the number of paths (out of the 2^n possible paths between 0 and n) which reach the level $y = -c$ at some stage. First of all, we consider those which terminate beyond that level, $Y(n) < -c$, and we note that there are exactly the same number for which $Y(n) > -c$, since any path in this latter category can be obtained in one and only one way from one of the paths in the former category by reflecting (in the straight line $y = -c$) the final segment starting from the instant $t = k$ at which the level $y = -c$ is reached for the first time. In other words, we use the reflection

$$Y^*(t) = -c - (Y(t) + c) \qquad (k \leqslant t \leqslant n).$$

Finally, we note that (if $n - c$ is even) there are some paths for which $Y(n) = -c$. In terms of the probability (the number of paths divided by 2^n), the first group contribute to $\mathbf{P}(Y_n < -c)$, and, because of the symmetry revealed by André's reflection principle, so do the second group; the final group contribute to $\mathbf{P}(Y_n = -c)$. Expressing the result in terms of c rather than $-c$, we have therefore

$$q_n = 2\mathbf{P}(Y_n > c) + \mathbf{P}(Y_n = c) = \mathbf{P}(|Y_n| > c) + \tfrac{1}{2}\mathbf{P}(|Y_n| = c). \tag{43}$$

The basic idea is illustrated most clearly in Figure 8.6; to any path which having reached $y = -c$ finds itself at $t = n$ above that level, there corresponds, by symmetry, another path which terminates below it (and, indeed, if the first path terminates at $-c + d$, the second will terminate at $-c - d$; the reader should interpret this fact). Essentially, we could say that reflection corresponds to exchanging the rôles of Heads and Tails (from the instant of ruin onwards), a device that we have already come across (in example (*D*) of Chapter 7, 7.2.2).

A further principle—similar to that of Desiré André—was introduced by Feller (Vol. I, p. 20) under the heading of the duality principle; we prefer to call it the *reversal principle*, because the other term suggests connections, which do not actually exist, with other, unrelated, notions of 'duality'. The idea is that one reverses the time order; i.e. the ordered events $E_1 \ldots E_n$ become $E_n \ldots E_1$, by setting $E_i = E_{n-i+1}$. The reversed gain is then

$$Y^*(h) = Y(n) - Y(n-h),$$

and the path is reversed (i.e. rotated by 180°) with respect to the central point $(\frac{1}{2}n, \frac{1}{2}Y(n))$.

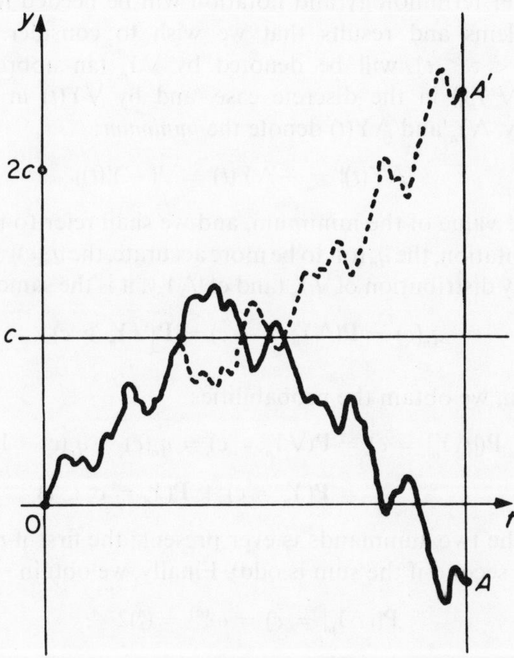

Figure 8.6 Desiré André's argument in the case of a single barrier. The paths which, after having reached level c, are below it at the end of the interval of interest (point A), correspond in a one-to-one manner, by symmetry, to those terminating at A' (symmetric with respect to the barrier $y = c$). It follows (in a symmetric process) that the probability of ending up at A', or at A, having reached level c, is the same.

The point $2c$ on the y-axis has been marked in because it corresponds to the 'cold source' in Lord Kelvin's method (Section 8.6.7)

The argument and the result given above have a far more general range of application. Indeed, the only facts about the distribution of the increments that we have used are its symmetry ($Y(t_0 + t) - Y(t_0)$ has equal probability of being $> a$ or $< -a$, and, in particular, of being $\gtrless 0$), and the fact that the level $y = -c$ cannot be exceeded by 'skipping' it (i.e. anything exceeding it must actually pass through it). This holds for the Heads and Tails process if c = integer,† and for the Wiener–Lévy process (assuming it to be continuous) for arbitrary c. For other cases, it may have approximate, or asymptotic, validity (as in the *Remark* of Section 8.6.1), provided that the jumps in the direction in which the fixed level must be exceeded are small, or, at any rate, the large jumps are relatively rare (this brief comment will suffice; we do not wish to complicate matters by going into all the details).

Some further terminology and notation will be needed for certain more general problems and results that we wish to consider. The *maximum* of $Y(\tau)$ in $0 < \tau \leqslant t$‡ will be denoted by $\vee Y_n$ (an abbreviated form of $Y_1 \vee Y_2 \vee \ldots \vee Y_n$§) in the discrete case, and by $\vee Y(t)$ in the continuous case; similarly, $\wedge Y_n$ and $\wedge Y(t)$ denote the *minimum*;

$$|\wedge Y(t)| = -\wedge Y(t) = \vee(-Y(t))$$

is the absolute value of the minimum, and we shall refer to it as |minimum|.

With this notation, the q_n (or, to be more accurate, the $q_n(c)$) of (43) determine the probability distribution of $\vee Y_n$ (and of $|\wedge Y_n|$; it is the same, by symmetry);

$$q_n(c) = \mathbf{P}(\wedge Y_n \leqslant -c) = \mathbf{P}(\vee Y_n \geqslant c).$$

By subtraction, we obtain the probabilities

$$\begin{aligned} \mathbf{P}(|\wedge Y_n| = c) &= \mathbf{P}(\vee Y_n = c) = q_n(c) - q_n(c-1) \\ &= \mathbf{P}(Y_n = c) + \mathbf{P}(Y_n = c+1) \end{aligned} \tag{43'}$$

(only one of the two summands is ever present; the first if n and c have an even sum, the second if the sum is odd). Finally, we obtain

$$\mathbf{P}(|\wedge Y_n| = c) = \omega_h^{(n)} = \binom{n}{h} 2^{-n},$$

where either $2h - n = c$, or $2h - n = c + 1$.

It is important to pay particular attention to the cases $\wedge Y_n = 0$ and $\wedge Y_n = -1$. They are not contained in the general case (which is based

† The set of levels which cannot be skipped may take various forms: either all c (positive, negative, or both) if $Y(t)$ varies continuously (non-decreasing, non-increasing, or completely general); or all multiples of some given k (positive, negative, or both) if all positive jumps are $= k$, and the negative ones are multiples of k (or conversely, or if they are all $\pm k$); in all other cases there are no such levels.

‡ We refer to $0 < \tau$ (instead of $0 \leqslant \tau$) in order to facilitate comparison with the discrete case (although the distinction loses its meaning in the continuous case); what is important is to stress that $\tau = t$ is to be included, and, indeed, we should stress that $Y(t)$ is to be understood as $Y(t + 0)$ (i.e. taking into account a possible jump occurring exactly at t).

§ The omission of Y_0 is irrelevant, except when it is useful to distinguish two cases which otherwise would both yield $\vee Y_n = 0$ (all $Y_n \leqslant 0$): with the convention adopted, we have, instead, $\vee Y_n = -1$, if $Y_1 = -1$, and the successive values are all $\leqslant -1$ (or, in general, stepping outside the example of Heads and Tails, $\vee Y_n$ can be an arbitrary negative value).

on the assumption $c > 0$), but they can easily be reduced to the appropriate form. For $c = 1$, we have $\mathbf{P}(\vee Y_n < 1) = u_n$ (i.e. $\mathbf{P}(Y_n = 0)$ or $\frac{1}{2}\mathbf{P}(|Y_n| = 1)$, according as n is even or odd), and the two cases $\vee Y_n = 0$ and $\vee Y_n = -1$ are equally probable if n is even (and they do not differ by very much if n is odd). More precisely, we have $\mathbf{P}(\vee Y_n = -1) = \frac{1}{2}u_{n-1}$ (the first step is -1, and we do not then go up by $+1$), and, by subtraction

$$\mathbf{P}(\vee Y_n = 0) = u_n - \tfrac{1}{2}u_{n-1}$$

($u_n = u_{n-1}$ if n is even; otherwise $u_{n-1} = u_n n/(n-1)$). In words (for n even): u_n is also the probability that Y_h remains *non-negative* in $0 < t \leqslant n$ (and a similar argument holds for the *non-positive* case).

Let us also draw attention to the following (interesting) interpretation of (43′).

The probability of attaining $y = c$ $(c > 0)$ as the maximum level for $t \leqslant n$ is precisely the same as that of attaining the same level c (or $c + 1$, according as we are dealing with the even or odd case) at $t = n$ (but not, in general, as the maximum level).

To put it another way: the probability $2\omega_h^{(n)}$ that $|Y_n|$ assumes the value $c = 2h - n$ $(h > n/2)$ splits into two halves for $\vee Y_n$; one half remains at c, and the other at $c - 1$. If this partial shift of one is negligible in a given problem (as it is, in any case, asymptotically) we may say that the distributions of the absolute value, $|Y_n|$, of the maximum, $\vee Y_n$, and of the |minimum|, $|\wedge Y_n|$, are all identical. Obviously, the maximum and the |minimum| are non-decreasing functions, and hence we can define their inverses. We denote by $T(y)$ $(y \geqslant 0)$ the inverse of $\vee Y(t)$, and by $T(-y)$ the inverse of $\vee(-Y(t))$ (they also have the same probability distribution as processes; however, $T(y)$ for $y \gtrless 0$ is not to be understood as a single process for $-\infty < y < +\infty$, but rather as a unified notation for two symmetric but distinct processes):

$T(y) =$ the minimum t for which

$$\vee Y(t) \geqslant y (y > 0), \qquad \text{or} \qquad \vee(-Y(t)) \geqslant -y (y < 0),$$

so that

$$(T(y) \leqslant t) = (\vee Y(t) \geqslant y)(y > 0) \vee (\vee(-Y(t)) \geqslant -y)(y < 0).$$

For every y, $T(y)$ is the random quantity expressing the instant (or, equivalently, the waiting time) either until ruin occurs, or until the first arrival at level (or point) y occurs, or until a particle is absorbed by a possible absorbing barrier placed at y, and so on.

With the present notation, we can express, in a direct form, the probabilities of ruin (or of absorption, etc.) as obtained in (43) on the basis of Desiré André's argument:

$$\begin{aligned} &\mathbf{P}(\vee Y(t) \geqslant y) = \mathbf{P}(\wedge Y(t) \leqslant -y) = \mathbf{P}(T(y) \leqslant t) = \mathbf{P}(T(-y) \leqslant t) \\ (44) \qquad &= 2\mathbf{P}(Y(t) > c) + \mathbf{P}(Y(t) = c) \\ &= \mathbf{P}(|Y(t)| > c) + \tfrac{1}{2}\mathbf{P}(|Y(t)| = c). \end{aligned}$$

We omit the term corresponding to $Y(t) = c$, which is only required in order to obtain exact expressions for the Heads and Tails case, and which is either zero or negligible in general (for the exact expressions of the Wiener–Lévy process, the asymptotic expressions for Heads and Tails, and for other cases). If we denote by $F^t(y)$ and $f^t(y)$ the distribution function and the density (if it exists) of $Y(t)$,† then the distribution function and the density of $|Y(t)|$, and therefore (either exactly or approximately) of $\vee Y(t)$ and $\vee(-Y(t))$, are given by

$$(45) \quad 2F^t(y) - 1 \qquad (\text{for } 0 \leqslant y < \infty), \qquad 2f^t(y) \qquad (\text{for } 0 \leqslant y < \infty).$$

8.6.4. Substituting the exact forms for the Heads and Tails case into (44) or (45), we would have

$$(46) \quad q_n = (\tfrac{1}{2})^{n-1}\sum_h \tbinom{n}{h}[0 \leqslant h < \tfrac{1}{2}(n-c)] + (\tfrac{1}{2})^n\tbinom{n}{(n-c)/2} \qquad (\text{if } n - c \text{ is even}).$$

It is more interesting, however, to consider the approximation provided by the normal distribution; this will, of course, be exact for the Wiener–Lévy process, and will hold asymptotically in the case of Heads and Tails, and for any other case with finite variance (which we shall always assume to be 1 per unit time). We have (for $y > 0$)

$$\begin{aligned} &q_y(t) = p_t(y) = \mathbf{P}(\vee Y(t) \geqslant y) = \mathbf{P}(T(y) \leqslant t) \\ (47) \qquad &= 2\mathbf{P}(Y(t) \geqslant y) = \sqrt{(2/\pi)}\int_{y/\sqrt{t}}^{\infty} \mathrm{e}^{-\frac{1}{2}x^2}\,\mathrm{d}x.\text{‡} \end{aligned}$$

Clearly, the form of (47), interpreted either as a function of y (with t as a parameter) or, alternatively, as a function of t (with y as a parameter), provides the distribution function of $|Y(t)|$ and $\vee Y(t)$, and that of $T(y)$ and

† This would also be valid for processes other than those which are homogeneous with independent increments (satisfying the conditions stated for Desiré André's argument) where there is less justification for writing F^t and f^t. In fact, however, we shall not be dealing with the general case.

‡ For y large (compared with $\sqrt{t}$), the approximation given by (20) (in Chapter 7, 7.5.4) can be used, and gives

$$(47') \qquad q_y(t) = p_t(y) \simeq K(\sqrt{t}/y)\mathrm{e}^{-y^2/2t}, \qquad K = \sqrt{(2/\pi)} \simeq 0{\cdot}8.$$

$T(-y)$. It is often useful to have this expressed in terms of both the parameters y and t; it can then be interpreted according to whichever is appropriate. With this notation, we shall denote it in all cases by $\mathbf{P}(\bigvee Y(t) \geqslant y)$ (even in the Heads and Tails context).

The distribution of the maximum (or |minimum|) $\bigvee Y(t)$ (or $|\bigwedge Y(t)|$), and of $|Y(t)|$, is clearly the *semi-normal* (the normal distribution confined to the positive real axis), whose density is given by

$$(48) \qquad f(x) = Kt^{-\frac{1}{2}}\, \mathrm{e}^{-\frac{1}{2}x^2/t}(x \geqslant 0), \qquad K = \sqrt{(2/\pi)} \simeq 0{\cdot}8.$$

This is half of the normal distribution with $\bar{m} = 0$ and $\bar{\sigma}^2 = t$; the mean and variance are given by $m = \sqrt{(2/\pi)}\bar{\sigma}$† and $\sigma = \sqrt{(1 - 2/\pi)}\bar{\sigma}$; i.e., numerically, $m \simeq 0{\cdot}8\bar{\sigma}$ and $\sigma \simeq 0{\cdot}6\bar{\sigma}$ (these should not be confused!).

On the other hand, $T(y)$ (or $T(-y)$) has density

$$(49) \qquad f(t) = Kyt^{-\frac{3}{2}}\, \mathrm{e}^{-\frac{1}{2}y^2/t}\ (t \geqslant 0), \qquad K = 1/\sqrt{(2\pi)} \simeq 0{\cdot}4.$$

This is the stable distribution with characteristic exponent $\alpha = \frac{1}{2}$ (which we mentioned in Section 8.4.4); in other words, it corresponds to jumps x whose density of intensity is $x^{-\frac{3}{2}}$. Since $\alpha < 1$, it has infinite prevision (in line with what we established directly in Section 8.6.2).

The fact that $T(y)$ had to have the stable distribution with $\alpha = \frac{1}{2}$ could have been deduced directly from the fact that

$$T(y_1 + y_2) = T(y_1) + [T(y_1 + y_2) - T(y_1)] = T(y_1) + T(y_2).$$

The time required in order to reach level $y_1 + y_2$ is, in fact, that required to reach y_1 plus that then required to proceed further to $y_1 + y_2$. However, given that, by the continuity of the Wiener–Lévy process, the level y_1 is reached (and not bypassed) at $T(y_1)$ with a jump, it is a question of going up by another y_2 under the same conditions as at the beginning. By virtue of the homogeneity, however, the distribution can only depend on y^2/t (and the density, in terms of y^2/t, could therefore only be—as, in fact, it actually is—a function of y^2/t divided by t).

We shall return (in Section 8.7.9) to the exact form for the Heads and Tails case, having encountered (for Ballot problems, in Section 8.7.1) an argument which enables us to establish it in a straightforward and meaningful way.

---

† $m = \bar{\sigma}[2/\sqrt{(2\pi)}] \int_0^\infty x \exp(-\frac{1}{2}x^2)\, \mathrm{d}x$;

it can easily be shown that the integral is equal to one.

8.6.5. In the case where we consider *two gamblers*, G_1 and G_2 (with initial fortunes c' and c'', where $c' + c'' = c^*$), Desiré André's argument still applies, but now, of course, in a more complicated form. If we denote by A and B passages through levels c' and c'', respectively, a path whose successive passages are $ABABAB\ldots$ signifies the ruin of G_1; if the sequence begins with B, it signifies the ruin of G_2. (It does not matter how many times A is followed by B or B by A, nor does it matter whether the sequence ends in an A or a B; successive passages through the same level are not counted; e.g. $ABBAAABAABB = ABABAB$.) Desiré André's argument (as applied in the one-sided case) does not directly enable one to count the paths $\{A\}$ which signify G_1's ruin, nor the paths $\{B\}$ which signify G_2's ruin, nor the paths $\{0\}$ which indicate that neither is ruined (all belonging to the 2^n paths in the interval $[0, n]$). It does, however, enable one to count those of 'type (A)', 'type (B)', 'type (AB)', 'type (BA)', 'type (ABA)', etc., where these refer to paths containing, in the sequence, the groups of letters indicated (which might be sandwiched between any number of letters).

Everything then reduces to the previous case of only one gambler; in other words, to

$$p_t(y) = p(y) = \mathbf{P}[\vee Y_n \geqslant y].$$

The probability of paths of 'type (A)' is, in fact, $p(c')$; that of paths of 'type (AB)' is $p(c' + c^*)$ (because to first reach $-c'$ and then to reach $-c''$ requires a zigzag path along $c' + (c' + c'')$; this amounts to reflecting the path with respect to $y = -c'$, starting from the instant it reaches this level and continuing up until when it reaches c''); for 'type (ABA)' we have $p(c' + 2c^*)$, and so on. Similarly, for 'types' (B), (BA), (BAB), ..., we have

$$p(c''), p(c'' + c^*), \quad p(c'' + 2c^*), \ldots,$$

and in this way we arrive at the required conclusion.

The paths $\{A\}$ are, in fact, those given by

$$(A) - (BA) + (ABA) - (BABA) + (ABABA) - \ldots$$

(i.e. we start with those reaching $-c'$ and we exclude those first reaching c''; in this way, however, we exclude those that reach $-c'$ prior to c''; and so on). The same thing holds for $\{B\}$; the $\{0\}$ are those remaining (i.e. neither $\{A\}$ nor $\{B\}$).

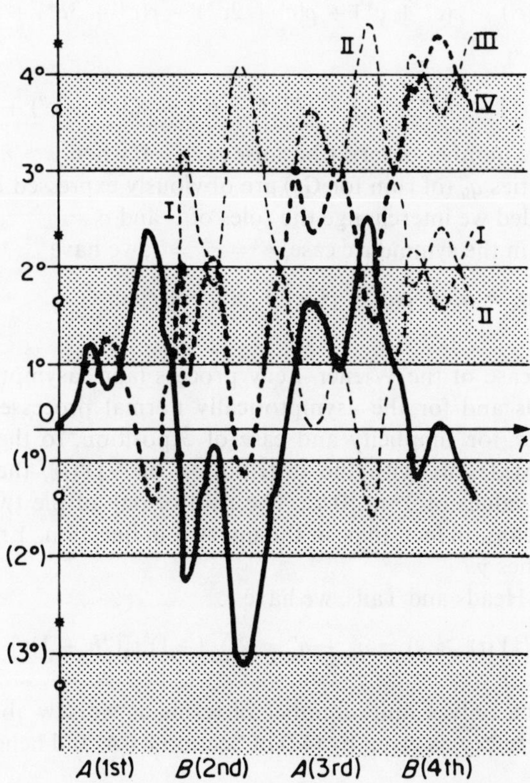

Figure 8.7 Desiré André's argument in the case of two barriers. The barriers are the straight lines bounding the white strip around the origin 0; the other strips are the *proper* images (white strips) and *reversed* images (dark strips) with the respective *hot* (black) and *cold* (white) sources (corresponding to Lord Kelvin's method; see Section 8.6.7). The actual path is indicated by the solid black line; its four successive crossings are denoted by A(1st), B(2nd), A(3rd), B(4th) (consecutive repeat crossings of A or B are not counted).

The final image of the path (obtained by repeated application of André's reflection principle) is indicated by the heavy broken line; it follows the same path up until A(1st) and is then given by the reflection (I) of it with respect to the 1st level. Then, after B(2nd), it is given by the reflection (II) of (I) with respect to the 2nd level, and so on. The continuations of the reflected paths (after the section in which they constitute the final image) are indicated by the lighter, dashed line. The image paths reaching the 1st, 2nd, 3rd levels, etc., correspond to paths of types A, AB, ABA, etc. (instant by instant); the same is true in the opposite direction (1st. 2nd. 3rd levels. etc.. in the negative half-plane) for paths of types B. BA. BAB. etc.

It follows that the probabilities of ruin within n tosses are, for G_1, given by

$$q'_n = p(c') - p(c'' + c^*) + p(c' + 2c^*) - p(c'' + 3c^*) + \dots \tag{50}$$

or (in terms of c''),

$$p(c^* - c'') - p(c^* + c'') + p(3c^* - c'') - p(3c^* + c'') + \dots \tag{50'}$$

(there are a finite number of terms because $p(y) = 0$ when $y > n$).

The probabilities q''_n (of ruin for G_2) are obviously expressed by the same formulae, provided we interchange the rôles of c' and c''.

In particular, in the symmetric case, $c' = c'' = c$, we have

$$q'_n = q''_n = p(c) - p(3c) + p(5c) - p(7c) + \dots \tag{51}$$

8.6.6. In the case of the Wiener–Lévy process (and, asymptotically, for Heads and Tails and for the asymptotically normal processes), we shall restrict ourselves, for simplicity and ease of exposition, to the symmetric case, which provides us with the distribution of $\vee|Y(t)|$, the maximum of the absolute values of Y in $[0, t]$. The ruin of one of the two gamblers within time t means, in fact, that, in the interval in question, Y reaches $\pm y$; i.e. that $|Y|$ reaches y.

In the case of Heads and Tails, we have

$$\mathbf{P}(\vee|Y(t)| \geqslant y) = q'_n + q''_n = 2\sum_h (-1)^h p[(2h + 1)y] \tag{52}$$

(there are, in fact, only a finite number of terms as we saw above). In the Wiener–Lévy case, $p(y)$ is given by (47) of Section 8.6.4, and hence

$$\mathbf{P}(\vee|Y(t)| \geqslant y) = 2\sqrt{(2/\pi)} \sum_{h=0}^{\infty} (-1)^h \int_{(2h+1)y/\sqrt{t}}^{\infty} \mathrm{e}^{-\frac{1}{2}x^2}\,\mathrm{d}x. \tag{53}$$

Differentiating with respect to y, we obtain the density

$$\begin{aligned} f(y) &= K \sum_{h=0}^{\infty} (-1)^h (2h + 1)\,\mathrm{e}^{-\frac{1}{2}[(2h+1)y]^2/t} \\ &= K(\mathrm{e}^{-y^2/2t} - \mathrm{e}^{-(3y)^2/2t} + \mathrm{e}^{-(5y)^2/2t} - \mathrm{e}^{-(7y)^2/2t} + \dots), \end{aligned} \tag{54}$$

$$K = 2\sqrt{(2/\pi t)}.$$

8.6.7. It is instructive to compare the present considerations, based on Desiré André's argument, with those, essentially identical, based on Lord Kelvin's *method of images*, which is often applied to diffusion problems. We have already observed (in Chapter 7, 7.6.5) that the Heads and Tails process can be seen, in a heuristic fashion, to approach a diffusion process, and that the analogy becomes an identity when we consider the passage to the limit which transforms the Heads and Tails process into the Wiener–Lévy process.

In order to use the method to formulate the problem of the ruin of *one gambler* (occurring when level $y = c$ is reached), it suffices to find the solution of the diffusion equation (given by (32) of Chapter 7, 7.6.5) in the region $y \leqslant c$ (where we assume $c > 0$), satisfying the initial condition of concentration at the origin, *together with the condition that it vanishes at* $y = c$. By virtue of the obvious symmetry (giving a physical equivalent to Desiré André's argument), it suffices to place initially, at the point $y = 2c$, a mass equal and opposite to that at the origin (the 'cold source'); this gives a process which is identical to the other, but opposite in sign, and symmetric with respect to the line $y = 2c$ instead of to $y = 0$. On the intermediate line, $y = c$, the two functions therefore cancel one another out, and their sum provides the desired solution. The probability of ruin at any instant can be interpreted in terms of the flux of heat out past the barrier, together with an incoming cold flux; and so on.

In the case of two gamblers (i.e. barriers at $\pm c$, and the initial position of the mass at $y = 0$, or at $y = a$, $|a| < c$†), if we are to use the same trick we have to introduce an *infinite* number of sources, alternatively hot and cold—like alternate images of the face and the back of the head in a barber's shop with two mirrors on opposite walls. We have an infinite number of images of the mirrors (lines $y = (2k + 1)c$, k being an integer between $-\infty$ and $+\infty$), and in between them an infinite number of strips (proper and reversed images of the shop), and inside each of these strips the image ('hot' or 'cold') of the source. If the source is at the centre ($y = 0$), the others are at $y = 2kc$ (hot if k is even, cold if k is odd); otherwise—if it is at $y = a$—the hot sources are at $2kc + a$, and the cold ones at $2kc - a$ (still with hot corresponding to k even, cold to k odd).

Using the techniques of this theory (Green's functions, etc.) one can obtain solutions to even more complicated problems of this nature (e.g. those with curved barriers), where this kind of intuitive interpretation would necessitate one thinking in terms of something like a continuous distribution of hot and cold sources.

8.7 BALLOT PROBLEMS; RETURNS TO EQUILIBRIUM; STRINGS

8.7.1. We turn now to what we referred to in Section 8.5.2 as the second group of questions concerning random processes; those in which the study of the process reduces to an examination of certain segments into which it

† This is just a more convenient way of saying that one starts from 0 but places barriers at $c - a$ and $c + a$.

may be useful to subdivide it. More precisely, we shall consider the decomposition into *strings*; i.e. into the segments between successive returns to equilibrium (i.e. between successive *zeroes* of $Y(t)$).

Over and above their intrinsic interest, these questions often point the way to the formulation, understanding and solution of other problems which entail *recurrence* in some shape or form. In other words, problems relating to processes which, after every repetition of some given phenomenon (in this case, the return to equilibrium), start all over again, with the same initial conditions (or with modifications thereof which are easily taken into account).

The simplest scheme involving the notion of recurrence is that of events E_h forming a *recurrent sequence*† (such as the $E_h = (Y_h = 0)$ in our example), for which, when the outcomes of the preceding events are known, the probabilities depend on the number of events since the last success. In other words, the index (or 'time') $T^{(k)}$ of the kth success is the sum of the k *independent* waiting times $T_1, T_2, \ldots, T_k$: the T are integers in this case, but this is merely a special feature of this simple scheme.

We now provide a brief account of the most important aspects of the theory (for a fuller account, cf. Feller, Vol. I, Chapter XIII). We denote by f_h the probability that E_h is the first success (or, equivalently, that, following on from the last success obtained, the first success occurs in the hth place); in other words, $f_h = \mathbf{P}(T = h)$, where T = waiting time. It follows immediately that either $f = \sum f_h = 1$, or it is <1 (if the probability of a success does not tend to certainty as the number of trials increases indefinitely). We adopt the convention of denoting the difference $1 - f$ by f_∞ (the probability that the waiting time is infinite).

By convolution, we obtain the probabilities $f_h^{(2)}$ of E_h being the second success, and so on for $f_h^{(3)}$, etc. In general, we have

(55) $$f_h^{(r)} = f_1 f_{h-1}^{(r-1)} + f_2 f_{h-2}^{(r-1)} + \ldots + f_{h-1} f_1^{(r-1)}.$$

Summing over r, we obtain the probability u_h of E_h being a success (without taking into account whether it was the first, second, ..., or whatever; as above, this holds also for a success at the hth place following on from some success already obtained, assuming that nothing is known about successes for subsequent events):‡

(56) $$u_h = f_h + f_h^{(2)} + f_h^{(3)} + \ldots + f_h^{(h)} \text{ (obviously, } f_h^{(r)} = 0 \text{ for } r > h).$$

† These are usually called 'recurrent events', but this terminology does not fit in with ours (nor, in a certain sense, with the point of view we have adopted).

‡ I would prefer to write $\mathbf{P}(E_h) = p_h$ (instead of u_h) as usual, in order not to make it appear that we are dealing with a special case, and in order to avoid confusion with the standard use of u_n (as the maximum probability for Heads and Tails). The reason we have used u_h is for the convenience of the reader who wishes to pursue this topic (which we are only scratching the surface of) using Feller: however, it only occurs in this section, so that there should be no cause for any confusion.

The sum of the $f_h^{(r)}$ gives f^r ($=1$ or <1, the same as f), and provides the probability that an rth repetition takes place. The sum

$$f + f^2 + f^3 + \ldots + f^r + \ldots$$

therefore gives the prevision of the total number of successes (finite or infinite, according as $f < 1$ or $f = 1$). The same prevision can be expressed in a different way, however, by the sum of the u_h; we therefore obtain

(57) $$u = u_1 + u_2 + \ldots + u_h + \ldots = f/(1 - f),$$

(57′) $$f = u/(u + 1).$$

If $f = 1$, $u = \infty$, the events E_h are called *persistent*; in the opposite case, $f < 1$, $u < \infty$, they are called *transient*. In the case of persistent events,† if we denote the prevision of the waiting time by τ, i.e.

(58) $$\tau = f_1 + 2f_2 + 3f_3 + \ldots + hf_h + \ldots \qquad \text{(which could be } \infty\text{)},$$

then, as h increases, the probability u_h of success tends to the limit

(59) $$\bar{u} = 1/\tau \qquad \text{(in particular, } u_h \to \bar{u} = 0 \text{ if } \tau = \infty\text{)}.$$

Let us now return to the central topic of this section and explain how the 'Ballot problem' enters into the picture. This is simply a traditional way of referring to, and interpreting, a set of problems similar to those encountered under the heading of 'gambler's ruin', but relating now to drawings from an urn without replacement (which provides a model for the process of counting votes). The results also find application in statistics, where they form the basis of certain criteria (due to Kolmogorov and Smirnov) for considering the deviation of an empirical distribution function from a given hypothetical theoretical distribution. Anyway, although we shall refer to Ballot problems as a convenient aid to intuition, we shall think of this new case always in the context of Heads and Tails.

In fact, we shall study problems of Heads and Tails conditional on the knowledge of the number of successes, H, occurring in the N trials with which we are concerned (we have seen this argument before in Chapter 7, 7.4.3, where we used it to derive the hypergeometric distribution). Proceeding in this way, it is evident that, among other things, the graphical representation of this problem consists simply of the rectangular portion of the Heads and Tails lattice having opposite vertices at the origin (the starting point) and at $[N, H] = (N, 2H - N)$ (the point where the process terminates).

† The complication of 'periodic events' (E_h only possible for a multiple of some 'period' λ) can be avoided (and we assume this done) by confining attention to events $E_{h\lambda}$. Of course, this might not always be convenient in practice (e.g., if a sequence E_n, defined in terms of another sequence A_n, which is not periodic, turns out to be periodic: cf. Heads and Tails; returns to zero are only possible for n even).

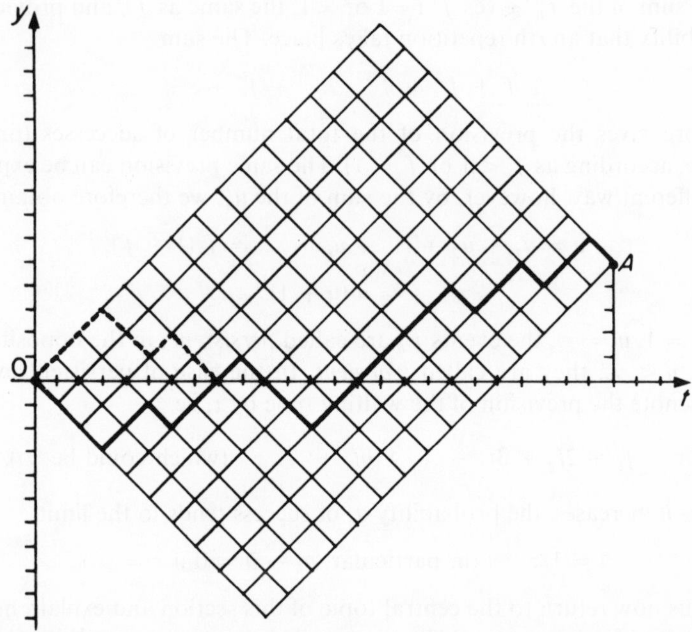

Figure 8.8 Desiré André's argument: the Ballot problem (i.e. the hypergeometric distribution). Paths from 0 to A which begin with a downward step correspond in a one-to-one fashion to those which start off upwards and then subsequently touch the t-axis (by symmetry in the interval before the t-axis is first reached)

There are $\binom{N}{H}$ possible paths joining these two points (see Figure 8.8); in order to fix ideas, we assume that $Y_N = 2H - N \geqslant 0$, i.e. $H \geqslant N - H$. For $0 < y < Y_N$, all paths either cross or touch each barrier of the form $y =$ constant. For the case $y = 0$ (and $y = Y_N$), we wish to consider how many paths touch it (either crossing it or not) after the initial $Y_0 = 0$ (or before the final $Y_N = N$, respectively). The same question (without the qualifications at the end-points) also arises for levels $y = Y_N + c$ $(c > 0)$, or, equivalently,† for $y = -c$.

(A) The case $y = 0$, the Ballot problem, is the simplest one (we restrict ourselves to $Y_N > 0$ for now, postponing the case of $Y_N = 0$ to Section 8.7.3). All paths whose first step is downward must cross $y = 0$ again, and, by reflecting the initial segment up to the first crossing, one obtains (with a one-to-one correspondence) all those having an upward first step which subsequently touch $y = 0$. But the first step has (like any other step) probability $(N - H)/N$ of being one of the $N - H$ downward steps; the probability

† This is an obvious application of the reversal principle; cf. Section 8.6.3.

of the eventual winner not being in the lead at some stage during the counting is therefore equal to twice this value, $2(N - H)/N$, and the probability that he is *always in the lead* during the counting is given by

$$1 - 2(N - H)/N = (2H - N)/N = Y_N/N. \tag{60}$$

(*B*) If we turn to the case $y = Y_N + c$ (or $y = -c$), $c > 0$, the principle of reflection (Desiré André) shows immediately that there are $\binom{N}{H+c}$ paths which touch this barrier in $0 \leqslant t \leqslant N$ (either crossing it or not, as the case may be). This is the number which finish up at the point whose ordinate is $(2H - N) + 2c = 2(H + c) - N$, the symmetric image of the given final point with respect to the barrier. In fact, these paths are obtained from the others (in a one-to-one onto fashion) by reflecting, with respect to the barrier, the final portion, starting from when the barrier is first reached (for $t = h$; $Y_h = Y_N + c$).

The probability s_c of reaching (and possibly going beyond) level $y = Y_N + c$ (or $y = -c$) is therefore given by

$$s_c = \binom{N}{H+c}\Big/\binom{N}{H} = \frac{(N - H)(N - H - 1)(N - H - 2)\ldots(N - H - c + 1)}{(H + 1)(H + 2)(H + 3)\ldots(H + c)}. \tag{61}$$

The explicit expression is particularly useful when c is small (note that for $c = 1$ we have $s_1 = (N - H)/(H + 1))$, and is instructive in that it shows how the successive ratios $(N - H - c + 1)/(H + c)$ give the probability of reaching the required level $(Y_N + c)$ given that one knows that the level immediately below it (i.e. $Y_N + c - 1$) has been reached. The complementary probability is

$$(2H - N + 2c - 1)/(H + c),$$

and so the probability r_{c-1} that the *maximum* level reached is $Y_N + c - 1$ is given by the formula for s_c with $(2H - N + 2c - 1)$ replacing

$$(N - H - c + 1)$$

in the final factor; this is also the probability that the *minimum* level is $-(c - 1)$ (alternatively, one can obtain this by noting that $r_{c-1} = s_{c-1} - s_c$). Observe that $s_c = 0$ for $c \geqslant N - H + 1$ (why is this so?).

(*C*) *The case of two barriers.* In the case of two barriers at levels $y = -c'$ and $y = Y_N + c''$ (c' and c'' positive), by performing successive reflections, as in the previous case, one obtains the paths terminating at the image points of the given final point ($Y_N = 2H - N$) with respect to the two barriers (thought of as parallel mirrors: there are an infinite number of images, but only those with ordinates lying between $\pm N$ can be reached). Setting

$c^* = Y_N + c' + c''$, the distance between the barriers, the ordinates of the images are given by

$$(2k + 1)c^* - c' \pm c''.$$

(N.B.: for $k = 0$, we have $c^* - c' - c'' = Y_N = 2H - N$, the given final point, and

$$c^* - c' + c'' = Y_N + 2c'' = 2(H + c'') - N,$$

the unique image in the case of a single barrier $y = Y_N + c''$; in that case, we used c instead of c''.)

It follows that the probability of the lower barrier being reached first is given by

$$(62) \qquad q'_N = (1/\tbinom{N}{H})[\tbinom{N}{H+c'} - \tbinom{N}{H+c^*} + \tbinom{N}{N+2c^*+c'} - \tbinom{N}{H+3c^*} + \dots]$$

(where we argue as in Section 8.6.5): the result for q''_N is similar (with c'' in place of c'). The sum, $q_N = q'_N + q''_N$, gives the probability of reaching a barrier (not distinguishing which, or which was reached first), and $1 - q_N$ that of not reaching either of them.

8.7.2. When studying a random process, it is often useful to consider it as subdivided into successive *strings*; i.e. into segments within which it retains the same sign. In our case,† this means those segments separated by successive *zeroes*, $Y_t = 0$, and necessarily having even length (since Y_t can only vanish for t even). Strings are either positive or negative (i.e. paths on the positive or negative half-plane, $Y_t > 0$ or $Y_t < 0$; cf. the second footnote in Chapter 7, 7.3.2), and between any two strings the path has a zero at which it either *touches* the t-axis or *crosses* it, according as the two strings have equal or opposite signs.

If one thinks in terms of gains, i.e. of the excess of the number of successes over the number of failures, a zero represents a *return to equilibrium* (equal numbers of successes and failures; gain zero), and the string represents a period during which one of the players has a strict lead over the other. Omitting the word 'strict' and including the zeroes, we obtain periods in which one or other player does not lose the lead (i.e. the union of several consecutive strings having the same sign). One might also be interested in knowing the length of time, within some given period up to $t = N$, during which either player has had the lead. If one thinks of a random walk, the zero is a *return to the origin*, and a string is a portion of the walk between two returns to the origin.

† That of processes with jumps of ± 1, with paths on a lattice, and (for convenience) starting at the origin, $Y_0 = 0$. In other cases, one could have changes in sign without passages through zero occurring (and one could even have, in continuous time, a discontinuous process $Y(t)$ with an interval in which changes of sign occur within neighbourhoods of each point).

The above discussion, together with the results obtained so far, leads us directly into this kind of question, either with reference to the special case of Heads and Tails, or to that of the Ballot problem (which reduces to the former, if one thinks of $Y_N = 2H - N$ as being known).

8.7.3. *The Ballot problem in the case of parity*: $Y_N = 0$, *i.e. there are equal numbers*, $H = N - H = N/2$, *of votes for and against. What is the probability that one of the two candidates has been in the lead throughout the count*? In our new terminology, this means that the process forms a single string; i.e. there are no zeroes except at the end-points (if we are thinking in terms of a particular one of the candidates, the string must be of a given sign and the probability will be one half of that referred to in the question above).

This can easily be reduced to the form of case (A) as considered in Section 8.7.1. In terms of the candidate who is leading before the last vote is counted, we must have $Y_{N-1} = 1$ (since we know that at the final step the lead disappears, and we end up with $Y_N = 0$); it follows that the required probability is given by

$$Y_{N-1}/(N-1) = 1/(N-1).$$

This is the probability that one of the two candidates (no matter which) remains strictly ahead until the final vote is counted; the probability of this happening for a particular candidate is $1/2(N-1)$.

8.7.4. *What is the probability that in a Heads and Tails process—or, more generally, in an arbitrary Bernoulli process—the first zero (return to equilibrium, passage through the origin) occurs at time* $t = n$ (where n, of course, is even)? For this to happen, it is first of all necessary that $Y_n = 0$; the problem is then that considered in Section 8.7.3. The probability that if a zero occurs it is the first one is therefore equal to $1/(n-1)$. The probability of the first zero at $t = n$ is thus $\mathbf{P}(Y_n = 0)/(n-1)$. In other words, this is the probability that the first *string* (and hence any string, since the process can be thought of as starting all over again after every zero) has *length n*.

(A) In the case of Heads and Tails, we have $\mathbf{P}(Y_n = 0) = u_n$, and so the probability of the first zero occurring at $t = n$ is given by

$$u_n/(n-1) \simeq (0{\cdot}8/\sqrt{n})/n = 0{\cdot}8n^{-\frac{3}{2}}.$$

The probability that the string is of length n and has a given sign (i.e. is to the advantage of a particular one of two gamblers) is one half of this.

More precisely, we have (setting $n = 2m$)

$$\frac{u_n}{n-1} = \tbinom{2m}{m}/2^{2m}(2m-1), \text{ and also } \frac{u_n}{n-1} = \frac{u_{n-2}}{n} = u_{n-2} - u_n.$$

Since

$$u_n \cdot 2^n = \binom{2m}{m} = \frac{2m!}{m!m!} = \frac{2m(2m-1)(2m-2)!}{m \cdot m \cdot (m-1)!(m-1)!}$$

$$= \frac{4(n-1)}{n}\binom{2m-2}{m-1} = \frac{4(n-1)}{n} u_{n-2} \cdot 2^{n-2},$$

we can verify at once that $u_n = u_{n-2}(n-1)/n$.

This establishes the following important conclusions:

(*a*) u_n is also the probability that there are no zeros up to and including $t = n$ (this is true for $u_2 = \frac{1}{2}$, and hence is true by induction, since $u_{n-2} - u_n$ is the probability of the first zero occurring at $t = n$);†

(*a*′) as in the footnote;

(*b*) since $u_n \to 0$, the probability that (as the process proceeds indefinitely) there is at least one return to equilibrium tends to 1 (and the same is therefore true for two, three or any arbitrary k returns to equilibrium);

(*c*) the form u_{n-2}/n tells us that $1/n$ is the probability that the string terminates at $t = n$ (since Y_n becomes 0), assuming that it did not terminate earlier (since u_{n-2} is the probability that $Y_t \neq 0$ for $t = 1, 2, \ldots, n-2$, and this is necessarily so for $t = n - 1 =$ odd);

(*d*) from this, we can deduce that u_n can be written in the form

$$(63) \qquad u_n = (1 - \tfrac{1}{2})(1 - \tfrac{1}{4})(1 - \tfrac{1}{6})(1 - \tfrac{1}{8}) \ldots (1 - 1/n);‡$$

in other words,

$$u_{n+2} = u_n\left(1 - \frac{1}{n+2}\right) = \left(u_n \frac{n+1}{n+2}\right)$$

† It follows that the probability of Y_t $(0 < t \leqslant n)$ being *always positive* (or *always negative*) is $u_n/2$. If, instead, one requires only that (*a*′) it is non-negative (or non-positive), the probability is double: i.e. it is still u_n (as can be seen from Section 8.6.3; special cases of (43′) for $c = 0$ and $c = -1$).

‡ We note that this enables us to establish Wallis's formula; from

$$u_{2m} = \frac{1}{2}\cdot\frac{3}{4}\cdot\frac{5}{6}\cdot\frac{7}{8}\cdots\frac{2m-1}{2m} \simeq \frac{\sqrt{(2/2m)}}{\sqrt{\pi}},$$

we obtain

$$\sqrt{\frac{\pi}{2}} = \lim_{m\to\infty} \frac{2.4.6\ldots 2m}{3.5.7\ldots(2m-1)}\cdot\frac{1}{\sqrt{(2m)}},$$

where π has its usual meaning, given by the double integral of Chapter 7, 7.6.7:

$$\int e^{-\frac{1}{2}\rho^2}\, dx\, dy = \int e^{-\frac{1}{2}\rho^2} \rho\, d\rho\, d\theta = 2\pi.$$

(as a product of complementary probabilities; a demographic analogy goes as follows: the probability of being alive at age n can be expressed as a product of the probabilities of not dying at each previous age);

(*e*) a further meaningful expression for u_n is given by

$$(64) \qquad u_n = \sum_k \frac{u_k}{k-1} u_{n-k} \text{ (the sum being over even } k,\ k \leqslant n);$$

observe that, in fact, each summand expresses the probability that the first zero is at $t = k$, and that there is another zero at $t = n$ (i.e. after time $n - k$); for $k = n$ (the final summand), we must take $u_0 = 1$, and hence $u_n/(n-1)$, a term which can be taken over to the left-hand side to give the explicit expression $u_n = [(n-1)/(n-2)] \sum'$ (where $\sum'$ denotes the same sum, but without the final term).

We shall encounter further properties of u_n and $u_n/(n-1)$ later.

(*B*) In the general case (the Bernoulli process with $p \neq \frac{1}{2}$), we have instead

$$\mathbf{P}(Y_n = 0) = \binom{2m}{m}(p\tilde{p})^m = u_n(4p\tilde{p})^m = u_n[2\sqrt{(p\tilde{p})}]^n, \qquad 2\sqrt{(p\tilde{p})} < 1.$$

The probability of the first zero at $t = n$ is, therefore,

$$(65) \qquad [u_n/(n-1)][2\sqrt{(p\tilde{p})}]^n < u_n/(n-1) \qquad (n \text{ even}),$$

and the sum of such probabilities is <1.

The remaining probability, P, given by $P(x) = 1 - \sum_n [u_n/(n-1)](1-x)^n$ with $x = 1 - 4p\tilde{p}$, is the probability that a string has infinite length (which, with probability $= 1$, will be to the advantage of the favourite; i.e. the player with $p > \frac{1}{2}$). A the beginning of each new string, there is probability $(1 - P)$ of a finite string, advantageous to one or other of the players, and probability P that the favourite embarks on an infinite string. The probability that the kth string turns out to be infinite is given by $P(1-P)^{k-1}$.

Note, in particular, that property (*b*) only holds in the case of Heads and Tails (i.e. *only* if we have $p = \frac{1}{2}$). Otherwise, it is not at all *asymptotically certain* that a return to equilibrium takes place (and even less is it certain that such a return to equilibrium takes place an arbitrarily large number of times). On the contrary, it is asymptotically certain that the *favourite* (the player with $p > \frac{1}{2}$) maintains his lead from some given time onwards.

Remark. We have introduced the phrase '*asymptotically certain*' to mean that some given fact—e.g., in the case under consideration, the occurrence of a return to equilibrium, or of k returns to equilibrium—has a probability tending to 1 of occurring in a random process, provided the process goes on indefinitely; i.e. if p_N is the probability that it occurs before time N, then $p_N \to 1$ as $N \to \infty$.

We note that if some given fact is asymptotically certain, then its occurrence k times (k arbitrary) is also asymptotically certain, provided (as in our case) that each time it occurs we find ourselves with the same initial conditions.† Without this latter stipulation, the conclusion of (b)—'and the same is therefore true for two, . . . '—no longer holds (this is obvious, but was not mentioned explicitly in (b) for the sake of brevity).

We note also that 'asymptotically certain' in no way (logically) implies 'certain' provided the process continues indefinitely (not even if we were to assume that we could examine the process in its entirety, placing ourselves beyond the end of time). It is even more important to note that the fact that the occurrence of an event k times (with k arbitrarily large) is asymptotically certain *does not imply* that, in a process of infinite duration, its occurrence *an infinite number of times* is certain (necessary), nor even that it has probability 1 (nor even that it is probable, or even possible). We can only say that this number of repetitions N (assuming that it makes sense to speak about it) is a random quantity (either integer or $+\infty$), which has probability 0 of taking on any individual finite value, and hence of belonging to any given finite subset of integers, such as those less than some preassigned integer k. It could, however, be certainly finite, like an 'integer chosen at random' (cf. Chapter 4, 4.18.3).

8.7.5. *What is the prevision of the length L of a string (i.e. of the waiting time $t = L$ until the first zero)*? In the case of Heads and Tails, we see immediately that $\mathbf{P}(L)$ is infinite. In fact, $n(u_{n-2}/n) = u_{n-2}$; i.e. the contribution to the prevision corresponding to $L = n$ tends to zero like $n^{-\frac{1}{2}}$, and the sum diverges.

As we remarked above (cf. the discussion following (65)), if $p \neq \frac{1}{2}$ this no longer happens (because of the presence of the factor $2\sqrt{(p\tilde{p})} < 1$ in the geometric progression): the (finite) strings have, in prevision, finite length. However, the prevision becomes infinite if we take into account the fact that each string could be the *final* one, of *infinite*‡ length (but, if we distinguish between the two players, this only happens for the favourite: the first player if $p > \frac{1}{2}$).

Remark. The result which we are interested in (for Heads and Tails) is more conveniently expressed in the following form (which also gives us the opportunity to make an observation of a more general character). Each string consists, in prevision, of $\frac{1}{2}$ a string of length 2, of $\frac{1}{8}$ a string of length 4, . . . , of $u_n/(n-1)$ a string of length n, and so on. In terms of the prevision of length, we have, in contrast, $2/2 = 1$ for strings of length 2, $4/8 = \frac{1}{2}$ for strings of

† We are referring, therefore, to a recurrent sequence of events (cf. Section 8.7.1).

‡ It seems out of place here to complicate such expressions in order to repeat critical comments of the kind mentioned in the previous 'Remark' (following (B)).

length 4, ..., $n(u_{n-2}/n)$ for strings of length n, and so on. Observe, in particular, that, for an individual string, the prevision of long strings is negligible (i.e. the prevision of strings $>n$ is less than any given $\varepsilon > 0$, provided n is taken sufficiently large), whereas, for the prevision of length (which is infinite), the prevision of the length of short strings (i.e. of those less than some preassigned, arbitrary, finite n) is negligible. It makes no difference if one makes the same statements but multiplies by 1000 (or a million, or whatever): 'out of 1000 strings, in prevision 500 have length 2, and their total length is, in prevision, 1000', and so on. Usually, one says 'on average'. We shall see later (in Section 8.8.4) that there are dangers in using this form of expression.

8.7.6. *For the Ballot problem in the case of parity, what is the probability that one of the two candidates has never been behind during the count?* This is almost the same question as we asked in Section 8.7.3, except that we are here asking for something less: we admit the possibility that at some stages during the count the votes for the two candidates may also have been equal. It is sufficient that the lead of the candidate in question never becomes negative; i.e. that it never reaches the level $y = -1$. As in Section 8.7.3, we assume $Y_{N-1} = 1$; we are then back in case (*B*) of Section 8.7.1, and we can apply s_c for the case $c = 1$, which we have already given explicitly:

$$s_c = (N - H)/(H + 1),$$

where we have to put $N - 1$ in place of N, and $N/2$ in place of H (since $Y_{N-1} = 2H - (N - 1) = 1$). We hence obtain

$$(N - 2)/(N + 2) = 1 - 4/(N + 2)$$

for the probability that level $y = -1$ is reached, and $4/(N + 2)$ is then the required probability. If we are thinking in terms of one particular candidate, the probability that the lead is never negative is one half of this; i.e. $2/(N + 2)$.

Since the probability of the lead always being positive is $1/2(N - 1)$ (case (*A*) of Section 8.7.1), we obtain, by subtraction, the probability that the lead is always non-negative, but sometimes zero: this probability is equal to the previous one multiplied by

$$3(N - 2)/(N + 2).$$

In a different form, assuming that the lead is always non-negative, the probability that it is always positive is $(N + 2)/4(N - 1)$, and the probability that it is zero is $3(N - 2)/4(N - 1)$ (i.e., for large N, they are practically equal to $\frac{1}{4}$ and $\frac{3}{4}$, respectively).

Remark. We have seen that $2/(N + 2)$ is the probability of a given candidate being ahead (whether strictly or not) either *always* or *never* (i.e. for N or 0 steps; paths either all in the positive or all in the negative half-plane). The possible values for the number of steps when he is in the lead are

$$0, 2, 4, \ldots, N - 2, N,$$

and since there are $(N + 2)/2$ possible values, their average probability must be $2/(N + 2)$. But this is the probability in the two extreme cases we have considered, and, moreover, by an obvious symmetry, the probabilities for h and $N - h$ are equal. It follows that either they are all equal, or they have a strange wavy behaviour—with at least three turning points. Actually, they are all equal: in other words, *we have a discrete uniform distribution over the steps spent in the lead*. The proof is not as straightforward as the statement, and will be omitted,† in order not to interrupt the discussion and overcomplicate matters. A further justification for this is that the previous considerations have already made the conclusion highly plausible.

8.7.7. *What is the probability that in the Heads and Tails process* (*or, more generally, in any Bernoulli process*) *the first crossing of* $y = 0$ (i.e. the first zero where the path does not simply touch the axis) *occurs at time* $t = n$ (*where* n, *of course, is even*)? In other words, we are asking for the probability that the duration of the initial period during which one of the players *never falls behind* is equal to n; i.e. that this is the sum of the lengths of the initial consecutive strings whose sign is that of the first string. In order for this to happen, it is first of all necessary that $Y_n = 0$, and that no crossings have taken place for $t < n$; in addition, we require that the first toss after $t = n$ (i.e. the $(n + 1)$st) is opposite in sign to the very first toss (and thus to the strings already obtained). The probability we seek is therefore given by

$$\mathbf{P}(Y_n = 0)\,.\,[4/(n + 2)]\,.\,2p\tilde{p},$$

where $4/(n + 2)$ is the probability of no crossing occurring (as determined in Section 8.7.6), and $2p\tilde{p}$ is the probability that the first and $(n + 1)$st tosses have opposite signs. In the case of Heads and Tails, $p = \tilde{p} = \frac{1}{2}$, this reduces to $2u_n/(n + 2)$ (or to $u_n/(n + 2)$ if one specifies which of the two players is to be ahead initially). In the general case ($p \neq \frac{1}{2}$), the probabilities (for each finite n) are smaller, and one has the residual probability of the lead being maintained indefinitely (by the favourite, as for the strings). Apart from a comment of this kind—made for comparative purposes—we shall restrict ourselves to the case of Heads and Tails.

If we compare the result obtained for the length L of a string with that for the lead, V say, we have

$$\mathbf{P}(L = n) = u_{n-2}/n, \qquad \mathbf{P}(V = n) = 2u_n/(n + 2);$$

from this it is clear that

$$\mathbf{P}(V = n) = 2\mathbf{P}(L = n + 2),$$

and hence that

$$\mathbf{P}(V \geqslant n) = 2\mathbf{P}(L \geqslant n + 2) = u_{n+2}\,.$$

---

† We merely note that it follows from (64) of Section 8.7.4(*e*), and that the arguments are similar to those mentioned in Section 8.7.10 (for the arc sine distribution).

It is instructive to consider the implications of this; on the one hand for the first few values (small values, corresponding to short strings), and, on the other, asymptotically (large values, corresponding to long periods of lead).

In the case of the first few possible (even) values, we have:

$$n = 2 \quad 4 \quad 6 \quad 8 \quad 10 \ldots$$

$$u_n = \tfrac{1}{2} \quad \tfrac{3}{8} \quad \tfrac{5}{16} \quad \tfrac{35}{128} \quad \tfrac{63}{256} \cdots$$

and hence

$$\mathbf{P}(L = n) = u_n/(n-1) = \tfrac{1}{2} \quad \tfrac{1}{8} \quad \tfrac{1}{16} \quad \tfrac{5}{128} \quad \tfrac{7}{256} \cdots$$

and

$$\mathbf{P}(V = n) = 2u_n/(n+2) = \tfrac{1}{4} \quad \tfrac{1}{8} \quad \tfrac{5}{64} \quad \tfrac{7}{128} \quad \tfrac{21}{512} \cdots$$

As we know from the last remark, the values in the final line are twice those of the penultimate line each shifted one place to the left (except for the final one, which equals $\frac{1}{2}$; doubling the remaining values, whose sum is $\frac{1}{2}$, we obtain again the total probability 1). This direct comparison shows that for $n = 2$ the probability for L is greater (as is obvious: in order to have $V = 2$, we must have $L = 2$, and, moreover, the subsequent string must be of opposite sign). For $n = 4$, they are equal, and subsequently the probabilities for V become greater ($\frac{1}{16} = \frac{4}{64} < \frac{5}{64}$; $\frac{5}{128} < \frac{7}{128}$; $\frac{7}{256} = \frac{14}{512} < \frac{21}{512}$). All this could be seen directly, by simply noting that the ratio $2(n-1)/(n+2)$ is equal to $2 - 6/(n+2)$.

For large values, this ratio is (asymptotically) equal to 2, and, in any case, $\mathbf{P}(V \geqslant n) \simeq 2\mathbf{P}(L \geqslant n) \simeq 2(0{\cdot}8/\sqrt{n}) = 0{\cdot}8/\sqrt{(n/4)}$. This can be expressed by saying that, in a sense, long periods in the lead are four times as long as long strings (more precisely, this is true in the sense that V has the same probability of reaching some (long) length n as L has of reaching length $n/4$).

8.7.8. *For the Ballot problem in the case of parity, what is the probability distribution of the maximum lead attained during the count by a particular candidate? What is the probability distribution of the absolute value of the lead? And what does it become conditional on the fact that a given candidate never lost the lead throughout the count? Or was strictly in the lead throughout the count?* Clearly, if we drop the reference to the Ballot problem, we see that we are dealing with the most general question of the probability distribution of $\vee Y_N$, or of $\vee|Y_N|$ (either for Heads and Tails, or for any Bernoulli process), assuming $Y_N = 0$, and possibly, also, $Y_t \geqslant 0$, or even $Y_t > 0$, for $0 < t < N$. This last assumption is the most restrictive, and it amounts to seeking the probability distribution of the maximum *in a single string*. Under the next to last assumption, we could be dealing with a segment composed *either by a single string, or by several consecutive strings all having the same sign.* In the general case, on the other hand, the segment might consist of a single string or of several, with arbitrary signs: it is only in this latter case that we need to distinguish between $\vee Y_N$ and $\vee |Y_N|$.

In fact, these are simply variants of problems (A) and (B) of Section 8.7.1. We shall consider them separately.

(a) The only assumption is that $Y_N = 0$, and we seek the distribution of the maximum of Y_t. From (B) of Section 8.7.1, we see, taking $H = N/2$, that $s_0 = 1$, and, for $c \geqslant 1$,

$$(66) \qquad s_c = \mathbf{P}(\vee Y_N \geqslant c) = \frac{N(N-2)(N-4)\dots(N-2c+2)}{(N+2)(N+4)(N+6)\dots(N+2c)}$$

$$(s_c = 0 \text{ for } c \geqslant (N+2)/2),$$

$$(67) \qquad r_c = \mathbf{P}(\vee Y_N = c) = s_c - s_{c+1} = \frac{4c+2}{N+2c+2} s_c$$

(in particular, $r_0 = 2/(N+2)$, as we already know). Applying Stirling's formula as given in (30) of Chapter 7, 7.6.4, we have, approximately (for c large, but $2c/N$ small; i.e. N much larger still), $s_c \simeq \mathrm{e}^{-2c^2/N}, r_c \simeq (4c/N)\,\mathrm{e}^{-2c^2/N}$.

(b) Continuing with $Y_N = 0$ as the only assumption, we seek the distribution of the maximum of $|Y_t|$. Arguing as in (C) of Section 8.7.1, but also taking into account the symmetry ($N = 2H$; i.e. $Y_N = 0, c' = c'' = c, c^* = 2c$), we find that the probability $\bar{s}_c$ of either reaching or crossing $\pm c$ (which was denoted in (C) by $q_N = q'_N + q''_N$, here $q'_N = q''_N$) is given by

$$(68) \qquad \bar{s}_c = (2/\tbinom{2H}{H})[\tbinom{2H}{H+c} - \tbinom{2H}{H+2c} + \tbinom{2H}{H+3c} - \tbinom{2H}{H+4c} + \dots].$$

Expressed more simply, using the s_c from (a) above, we have

$$(68') \qquad \bar{s}_c = s_c - s_{3c} + s_{5c} - s_{7c} + \dots,$$

and similarly

$$\bar{r}_c = r_c - r_{3c} + r_{5c} - r_{7c} + \dots.$$

Asymptotically, we therefore see from the previous expression that

$$(68'') \qquad \bar{s}_c = 2\sum_{k=1}^{\infty}(-1)^{k+1}\,\mathrm{e}^{-2k^2c^2/N}, \qquad \bar{r}_c = (8c/N)\sum_{k=1}^{\infty}(-1)^{k+1}k\,\mathrm{e}^{-2k^2c^2/N}.$$

(c) Let us now assume, in addition to $Y_N = 0$, that Y_t has not changed sign throughout $0 \leqslant t \leqslant N$, and, in order to fix ideas, let us assume that it is non-negative. It therefore makes no difference whether we talk in terms of the maximum of $|Y_t|$ or of Y_t (or of $-Y_t$, had we made the opposite assumption). We again argue as in (C) of Section 8.7.1, but now with $c' = 1$, $c'' = c$, in order to obtain the probability that Y_t always remains strictly between -1 and c. Dividing this by $2/(N+2)$ (the probability that $0 \leqslant Y_t$, i.e. that $-1 < Y_t$), we obtain the probability of $\vee Y_t < c$ conditional on the given hypothesis; i.e. the $1 - \bar{s}_c$ of the present case. If we use s_c to denote the same probability as in case (*a*), we have, therefore,

$$(69)\ \bar{s}_c = 1 - \frac{N+2}{2}$$

$$\times (s_1 + s_c - s_{c+2} - s_{2c+1} + s_{2c+3} + s_{3c+2} - s_{3c+4} - s_{4c+3} + \dots), \text{ etc.}$$

(*d*) This proceeds similarly: under the assumption of having been strictly in the lead ($Y_t \geqslant 1, 0 < t < N$), we can reduce to the previous case by taking the axis $y = 1$ as the base line on the interval from $t = 1$ to $t = N - 1$ ($Y_1 = Y_{N-1} = 1$); for large N, the difference is very small.

8.7.9. *Similar problems for an arbitrary segment of the Heads and Tails process* (i.e. for a segment $0 \leqslant t \leqslant n$, where we do not assume, as in the previous examples, that $Y_n = 0$). The segment could consist of one string, or several strings, or none, and, in general, it will end in an incomplete string. We shall give a brief review of certain problems and their solutions, in order to draw attention to various of the points which need considering.

(*a*) So far as *periods in the lead* are concerned (cf. Section 8.7.6 and the *Remark*), we know, from Section 8.6.3, that u_n is the probability of Y_t remaining non-negative (in $0 \leqslant t \leqslant n$); i.e. that the lead is maintained for n steps out of the n (and the same holds true, obviously, for 0 out of n). Assuming n to be even, the number of steps spent in the lead can be any of

$$0, 2, 4, \dots, n-2, n.$$

We therefore have $(n + 2)/2$ possible values, and the average probability is $2/(n + 2)$. However, the extreme cases have probabilities $u_n > 2/(n + 2)$,† and so it is likely (by the same kind of argument that in the *Remark* of Section 8.7.6 led us to believe that in that case, a segment consisting of complete strings, the probabilities were equal) that, in the general case, there is a very small probability that a subdivision into periods of lead will consist of nearly equal lengths, and much greater probabilities for the less equal subdivisions. There is another consideration which makes this plausible. We already know that for the segment leading up to the last zero we have equal probabilities for all subdivisions, and that, from the last zero on, the lead does not change hands. In any case, the fact is that this turns out to be true, and in precise terms, we obtain $p_h = u_n u_{n-h}$ (h and n even) for the probability of being in the lead for h out of the n steps. The proof is much more difficult than one might expect from the simplicity of this formula, and we shall omit

† Since it is the maximum of $n + 1$ probabilities $\omega_h^{(n)}$ ($h = 0, 1, \dots, n$), u_n is certainly $> 1/(n + 1)$. It is, in fact, much greater than this, becoming more and more so as n gets larger: asymptotically, $u_n \simeq (2/(n + 2)) \,.\, 0{\cdot}4\sqrt{n}$.

it.† We shall restrict ourselves to a consideration of how this probability behaves.

Recalling that $u_{h+2}/u_h = (h+1)/(h+2)$, we see that the ratio

$$p_{h+2}/p_h = (h+1)(n-h)/(h+2)(n-h-1)$$

is less than, equal to, or greater than 1, according as $h + 1 \lesseqgtr n/2$: taking the asymptotic expression for u_n, we have $p_h = u_h u_{n-h} = (2/\pi)\sqrt{[h(n-h)]}$. In the limit, we can say that the proportion of time during which a gambler is ahead, in a long period of play, is a random quantity X, whose probability distribution has density $f(x) = 1/(\pi\sqrt{[x(1-x)]})$. This is the *'arc sine'* distribution, so-called because the distribution function, $F(x) = \int f(x)\,\mathrm{d}x$, is equal to $(2/\pi)\sin^{-1}(\sqrt{x})$, which might be better written as

$$(1/\pi)\cos^{-1}(2x-1).$$

We shall come back to this again and again.

(*b*) We know that the *probability distribution of* Y_n is the *Bernoulli* (or binomial) distribution ($p_h = \omega_h^{(n)}$, $h = 0, 1, \ldots, n$), provided we know only that $Y_0 = 0$ (this holds similarly if we are given certain values, of which the last one is $Y_k = y$ with $k < n$; we then have $p_{y+h} = \omega_h^{(n-k)}$). The distribution is the *hypergeometric* if, in addition to Y_0, we are also given a value Y_N, $N > n$ (and similarly if two arbitrary values are known, one before and one after; $Y_{t'}$ and $Y_{t''}$, say, with $t' < n < t''$).

In general, we can say that any change in the state of information produces a change in the probabilities. In particular, if, in addition to knowing the initial value $Y_0 = 0$ (and, possibly, a subsequent value, or two arbitrary values, one before and one after), one also knows that Y_t has remained non-negative throughout $0 \leqslant t < n$, we have a range of possibilities as discussed above. The same holds if we have non-negativity throughout the entire process, $0 \leqslant t \leqslant N$, or throughout $t' \leqslant t \leqslant t''$ in the case where we know the values at the two points on either side, or even only for $t' \leqslant t < n$, or $n < t \leqslant t''$. Different probabilities also result if we assume the process to be strictly positive or strictly negative, or above or below some given level, or in between two levels, and so on. All this is obvious, but it needs emphasizing, and should be borne in mind.

(*c*) *The probability that level* $y = c$ *is reached for the first time at* $t = n$ ($Y_n = c, \wedge Y_{n-1} < c, c > 0$; symmetrically if $c < 0$) *is equal* (by the principle of reversal the problem is unchanged) *to the probability that* $Y_n = c$ *without there being any zeroes beforehand* (i.e. all Y_t, $0 < t < n$, have the same sign as $c = Y_n$). This probability is given by $\mathbf{P}(Y_n = c)$ multiplied by c/n, where c/n is *the probability that*, the value c of Y_t at $t = n$ being known, *this level c*

---

† Cf. the notes in Section 8.7.10.

has been reached for the first time at that point, and also the probability that the starting level ($y = 0$) has not been reached again; i.e. that Y_t *has always had the same sign as c*.

If we denote by H one or other of the two hypotheses mentioned, we have that $\mathbf{P}(H) = u_{n-1}/2$ (Section 8.6.3) and that $\mathbf{P}(H|Y_n = c) = c/n$ (the Ballot problem: case (A) of Section 8.7.1). The probability that we seek (which can also be obtained as the probability of ruin occurring precisely at the nth toss, $p_n(c) = q_n(c) - q_{n-1}(c)$, cf. Sections 8.6.1 and 8.6.3) therefore has the value stated:

$$\text{(70)}\qquad \mathbf{P}[(Y_n = c).H] = (c/n)\mathbf{P}(Y_n = c) = (c/n)\omega_h^{(n)} \qquad (c = 2h - n, c > 0).$$

The probability distribution of Y_n conditional on H (one of the two hypotheses) is therefore proportional to $c\omega_h^{(n)}$; in fact, we have

$$\text{(71)}\qquad \mathbf{P}(Y_n = c|H) = \mathbf{P}[(Y_n = c).H]/\mathbf{P}(H) = (2/nu_{n-1})c\omega_h^{(n)} \qquad (c = 2h - n > 0).$$

Expressed in words: the probabilities of the possible values c for Y_n are altered (by the condition H) in a manner proportional to c (those for $c \leqslant 0$ are clearly 0), and the normalization factor K has been found explicitly in passing. The same result (and proof) goes through for the opposite hypothesis, and provides the probability that $Y_n = c$ given that Y_n is greater than any value obtained previously (for $0 \leqslant t < n$, all Y_t are $< Y_n$; in this case, of course, we do not exclude the possibility of negative values). Asymptotically, we have a distribution of the form $f(x) = Kx\,\mathrm{e}^{-x^2/2}\ (x \geqslant 0)$.

(d) We now restrict ourselves to the special case of knowledge of one value before and one after, $Y_0 = 0$ and $Y_N = 0$, together with the condition that $Y_t > 0$ throughout the given interval. The distribution of Y_n (for any integer n, $0 < n < N$) is obtained in a similar way, by observing that the probability that $Y_n = c$ ($c = 2h - n > 0$), and that the given conditions hold, is, setting $N = 2H$, given by

$$\text{(72)}\qquad \left[\binom{H}{h}\binom{H}{n-h}\Big/\binom{2H}{n}\right].(c/n).(c/(N-n))$$

(the product formed by taking $\mathbf{P}(Y_n = c)$ from the hypergeometric case and multiplying it by the probability that the process does not vanish in the passage from 0 to c during the first n and subsequent $N - n$ steps). We note, however, that this probability can also be thought of as the product of the p_h we are after, multiplied by the probability of the hypothesis that at time $t = N$ we obtain the first zero, the sign having previously been positive. This latter probability is equal to $u_n/2$, and so we obtain

$$\text{(73)}\qquad p_h = \left[\frac{2}{u_n n(N-n)\binom{N}{n}}\right]c^2\binom{H}{h}\binom{H}{n-h} = K\,.\,c^2\,.\,\overline{\omega}_h^{(n)}$$

(we have placed a bar over the ω in order to stress the fact that it is the ω of the hypergeometric rather than the Bernoulli process, as above). One should note, however, the meaningful analogy between the two (every condition of positivity, on the left and on the right, entails a modification proportional to c). Asymptotically, we have a distribution of the form $f(x) = Kx^2\,\mathrm{e}^{-x^2/2}$ $(x > 0)$.

8.7.10. *Remark.* It is instructive to consider in more detail some of the problems which lead (asymptotically) to the arc sine distribution (cf. Section 8.7.6, and the *Remark* in Section 8.7.9 (*a*)). We have omitted the proofs, and we suggest that the reader refers to the third edition of Feller, Vol. I (1968). Comparison with earlier editions will reveal the simplifications in formulation which took place from one edition to the other, partly as a result of greater insight into the problems and their connections with one another, and partly for purely fortuitous reasons (cf. Chapter III, Section 4 of Feller, 'Last visit and long leads', and, in particular, the historical notes on pages 78 and 82).

The arc sine distribution was considered by P. Lévy (1939) in connection with the Wiener–Lévy process (cf. Section 8.9.8).† The application to Heads and Tails and other cases (obvious in the asymptotic case) seemed to require 'mysterious' forms of explanation, until their combinatorial character was revealed (by Sparre Andersen, in 1953). The methods used were quite complicated, and still were in the first edition of Feller (where they were due to Chung and Feller); the new feature, which arises in the passage from the second to the third edition, lies in the preliminary statement of a simpler theorem, which, from a qualitative viewpoint, begins to explain the weighting towards very unequal subdivisions of periods in which the lead does not change hands.

We can prove this in just a few lines. *The probability that in $2m$ tosses at Heads and Tails the final return to zero, $Y_t = 0$, occurs at $t = 2k$ is given by $u_{2k}u_{2m-2k}$* (which is the discrete version of the arc sine distribution). In fact, the probability that $Y_{2k} = 0$ is u_{2k}, and the probability of no zeroes in the $2m - 2k$ subsequent tosses is u_{2m-2k} (cf. Section 8.7.4(*a*)). This is trivial, and yet the theorem is new (according to Feller); moreover, it was discovered by chance, experimentally, on the basis of observed statistical properties of random sequences produced by a computer. These were detected by capable mathematical statisticians, who then simply pointed out, and subsequently proved, that the distribution was symmetric (without realizing that it was the arc sine distribution).

This is by no means intended as in any way disrespectful to the number of authors who have made valuable contributions to this topic. It merely goes to show that tucked away in the vast rock-pile of problems there is the odd nugget lying unobserved; once noticed, of course, it appears obvious.

The following little calculation, which I made simply out of curiosity, may be new, and possibly of some interest. I observed that it was not appropriate to call one

† I (vaguely) remember that an obvious property of the arc sine distribution—the density taking its maxima at the end-points—was considered paradoxical, even in cases where it was natural, as for observations of periodic phenomena (for example, in the case of a river flooding, the level remains around the maximum longer than it does around intermediate values, which are passed through more rapidly, both when the level is increasing and decreasing): see Figure 8.9. Of course, when periodicity is crude (for example, seasonal temperature changes, where maxima and minima vary from year to year) there are smoothed peaks, or sequences of peaks.

gambler 'luckier' simply because he has led for most of the game so far (see the footnote to Section 8.8.1): it might well turn out that his luck runs out at the end of the game—'he who laughs last laughs longest'. The probability of this happening is given by

$$1/\sqrt{3}\pi = 0{\cdot}184 = \int_{\frac{1}{2}}^{1} (2\pi t\sqrt{[t(1-t)]})^{-1}\,dt. \tag{74}$$

To see this, let t denote the time when the final zero occurs (taking $[0, 1]$ as the whole interval). If $t < \frac{1}{2}$, the one who is leading at the end is also the one who has spent most time in the lead. If $t > \frac{1}{2}$, in order for the one who is leading at the end to have spent most time in the lead, he must previously (i.e. before time t) have spent at least an additional time $t - \frac{1}{2}$ in the lead. Because of the uniformity of the distribution of the lengths in an interval between two zeroes, this, for given t, has probability $(t - \frac{1}{2})/t = 1 - \frac{1}{2}t$. This leaves a probability $\frac{1}{2}t$ for the opponent, conditional on t (having the arc sine distribution) being greater than $\frac{1}{2}$; we thus obtain (74).

8.8 THE CLARIFICATION OF SOME SO-CALLED PARADOXES

8.8.1. We have already found (and will do so again) that certain of the conclusions we have arrived at have had a paradoxical air, or, at any rate, have been easy to misinterpret. We have discussed various examples where such misinterpretation arises, and, in so doing, we have attempted to clarify the issues involved. In particular, we recall the laws of large numbers, and, in the gambling context, the long expected time to ruin. The topics we have just been considering also lend themselves to discussions of this kind. Indeed, it is hard to decide whether their main value lies in the knowledge they provide, and the light they throw on a number of important theoretical and practical questions, or in the opportunity they give one to clear up a number of misconceptions and confusions, which otherwise could make one rather wary of entering into the probabilistic domain at all.

In those aspects of the Heads and Tails process which we have just been studying, it surely seems rather strange and mystifying that some kind of 'stationarity' or regularity does not hold. In particular, why is there not a tendency for the periods of unbroken lead to be equally distributed in the two opposite directions (all the more so after having seen that the process can be considered as an indefinite sequence of *strings*, at the end of each of which the process begins all over again under identical conditions)?

In particular, since the alternation of strings in the two directions (i.e. in the positive and negative half-planes) is itself a Heads and Tails process when the strings are considered as 'tosses', it seems obvious that the balancing of periods in the lead should hold by analogy with the balancing up of the frequencies of Heads and Tails. In actual fact, this conclusion is true if one considers the *number* of strings giving the lead in one or other of the two directions, but it is *not* true if one wishes to consider the respective total *durations* of periods in the lead. In fact, we have seen (cf. the *Remark* in Section 8.7.6) that in an interval formed by complete strings (i.e. those ending in a zero) all durations are equally probable, instead of, as one might have expected, those of intermediate length being more probable (i.e. we have a distribution into almost equal periods). In the general case (where the final string may not be complete, i.e. the interval does not end in a zero; cf. the *Remark* in Section 8.7.9), the situation is, in fact, the very opposite; it is the most unbalanced distributions which are most likely.

Both the form of the density, $f(x) = K/\sqrt{[x(1-x)]}$, and that of the distribution function, $F(x) = K\cos^{-1}(2x-1)$, show clearly that the extremely asymmetric values are favoured (although in a symmetric fashion so far as the two opposite directions are concerned). The best way to visualize the result is to note that the splitting of the total duration of a long game into the fractions x and $1-x$ of the duration in which one or other of the two gamblers is ahead can be thought of as brought about by choosing 'at random' (i.e. with uniform density) a point on the circumference of the semi-circle having the segment $[0, 1]$ as diameter, and then obtaining x by projecting that point onto this diameter. In other words, if the circumference is divided into an arbitrary number of equal arcs, their projections onto the diameter (which will clearly be smaller the nearer they are to the end-points) are equally probable (because they contain the point dividing the two parts x and $1-x$).

The reader should now examine Figure 8.9(c), (b), (a) (in reverse order, from bottom to top), together with the notes which accompany it.

Feller (Vol. I, Chapter III) provides numerical data which illustrate this phenomenon and make clear why it is not, in fact, surprising. Imagine that two gamblers play continuously for a year (making a toss every hour, minute or second, it does not matter which). It turns out that there is only a 30% probability of both being ahead for more than 100 days (about 28% of the total time), whereas there is a 50% probability that one of them remains ahead for less than 54 days (15% of the time), 20% that he remains ahead for less than 9 days (2·4% of the time), 10% for less than $2\frac{1}{4}$ days (i.e. less than 0·6% of the time—more than 99·4% for his opponent!).† Feller also provides the details of the behaviour resulting from a computer experiment.

† Sometimes people refer to 'the less fortunate' player. This is not quite right, however, since it is possible (although not very probable) that the one who has been in the lead for most of the time finds himself behind at the end (cf. Section 8.7.10).

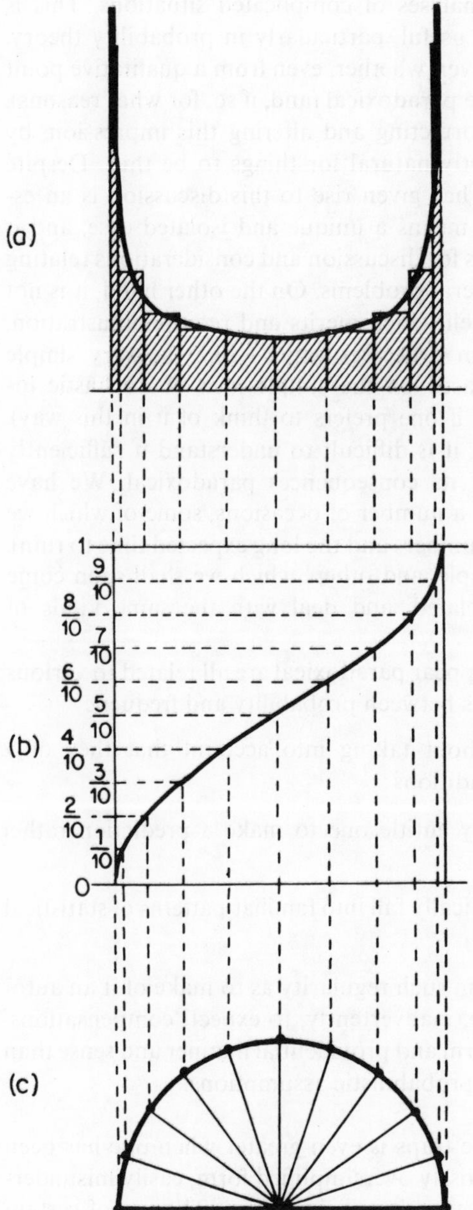

Figure 8.9 These should be read in reverse order (i.e. (c), then (b) and finally (a)).

(a) The density of the arc sine distribution. The histogram shows the average density in each of the intervals between the deciles. The graph shows the density, whose equation, taking the interval to be [0, 1], is

$$f(x) = K/\sqrt{[x(1-x)]},$$

and is infinite at the end-points.

(b) The distribution function of the arc sine distribution (obtainable using the device shown in (c)).

The abscissae shown correspond to the 'deciles' (cf. Chapter 6, 6.6.6) since they are obtained from the corresponding ordinates.

The ten intervals between the deciles are equally probable (with probability $\frac{1}{10}$). Note how much more dense the probability is near the end-points.

(c) The probability distribution of the projection (onto the diameter) of a point 'chosen at random' (with uniform density) on the circumference.

This distribution occurs, for example, if one measures, at a 'random' instant, the position (or velocity) of a point performing harmonic oscillations.

The division of the circumference into 10 equal parts (18°) gives the deciles

8.8.2. It is not really surprising that these numbers are not what we would imagine intuitively. Intuition cannot guide us—not even roughly sometimes—in foreseeing the results from analyses of complicated situations. This is precisely why mathematics is so useful, particularly in probability theory.

We should ask ourselves, however, whether, even from a qualitative point of view, the above conclusions are paradoxical (and, if so, for what reasons), and how one might set about correcting and altering this impression, by showing that it is, in fact, perfectly natural for things to be thus. Despite the fact that the example which has given rise to this discussion is an especially striking one, it is by no means a unique and isolated case, and it provides us with an excellent basis for discussion and considerations relating more or less directly to more general problems. On the other hand, it is not so much the individual result itself which merits and requires illustration, but rather the nature of random processes which—like the very simple case of Heads and Tails—are based on the simple idea of stochastic independence (or lack of memory, if one prefers to think of it in this way). Although this is a simple notion, it is difficult to understand it sufficiently well to avoid finding certain of its consequences paradoxical. We have already commented upon this on a number of occasions, some of which we recalled above (the laws of large numbers and the long expected time to ruin). These are—like the present example, and others which we shall soon come across—topics which are interrelated, and deal with the same kinds of questions.

The reasons why these results appear paradoxical are all related to various kinds of distortion of the relations between probability and frequency:

by assuming connections without taking into account that they only exist under certain restrictive conditions;

by thinking that they virtually entitle one to make a prediction rather than a prevision;

by assuming that they systematically fall into familiar patterns of statistical 'regularity';

by having such a strong belief in such regularity as to make of it an autonomous principle; this leads one, inadvertently, to expect 'compensations' to take place in a more extreme form and providential manner and sense than can be derived legitimately from probabilistic assumptions.

The danger of falling into these traps is even greater when one has been taught statistical concepts in a grossly oversimplified form, easily misunderstood and without the necessary warnings being given. The use of certain forms of terminology—for example, saying that something occurs *on average* a given number of times per unit time (instead of saying *in prevision*)—can lead one to regard such 'regularities' as certain, instead of merely being probable; that is, as predictions instead of previsions.

If 'regularity' is assumed as an 'article of faith' (and there is a book on statistics, inspired by this outlook, which is entitled 'Gleichformigkeit der Welt'), how can it be, one might ask, that phenomena like returns to equilibrium and the distribution of leads could violate this regularity, thus challenging the supreme dictate of the ordered universe? If one thinks of returns to equilibrium (which are practically certain) as revealing a tendency towards, or desire for, such 'regularity', one would expect that a particle, having gone a long way below the origin while performing a random walk, would make an about turn in order to return to the fold. On the contrary, it has no memory and there is no fold to return to. It might carry on until it is twice as far from the origin before it turns back towards it, or it might have only gone half as far. If it does end up by going back to the origin with certainty, this is simply because, being on a random walk, it will sooner or later pass through all the points (but without any possibility of recognizing the point we have labelled the 'origin', nor any desire to do so).† One would have a stronger case (and, indeed, a valid one were it not for omitting to point out the fact that the expected duration is infinite, or, at least, for not taking it into account) if one were to argue that the phenomenon should reproduce itself 'regularly' because after each return to the origin a new string begins under precisely the same conditions.

If we attempt to identify and explain those reasons which we assume to underlie the tendency to talk in terms of 'paradoxes', we find the answers staring us in the face. The probability–frequency relation as it occurs in the law of large numbers should not be assumed to hold, since the successive Y_n are not independent. They depend upon a 'cumulative effect', which tends to be dominant; deviations take place only slowly, and returns to equilibrium and changes of sign, i.e. of 'lead', only seldom. We have already mentioned above the idea of some kind of restoring force causing returns to equilibrium. However, only the last point is really important, because it pin-points a subtle and basic difference (whereas the other points simply caution one against the possibility of trivial and rather absurd misunderstandings).

The very fact that the probability of a return to the origin at time $t = n$ *tends to zero* (for n even, $u_n = 0{\cdot}8/\sqrt{n}$) should be sufficient to rule out the 'regularity' or 'stationarity' of behaviour that is the implicit and unconscious assumption occasioning all the 'astonishment' at these 'paradoxes'. The

† I do not mean to imply that fallacious ideas of this kind are accepted statistical doctrine in some other approach differing from the one we follow. However, the environment created by a few introductory sentences followed by empirical clarifications, etc., does not seem to be sufficiently antiseptic to prevent the germs of these dangerous distortions from multiplying in the subconscious.

latter appear as such simply because they do not fit into that particular framework; a framework which has created, in its own image and likeness, psychological attitudes whose tendency to rise to the surface becomes, in the absence of any process of re-education, general and indiscriminate. The following points may serve to provide a better appreciation of just how 'sensational' are the consequences of this probability tending to zero (a fact which is so simple when it is accepted as it is, without further thought). In a consecutive sequence of k tosses (e.g. 100,000), the probability that Heads always occurs from some n onwards is very small, but nonetheless finite; and this is also true for the probability that, starting from n, the sequence 11001... (where 1 = Heads and 0 = Tails) represents, in the binary code, either the first 100,000 decimal places of π, or the text of the Divine Comedy, or that it reproduces the initial segment obtained with the first k tosses, or any other preassigned segment of fixed length. Something of this kind occurs, in prevision, once every 2^k tosses, and it is practically certain that it occurs at least once within every segment of length N, if N is considerably larger than 2^k (and also that it occurs at least 10 times, or at least 1000, etc., provided we take a long enough sequence; the details are straightforward, and we shall not bother with them here). On the other hand, the expected number of returns to equilibrium with a given number, N, of consecutive tosses starting from $t = n$ is, approximately,

$$(0{\cdot}8/\sqrt{n})\,.\,N/2 = 0{\cdot}4N/\sqrt{n} \to 0,$$

and the probability of at least one is even smaller. This means that, if one proceeds far enough to have an interval sufficiently long to give a non-negligible probability of containing a return to equilibrium (e.g. 1% or 10%), then one has an interval which almost certainly (e.g. with probability 90% or 99%) contains the Divine Comedy at least once, and, if one carries on, at least 10 times, 1 million times, and so on, indefinitely.

8.8.3. Turning to a consideration of the values of Y_t (and not only at those instants for which $Y_t = 0$), a topic we shall be dealing with shortly, we can deduce an immediate and straightforward result about the extreme length of the strings at times far away from the starting time $t = 0$. We know that $|Y_n|$ has probability $\simeq 0{\cdot}8\, M/\sqrt{n}$ of being less than a preassigned M. For sufficiently large n, it is therefore almost certain that $|Y_n| > M$, in which case the string containing the instant $t = n$ *necessarily* has length $> 2M$. In fact, the length would equal $2M$ under the assumption that the increase from the previous zero to Y_n and then the decrease to the following zero take place in an unbroken sequence of M successes and M failures, respectively. From considerations made in the context of the ruin problem, however, we know that it is probable that the increase and the decrease take much longer.

But the above remarks only have an illustrative and introductory value: they help us to see what is happening, but they do not yet provide us with an explanation; neither do they resolve the confusion by going back to the source. At most, there is a restatement of the problem: instead of asking ourselves *why do the lengths of the strings become longer and longer* (despite the fact that they begin again from zero under the same conditions), we can ask why, given the same assumptions, do the ordinates Y_n become larger and larger (in absolute value); i.e. *why do the strings get further and further away from the axis* $y = 0$ (and it is clear that the two questions, even if not identical, are closely related).

Let us study then the history of each string, and we might as well consider the first one, starting at $t = 0$. The probability that its length is $n = 2m$, i.e. that it terminates at the nth toss (with a return to equilibrium), is

$$u_n/(n-1) = u_{n-2}/n.$$

But u_{n-2} is the probability that no zero has previously occurred, and so $1/n$ is the probability that the string terminates at time *n assuming that it does not terminate earlier* (for n even; otherwise, the probability is zero). It follows that a string has probability $\frac{1}{2}$ of terminating at the second toss, $\frac{1}{4}$ at the fourth (provided it did not terminate at the second), $\frac{1}{6}$ at the sixth (provided it did not terminate at the fourth), and so on. In demographic terms, a string can be thought of as an individual whose probability of dying decreases with age (this happens for new-born babies, the probability of survival increases each day they survive; the difference is that they grow old and the conditions become worse, whereas for the strings they continue to improve). We therefore see why the probability of 'the string enclosing the instant t' is smaller if t is smaller than if t is large. In the first case, it is necessarily 'young' (age at most t), and this bounds the past duration (the age that has been attained) and provides less favourable possibilities for the future (since an individual gets stronger with age): it has probability at least $1/(t+2)$ of terminating at the next even toss, and so on.

And what are the probabilities of the various possible values for Y_n? They are no longer those of the Bernoulli distribution that we had before. We are now in a different *state of information*, because we are discussing a string; in other words, we know (or assume) that Y_t has not vanished in the meantime, and that the probability we seek is that conditional on this hypothesis, H, as we have already seen (in Section 8.7.9(c)). This means that knowing that there are no zeroes modifies the distribution in favour of the larger values (as is natural); more precisely, it alters the probabilities in proportion to their sizes.

Every change in the state of information brings about a modification. For instance, if I knew the values of Y_n at every instant, then the probability of the string ending at the next toss—let us take n to be odd—is no longer

$1/(n+1)$, but is $\frac{1}{2}$ if $Y_n = \pm 1$, and zero otherwise. The situation would be different again if I knew just a *few* of the past values. If I knew Y_t at certain instants $t = n_1, n_2, \ldots, n_k$, the probabilities would be those conditional on the last value; i.e. on the hypothesis $H = (Y_{n_k} = c)$. But beware! This is only true if I have no knowledge, no clue whatsoever, relating to the results following the last known result. For example, if I obtain information every time $Y_t = 0$, I not only know the position of this last zero, but I also know that no other zeroes have occurred since (and I then have the distribution given above, for the present Y_n, whereas, otherwise, I would have the Bernoulli distribution). Yet another situation would arise if I knew it to be more probable for information to be available in the case of returns to the origin, or in the case of large values being attained, and so on, than in other cases: *absence of information can itself be informative*. In the cases just mentioned, it increases the probability of the non-occurrence of those things which, had they occurred, would probably have been reported.

Many of the mistakes which are made in the probabilistic treatment of problems and phenomena derive precisely from either ignoring, or forgetting, or giving insufficient weight to, the following fact: that everything depends upon the current, actual state of information (with all the attendant flexibility that attaches to this notion in practice). In interpreting deterministic laws, also, we need to keep circumstances of this kind in mind. An example is provided by the treatment of 'hereditary' phenomena, such as hysteresis, using integral or integral–differential equations. If one assumes that knowledge of the past enters the picture indirectly through the modification it produces in the structure determining the present state (which is itself not directly observable), then we see that certain information (in our case, 'the past') may or may not be 'informative' in so far as the effects we are interested in are concerned (here, *deterministic* prevision of the future—i.e. 'prediction'—rather than prevision), depending on whether certain other information is, or is not, available (here, complete information about the structure of the 'present state'). If this latter information is not available, the information concerning the past serves as a substitute. It may be a completely adequate substitute, or only partially so, according as the knowledge of the outside influences in the past are, or are not, considered sufficient to determine, completely and with certainty, the present unobservable situation. In the latter case, we are essentially back in the realm of probability, even if this remains obscured by the fact that the treatment deals only with macroscopic behaviour, neglecting the random aspects which are, in that context, negligible.

In the probabilistic field, however, information is always incomplete, and derives from the distinction—which, in any case, is never very clear-cut—between what one knows, or believes one knows, or definitely remembers, and what one does not know. A fundamental rôle is played by that certain something which is, in a sense, complementary to information, and which comes about by interpreting the reasons for the absence of information. We have illustrated this with the Heads and Tails process, and we shall return to it again and again, sometimes with illustrations which are particularly instructive because they are, at first sight, rather disconcerting.† For examples from more familiar fields, note that the knowledge of a person's age is, to some extent, a substitute for a medical examination in so far as the evaluation of the probability of death is concerned, and that, for a person who is insured, present age, together with the medical report dating back to when the policy was originally taken out, are taken as substitutes for a medical examination at the present time. In like manner, the fact of whether or not one receives direct news, or whether or not the newspapers carry reports of a certain situation, individual, firm, institution, etc., itself constitutes information (satisfactory or not, as the case may be). Any attempt—and these are still frequent—to base the theory of probability on some distinction between those things of which one is *perfectly certain* and others of which one is *perfectly ignorant* precludes, for the reasons we have given (and by not taking into account objections of principle), an understanding of the most meaningful aspects of problems requiring the use of probability theory.

Going back to the discussion of the questions we considered for the Heads and Tails process, and to the doubts expressed in this context ('*how can it be* . . . ?'), we see, therefore, that, in line with what has been said, the answer lies in making it clear that the situation—for example, the probability of a return to equilibrium at a given instant—does not alter merely because of the passage of time, or because time modifies something or other, but rather because our state of information changes. Initially, i.e. not conditional on any subsequent information, our state of information consists solely of the knowledge that $Y_0 = 0$. When we later place ourselves at times $t = n$, however, we have a *changed* state of information; one that consists in knowing that there was a passage through the origin *n steps ago* (without knowing whether or not this was the last zero, nor anything else which would lead us to depart from an identical state of information regarding the 2^n possible paths that could have been followed meanwhile).

8.8.4. Despite all this, the doubt might linger on, transplanted to the new ground opened up by the new information. How can it be that the variation in the state of information to which we referred continues to exert an influence even when we move indefinitely far away? There is, in fact, a case in which knowledge of the initial state, provided it is sufficiently remote, ceases to have any influence (this is the so-called *ergodic* case, which we shall

† For example, the equivalence of the Bayes–Laplace scheme to that of Pólya's 'contagion probabilities'; cf. Chapter 11, 11.4.4.

be concerned with briefly in Chapter 9, Section 9.1, when we deal with 'Markov Chains'). This is, however, something that only occurs under certain specified conditions, and the fact that it often crops up is merely an indication that these conditions are satisfied in many of the problems one considers, rather than because of some principle which permits one to use the idea indiscriminately.

Here, too, as in the case of belief in a tendency to equilibrium, it happens that a special circumstance is assumed as some kind of autonomous 'principle', rather than as a simple and direct consequence of conditions which may hold in some cases and not in others. In this way—partly by accident, and partly as a result of the usual obsession with replacing probability theory by something which is apparently similar, but which can, in fact, be reduced to the ordinary logic of certainty—one ends up by seizing upon the most fascinating results (like the laws of large numbers and the ergodic theorem) and raising them to the status of principles. When the applications of these principles to situations in which the theorems they misrepresent do not hold turn out to be contradictory, the results are then held to be paradoxical. As an analogy, it is as if, instead of the principle of conservation of energy, one took as a 'principle' the statement that a field of forces must be conservative, and then were faced with justifying the 'paradoxical' cases (like magnetic fields) where the 'principle' no longer holds.

As an example, it is often asserted—especially by philosophers—that the calculus of probability *proves* that the 'ergodic death' of the universe is inevitable. On the contrary, the calculus of probability (the logic of uncertainty) is completely neutral with respect to facts and behaviour relating to natural phenomena, and with respect to any other kind of 'reality' (in just the same way as the logic of certainty is). It is absurd to believe that the calculus of probability can itself rule out any particular belief or that it can force one to adopt it; whether it be a belief in 'ergodic death', or whatever. All that it does do is to rule out 'incoherent previsions', on the grounds that these are not previsions; in the same way as the logic of certainty precludes one making the assertion that a horse has three fore legs and four hind legs, making a total of five (whereas it would be admissible to say $3 + 4 = 7$, or $3 + 2 = 5$, or $1 + 4 = 5$). The point is that it must not be inconsistent; the question of whether or not the statements conform to what zoologists regard as admissible is not relevant. Ergodic death is very probable if one accepts, or at least assumes as the most plausible, that model of physical phenomena which regards things as deriving from the destruction of an initial state of order (as in the mixing of gases; kinetic theory). But the calculus of probability in no way precludes phenomena in which a new order is created (as in biology; in particular, the mechanisms of reproduction for DNA, and hence of cells, human beings, and—who knows?—new stars or galaxies);† on the contrary, its techniques provide the means of analyzing them.

Returning to our case, one could say that the 'ergodic principle' *no longer applies* (which is another way of saying that the 'ergodic theorem does not apply unless the necessary assumptions are satisfied'), because if, at time $t = n$, we know that $Y_0 = 0$, then, amongst other things, we already know that $|Y_n| \leqslant n$ with certainty (not to mention our knowledge of the distribution); this information is significant, although its significance varies a great deal with n. The opposite situation—the ergodic case—occurs if we think of the same random walk on an m-sided polygon (m odd; a step clockwise or anticlockwise, according as we get a Head or a Tail). It is clear that knowledge of the starting point is practically irrelevant for evaluating the probabilities of the m positions after n steps (for n large); these probabilities will all be practically equal (to $1/m$).

There is one thing that we have seen, however, which seems to contradict the (obvious) fact that the process begins again under identical initial conditions after every return to zero. If this is so, how is it that the first zeroes can be expected to be very close and then subsequently get further and further apart in the startling way brought home by our discussion of the interposed repetitions of the Divine Comedy? There is, in fact, no contradiction. Each time a zero occurs, it is to be expected that there will be several close to each other; the initial period is a special case of this. As a result, obviously, the groups of zeroes are even further from one another than the individual zeroes would be if we took the same number of them but assumed them approximately equidistant. For an arbitrary zero, for example, the kth, conveying *no information* about the length of adjacent strings (as would be the case if one said, for instance, 'the first zero after the nth toss', because it is likely that the nth toss will fall within a long string), the probability is always $\frac{1}{4}$ that the two adjacent zeroes are the minimal distance away ($=2$; in other words, that the two adjacent strings have the minimal lengths possible, i.e. 2); $(\frac{5}{8})^2 \simeq 0{\cdot}39$ that they are both not more than a distance of 4 away; $(\frac{193}{256})^2 \simeq 0{\cdot}75$ for not more than 10 away; and so on (in general, the probability is $(1 - u_n)^2 \simeq 1 - 2u_n \simeq 1 - 1{\cdot}60/\sqrt{n}$ that both adjacent strings have lengths not exceeding n). Briefly, there are a great number of short strings, but here and there we find long strings, some extremely long; when we count

† For a brief summary of how 'chance' comes to intervene continuously in thousands of complicated ways to bring about evolution (albeit, of course, according to our present conceptions), it is sufficient to read the two sections entitled 'The Development of Life' and 'A Chance Happening' in V. F. Weisskopf, *Knowledge and Wonder*, Heinemann, London (1962). As for the 'continuous creation of matter' and the formation of the galaxies, see the section entitled 'What happened at the beginning?', pp. 165–167; also see D. W. Sciama, *The Unity of the Universe*, Faber and Faber, London (1959); in particular, the section on the 'Steady State Model', pp. 155–157.

not the numbers of strings but the number of steps they contain, however, the proportionate contribution from short and long strings is inverted. This is what we saw (cf. the *Remark* in Section 8.7.6) when we compared the probabilities u_{n-2}/n with the previsions of the lengths $n(u_{n-2}/n) = u_{n-2}$. In the context of this conceptual discussion, it is convenient just to take up the final point (mentioned above) concerning the trouble caused by the expression 'on average', a notion inspired by the statistical formulation.

Once again, we are dealing with the attempt to replace a genuine and valid probabilistic concept, which applies under all circumstances, by a counterfeit notion, only partially valid and not always applicable (it does not apply here, for example). The probabilistic meaning is expressed perfectly—even though the expression may be rather unpalatable—by saying that each *individual* string has an infinite *expected length*, which is a result of the possible lengths 2, 4, 6, 8,... having probabilities $\frac{1}{2}, \frac{3}{8}, \frac{5}{16}, \frac{35}{128}, \ldots,$ respectively. It makes no difference (just as it makes no difference whether we quote an interest rate as 45 lire per 1000 lire, or as 0·045) if we say that in 1000 strings we have, in prevision, a total length of 1000 deriving from strings of length *two*, 375 from strings of length *four*, 312·50 from length *six*, 273·44 from length *eight*, and so on. There is no harm in this, and, indeed, it could be more expressive to consider that an expected length of 312·50 steps from 1000 strings derives from an expected number of strings of length six equal to 52·087. The trouble arises when one tries to interpret the phrase in a non-probabilistic sense, as if it were possible to state that in any 1000 strings *things will turn out in this way* (in some vague sense: no-one goes so far as to claim this to be logically true—i.e. in a definite necessary sense—but people omit to state that it is at most 'very likely'; it is as though the possibility of something else happening could be avoided by recourse to some hybrid notion of 'practically certain'). Anyway, this very fundamental objection, one which always applies, precludes one from making statements of this kind without qualifying them as being *almost certain*, where by *almost-certainty*—by the mere fact of it not being *certainty*— we mean simply a rather high degree of probability (the latter being always subjective).

But it is not enough merely to correct a conceptually and formally inadequate form of expression. We must also make clear that statements which assert that in a large number of trials (in our case, strings) the actual outcomes will very likely be close to the 'previsions' do not hold except under appropriate conditions. First of all, we need conditions like independence, and this holds in our case; so far as the lengths and the signs are concerned, the strings are stochastically independent. For this reason, we can conclude that we may be almost certain that the proportion of positive and negative strings is fifty–fifty; the sequence of strings thought of in terms of their signs is a Heads and Tails process. However, we cannot claim that the same is true for the number of steps made on the two half-planes: despite the independence, the conclusion fails to hold because the prevision of the length of a string is infinite. *A fortiori*, for precisely the same reason, we cannot make statements of almost certainty about the frequency distribution of the lengths of the strings (or of the number of steps, considered in terms of the length of the string to which they belong). On the contrary, it would be very difficult to formulate this problem (even if the expected length were finite, and the statement therefore essentially true), not least because there are an infinite number of cases (lengths) to be distinguished.

We have shown how even a fairly superficial examination of very simple cases, like that of the Heads and Tails process, can reveal a number of features which are both unsuspected and fascinating in their own right. The intrinsic interest of these results has interesting conceptual implications when one considers more deeply the reasons for the surprise and the air of paradox which they generate.

8.9 PROPERTIES OF THE WIENER–LÉVY PROCESS

8.9.1. In Section 8.3, we had a brief look at those properties of the Wiener–Lévy process which could be established immediately, and which served to enable us to make reference to the process. We now return to this topic, both in order to consider it in more depth, and to show how certain asymptotic properties, which hold in many cases of asymptotically normal processes, gain in simplicity and clarity when we observe that they are exact properties in the Wiener–Lévy case.

This is, paradoxically, both the simplest and the most pathological case. The disarmingly simple features we have seen already: all the quantities we consider, whether individually, in pairs, or n at a time, and under all the circumstances we have examined, are normally distributed (in either 1, 2 or n dimensions). What could be better?

There was one feature, however, which might perhaps have given us grounds for suspecting that troubles might lie ahead. We are referring to the property of projective invariance which, as we mentioned, enables us to reduce the study of asymptotic behaviour to that of the behaviour in the neighbourhood of the origin. At the time, we did not wish to frighten the reader by drawing attention to certain things which happen at infinity and cause awful trouble when one considers them concentrated near the origin. What is even worse is that everything which happens in the neighbourhood of the origin also happens in the neighbourhood of every other point of the curve $y = Y(t)$ (since the process is homogeneous with independent increments).

8.9.2. First of all, we recall how the Heads and Tails process (along with many others) provides—with an appropriate change of scale—a good approximation to the Wiener–Lévy process (to any desired degree of accuracy).

Let us consider the standardized Wiener–Lévy process ($m = 0$, $\sigma = 1$). A Heads and Tails process, in order to preserve these characteristics and to approximate the continuous process, must consist of a large number of small jumps (e.g. very frequent bets with very small stakes), and, instead of a single jump ± 1 per unit time, requires N^2 jumps of size $\pm 1/N$ per unit time. In this way, the standard deviation per unit time is, in fact, given by

$$\sigma\sqrt{n} = (1/N).\sqrt{N^2} = 1$$

as required.

If N is large—in the sense that the time intervals $\tau = 1/N^2$ and stakes $s = 1/N$ are small in comparison with the precision with which we wish, or are able, to measure intervals of time and amounts of money—this process is practically indistinguishable from the Wiener–Lévy process. In fact, if all the time intervals we wish, or are able, to consider contain a large number of small time intervals, τ, then the increments of $Y(t)$ are made up of the sums of a large number of independent increments, and are therefore approximately normally distributed.

If we think in graphical terms, we can say that if the graph of the Heads and Tails process (see Figure 7.1 of Chapter 7, 7.3.2) is collapsed by dividing the ordinates by $1/N$, and the abscissae by $1/N^2$, with N large enough to render imperceptible the segments of the broken line corresponding to the individual tosses, then we have the most exact obtainable representation of a Wiener–Lévy process. In a certain sense, this is precisely the same old process of approximation and idealization as is used when we consider changes in population (the number of inhabitants of some region, etc.) as a continuous graph: even though one wishes to consider them as drawn in, one chooses the scale in such a way as to render imperceptible the jumps which represent the individual births and deaths on which the behaviour of the curve actually depends.

8.9.3. Of course, as we remarked at the time, instead of starting from the Heads and Tails process in discrete time, we could start from that in continuous time (the Poisson variant, with jumps $\pm 1/N$, N^2 of them, *in prevision*, per unit time), or with any other distribution of jumps (e.g. normal, always taking the standard deviation to be $1/N$, etc.).

Conversely, the Wiener–Lévy process can be a useful representation (giving an excellent approximation on some given scale) of phenomena whose 'microscopic' behaviour may well be very different. Among other things, it provides a useful model of the *Brownian motion* of a particle (or, better, if we restrict ourselves to one dimension, the projection of its motion onto one of the axes). Of course, the scale must be chosen so that it no longer makes sense to attempt to follow the actual mechanism of the phenomena, with its free paths, collisions, and so on. We also note that P. Lévy often refers to the Wiener–Lévy process as the 'Brownian motion process' (the name Wiener–Lévy has come about in recognition of the two authors who made the greatest contributions to the study of this process; Bachelier also deserves a mention, however; he had previously discovered many of the properties and results, although in not such a rigorous manner).

8.9.4. We shall restrict ourselves in what follows to collecting together, as a survey, some of the more interesting facts about the process, but without providing proofs. In general, however, we shall be dealing with results which have already been proved implicitly—or at least made plausible—by virtue of results established for the Heads and Tails process.

Problems relating to the Wiener–Lévy process can be tackled in many different ways; in a certain sense, this reflects the various ways of looking at the normal distribution which we noted when the latter was first introduced (the Wiener–Lévy process can be considered, roughly speaking, as a particular form of extension of the normal distribution to the infinite-dimensional case). On the other hand, we get a better overall view if we say something briefly about each of the most important procedures.

Those procedures which derive basically from the Heads and Tails process, or something similar, are essentially rooted in combinatorial theory (together with whatever else may be required). The greater part of Chapters 7 and 8 are, in fact, devoted to this kind of procedure, and we have often pointed out how it might be used in the context of the Wiener–Lévy process. We shall shortly give the details of this.

Rather more direct procedures are derived from the properties of the normal distribution itself, together with the various techniques for dealing with distributions. This means that knowledge of the second-order characteristics (variances and covariances) are sufficient to determine the process. We have already given examples of this when introducing the preliminary properties of the Wiener–Lévy process.

The third kind of procedure will require a more thorough discussion. It involves the study of diffusion problems (the heat equation, and so on), and was briefly mentioned in Chapter 7, 7.6.5, and again in Chapter 8, 8.6.7. Despite the elegance and power of these methods, we shall not be able to say a great deal about them here. This is unfortunate, because, besides their power, they can be made very expressive in terms of the image of the spread of probability, considered as mass. However, we clearly cannot include everything, and this seemed a reasonable candidate for exclusion, since—from a conceptual viewpoint—it is more in the nature of an analogy than a genuine representation of the problems. This is in contrast to the other approaches, which, in their various ways, stick closely to the probabilistic meaning, and enable one to shed light on every aspect of it.

Anyway, we shall restrict ourselves here to illustrating, using the perspective provided by diffusion theory, some of the problems with which we are already familiar through other approaches.

The Wiener–Lévy process corresponds precisely to the basic case of diffusion starting from a single source. The so-called 'dynamic' considerations and conclusions, in which t is considered as the time variable (instead of just a constant), involve precisely this process (and not just the individual distributions at individual instants).

The gambler's ruin problem (in the version provided by the Wiener–Lévy process) requires the introduction of an absorbing barrier; the straight line $y = -c$, where $c =$ the initial capital. This problem can be solved, in the theory of heat transfer, by the *method of images* (due to Lord Kelvin). This involves placing an opposite (cold) source at the point $t = 0$, $y = -2c$ (a mirror-like image of the origin, a hot source, the image being taken with respect to the barrier). The resulting process, which, for reasons of symmetry, clearly has zero density on the barrier, gives, at each instant t, the density of the distribution of the gain. The missing part (the integral of the density is less than 1) is the mass absorbed by the barrier; i.e. the probability of ruin before the instant under consideration. It can be seen, without the need for any calculations, that this is twice the 'tail' which would go beyond the barrier (this tail is itself missing, and there is also the negative tail that has come in from the cold source). We note that this corresponds precisely to Desiré André's argument.

Similarly, in the case of the two-sided problem, the method of images leads to the introduction of an infinite number of hot and cold sources (images of the actual source, with an even or odd number of reflections in the absorbing barriers). This is the 'physical' interpretation of the formulae in Section 8.6.6.

8.9.5. Our survey of the results relating to the Wiener–Lévy process should begin, naturally enough, with those we gave in Section 8.3, and with those we came across subsequently. We shall only repeat those things which are required to make the survey sufficiently complete.

We begin with the results relating to the ruin problem (i.e. to the case with an absorbing barrier).

In the case of a single barrier (at $y = c$, say), the probability of ruin at or before time t, $F(c, t)$,† i.e. the distribution function for the time T spent by the process before ruin occurs, $F(c, t) = \mathbf{P}(T \leqslant t)$, is given by

$$F(c,t) = \mathbf{P}(|Y(t)| > |c|) = 2\mathbf{P}(Y(t) > |c|)$$

$$= \frac{2}{\sqrt{(2\pi t)}}\int_{|c|}^{\infty} e^{-y^2/2t}\,dy = \frac{2}{\sqrt{(2\pi)}}\int_{|c|/t}^{\infty} e^{-x^2/2}\,dx \tag{47}$$

† In Sections 8.6.4–8.6.5, this was denoted by $q(t)$ and $p(c)$ (or $q_c(t)$ and $p_t(c)$) because it was then convenient to think of one of the variables as fixed (i.e. included as a parameter).

(the above holds in a real sense, because the probability of $T < \infty$ is 1). The density has the form

$$\text{(49)} \qquad \frac{\partial F}{\partial t} = f_c(t) = \frac{|c|}{\sqrt{(2\pi)}} t^{-\frac{3}{2}} \mathrm{e}^{-c^2/2t} = \frac{|c|}{Nt} \cdot \frac{N}{\sqrt{(2\pi t)}} \mathrm{e}^{-c^2/2t}.$$

We recall that we are dealing with the stable distribution with index $\alpha = \frac{1}{2}$. The second form emphasizes the relationship with the Heads and Tails process, involving N^2 tosses per unit time, each involving a gain of $\pm 1/N$. The second factor expresses (asymptotically) the probability of a gain of $|c| = (N|c|)(1/N)$ in $N^2 t$ tosses, the first factor the probability that we are dealing with the first passage through level $y = |c|$ (cf. Section 8.7.9(*c*)).

When considered as a function of $|c|$ (and we shall write y rather than $|c|$), the distribution becomes the half-normal, with density

$$\text{(48)} \qquad f_t(y) = \sqrt{(2/\pi)}\, \mathrm{e}^{-y^2/2t} \ (y \geqslant 0) \text{ (i.e. zero for } y < 0).$$

We recall that, in addition, this holds for the following cases:

the absolute value of $Y(t)$
the absolute value of $\vee Y(t)$ (the maximum of $Y(\tau)$ in $0 \leqslant \tau \leqslant t$)
the absolute value of $\wedge Y(t)$ (the minimum of $Y(\tau)$ in $0 \leqslant \tau \leqslant t$)
the absolute value of $\vee Y(t) - Y(t)$
(the deviation from the maximum)
the absolute value of $\wedge Y(t) - Y(t)$
(the deviation from the minimum).

We now give the probability distributions of $Y(t)$ conditional on three different assumptions concerning the maximum of $Y(\tau)$ in $[0, t]$. The three assumptions are as follows:

that, with respect to some given $c > 0$, we have $\vee Y(t) \geqslant c$ (75);

that $\vee Y(t) \leqslant c$ (76);

that $\vee Y(t) = Y(t)$ (77).

In the first two cases we have:

$$\text{(75)} \qquad f(y) = K \exp\{-(c + |y - c|)^2/2t\}, \qquad 1/K = F(c, t),$$

$$\text{(76)} \qquad f(y) = K[\exp\{-y^2/2t\} - \exp\{-(2c - y)^2/2t\}](y \leqslant c), \qquad 1/K = 1 - F(c, t).$$

The first follows immediately from the reflection principle. Note that $c + |y - c|$ is equal to y (for $y \geqslant c$) or $2c - y$ (for $y \leqslant c$), and that the distribution is therefore the normal with the central portion (between $\pm c$) removed, and the two remaining tails attached to one another. For the second, it is sufficient to observe that, multiplying it and the first one by

$1 - F(c, t)$ and $F(c, t)$, respectively (i.e. by suppressing the K), and summing them, we must again obtain $K \exp(-y^2/2t)$.

Finally, suppose we assume that we know that either the value $Y(t)$ is greater than all those previously obtained, i.e. that $Y(t) = \vee Y(t)$ (without knowing anything more about the actual value), or that we know that $\wedge Y(t) = 0$, i.e. that the minimum is the initial zero, $Y(0) = 0$ (by the reversal principle, the two cases are equivalent). Conditional on either of these, the density of the distribution is like $\partial F/\partial t$ in (49), except that we now have to take $y = |c|$ as the variable rather than t. This changes K, and we obtain

$$f(y) = \frac{y}{t} \mathrm{e}^{-y^2/2t} \; (y \geqslant 0), \tag{77}$$

giving the distribution function

$$F(y) = [1 - \mathrm{e}^{-y^2/2t}] \; (y \geqslant 0). \tag{78}$$

The same thing holds, with the range of values reversed, when we take $Y(t)$ to be equal to the minimum rather than the maximum (or if $Y(0) = 0$ is the maximum).

This can be justified by considering (in a somewhat roundabout manner) the meaning of $\partial F/\partial t$, or, alternatively, by considering (71) of Section 8.7.9, which refers to the Heads and Tails process.

In the case of two barriers (the ruin problem for two gamblers), the distribution of $Y(t)$ conditional on the fact that neither has been ruined in $[0, t]$, i.e. conditional on

$$-c' \leqslant \wedge Y(t) \leqslant \vee Y(t) \leqslant c'' \qquad (\text{with } c' > 0 \text{ and } c'' > 0),$$

is given by

$$f(y) = K \sum_{h=-\infty}^{+\infty} [\exp\{-(y - 2hc^*)^2/2t\} - \exp\{-(y + 2c' + 2hc^*)^2/2t\}], \tag{79}$$

where $c^* = c' + c''$, and $-c' \leqslant y \leqslant c''$.

In the symmetric case, $c' = c'' = c$, $c^* = 2c$, this becomes

$$f(y) = K \sum_{h=-\infty}^{+\infty} (-1)^h \exp\{-(y - 2hc)^2/2t\} \qquad (-c \leqslant y \leqslant c). \tag{80}$$

Clearly, the probability, $1 - q(t)$, of neither barrier being crossed until time t is equal to $1/K$ (which is given by the integral of the $\sum$ between $\pm c$, or, in the general case, between $-c'$ and $+c''$). This $q(t)$ also appeared, in a slightly different form, in (53) (see Section 8.6.5). Note that, if we ignored the K, (79) would give $f(y)\,\mathrm{d}y =$ the probability that $Y(t)$ lies in $[y, y + \mathrm{d}y]$ *and* has never previously gone outside the interval $[-c', c'']$: the point is that we would be saying 'and', rather than 'assuming that'. Similar comments apply in all other cases of this kind.

The few remarks we have made about Lord Kelvin's 'method of images' (see Section 8.6.7) suffice to explain the result. If one so wished, one could verify it by checking that both the diffusion equation ((32) of Chapter 7, 7.6.5) and the boundary conditions, $f(y) = 0$ on the half-lines $y = -c'$ and $y = c''$ for $0 < t < \infty$, are satisfied.

8.9.6. In the case of the Wiener–Lévy process, we can provide complete answers to questions concerning the asymptotic behaviour of $Y(t)$ as $t \to \infty$. In principle, these answers are provided by a celebrated result of Petrowsky and Kolmogorov; 'in practice', i.e. in a less complete but more expressive way, they are given by a famous theorem of Khintchin, the so-called 'law of the iterated logarithm' (cf. the brief comment in Chapter 7, 7.5.4).

What we do is to compare $Y(t)$ with some function $\omega(t)$ (which we assume to be continuous, increasing and tending to $+\infty$), and then calculate the probability that the inequality $Y(t) < \omega(t)$ holds from some arbitrary time t onwards. More precisely, we examine the limit, as $t' \to \infty$, of the probability that the inequality holds from t' onwards. To be even more precise,† this latter probability is itself to be understood as the limit, as $t'' \to \infty$, of the probability that the inequality holds in (t', t''). The p of interest is thus given by

$$p = \lim_{t' \to \infty} [\lim_{t'' \to \infty} p(t', t'')];$$

the limit certainly exists, since $p(t', t'')$ increases as t'' increases and decreases as t' increases.

We can say a great deal more, however. By the Zero–One law, only the two values $p = 0$ or $p = 1$ are possible: either it is practically certain that $Y(t)$ remains below $\omega(t)$ from a certain time onwards, or it is practically certain that this does not happen; i.e. there will always be segments in which $Y(t)$ is greater than $\omega(t)$. The class of functions $\omega(t)$ can therefore be divided into two subclasses, which could be said to contain 'those which increase more (less) rapidly that the "large values" of $Y(t)$'. The general distinction (given by Petrowsky and Kolmogorov) is that $\omega(t)$ belongs to the upper or lower class according as the improper integral (from an arbitrary positive t_0 to $+\infty$) of

$$(81) \qquad \psi(t) \,.\, t^{-1} \exp\{-\tfrac{1}{2}\psi^2(t)\}\, dt, \text{ where } \psi(t) = \omega(t)/\sqrt{t},$$

converges or diverges.

---

† This further qualification is unnecessary if countable additivity is assumed. We recall similar caveats in the case of the strong law of large numbers (Chapter 7, 7.7.3), etc. For simplicity, we shall give an informal discussion here.

In terms of $\psi(t)$, the condition $Y(t) < \omega(t)$ can be written as $Y(t)/\sqrt{t} < \psi(t)$; i.e. in terms of a *standardized* function (for which we have made σ = constant = 1).

The more expressive distinction (that of Khintchin) simply considers the class of functions

$$\omega(t) = k\sqrt{(2t \log\log t)} \qquad (\text{i.e. } \psi(t) = k\sqrt{(2 \log\log t)}) \tag{82}$$

and asserts that these belong to the lower class when $k \leqslant 1$, and to the upper class when $k > 1$. The result can be strengthened by considering the functions

$$\omega(t) = \sqrt{[2t(\log\log t + k \log\log\log t)]}: \tag{83}$$

these belong to the lower class is $k \leqslant \frac{3}{2}$, and to the upper if $k > \frac{3}{2}$ (a generalization due to P. Lévy, which is proved by first using a direct approach, which leaves the cases $\frac{1}{2} \leqslant k \leqslant \frac{3}{2}$ undecided, and then removing the gap by using the Petrowsky and Kolmogorov result).

The proof of the general criterion is based on diffusion theory ideas (which relate to the theory of heat flow). For the law of the iterated logarithm, we recall the previous comments given for the case of Heads and Tails.

8.9.7. *Small-scale behaviour* is extraordinarily complicated and irregular. Not only do all the large-scale peculiarities reappear (shrunk by a factor of N^2 in the abscissa, corresponding to a factor of N in the ordinate), but, also, if we study the behaviour in the neighbourhood of a point—the origin, for instance—we find all the asymptotic properties corresponding to $t \to \infty$ reappearing in an inverted way. This can be seen most simply by observing that, if $Y(t)$ is given by a Wiener–Lévy process, then the same is true for the function

$$Z(t) = tY(1/t);$$

this has $m_t = 0$, $\sigma_t = t\sqrt{(1/t)} = \sqrt{t}$, the distribution is normal, and the correlation coefficient between $Z(t_1)$ and $Z(t_2)$ is the same as that between $Y(1/t_1)$ and $Y(1/t_2)$ (if $t_2 > t_1$, and hence $1/t_1 > 1/t_2$, it is equal to $\sqrt{[(1/t_1)/(1/t_2)]} = \sqrt{(t_2/t_1)}$); this is all we need.

It is practically certain, therefore, that in every neighbourhood of zero ($Y(0) = 0$) $Y(t)$ vanishes *an infinite number of times* (as in the case when $t \to \infty$), and that it touches infinitely often every curve

$$y = \omega(t) = k\sqrt{[2t \log\log(1/t)]}$$

with $k \leqslant 1$, but not those with $k > 1$ (which gives, *locally*, an almost certain 'modulus of continuity'; $|Y(t_0 + t) - Y(t_0)| < \omega(t)$ in a neighbourhood of t_0, with $0 < t < \varepsilon$). If, however, we want this to hold almost certainly for all the t_0 of some given interval simultaneously (still for all t between 0 and ε), we have to take

$$(84) \qquad \omega(t) = k\sqrt{[2t \log(1/t)]}, \qquad k > 1$$

(the simple rather than the iterated logarithm).

In order to be brief, the presentation of the results in these cases has been rather informal. We should point out, however—for reasons we shall see shortly—that there are grave dangers in treating these topics without sufficient care and attention. For every point t_0 at which $Y(t_0) = 0$, it is practically certain (probability $= 1$) that there are other roots (an infinite number of them) in every interval of the point, either to the left or to the right (and the same holds true at every other point if we consider crossings of the horizontal line $y = Y(t_0)$). On the other hand, between two roots there are always several intervals (and almost certainly a countable infinity) in which $Y(t)$ is either positive or negative, and hence there are isolated roots to the right or to the left—the end-points of such intervals. Since this can be repeated for all horizontal lines $y =$ constant (an *uncountably infinite* set), the points $y = Y(t)$ which are isolated (on at least one side) from points of the curve at precisely the same level y, form, in every interval, an uncountably infinite set, and among them there are always an infinite number of points isolated on either side (at least the maxima and minima).

This having been said, the length of the segment starting from the origin, where we assume $Y(0) = 0$ (or starting from some arbitrary t' at which we know that there is a root, $Y(t') = 0$), and containing no roots, is a random quantity X which has probability 1 of being precisely zero (if 0, or, in general, t', is a root which is adherent on the left to the set of roots). Indeed, such a random quantity, $X(t')$, can be considered, without changing the problem, for any arbitrary t'—even if $y' = Y(t')$ is not zero—as the length of the interval on the left of t' not containing points t at which $Y(t)$ again takes the value y'. In any case, we know that we necessarily have $X(t') > 0$ for an uncountable infinity of points in any arbitrarily small interval, and it can be shown that, *assuming* the length X to be greater than some given $x_0 > 0$, the probability of it being greater than some $x \geqslant x_0$ is $\sqrt{(x/x_0)}$. In other words, conditional on the hypothesis $X \geqslant x_0$ $(x_0 > 0)$, we can say that X has distribution function and density given by

$$(85) \qquad F(x) = 1 - K\sqrt{x},$$

(86) $$f(x) = \tfrac{1}{2}K/\sqrt{x},$$

where $K = 1/\sqrt{x_0}$ (so that $F(x_0) = 0$); as $x_0 \to 0$, we also have $K \to 0$.†

The result for Heads and Tails (where $X \geqslant n$ has probability $u_n \simeq 0{\cdot}8/\sqrt{n}$‡) corresponds to the case $K \neq 0$ (because, clearly, in discrete time there is no way for 'peculiarities on the small scale' to occur).

8.9.8. If, instead, we begin by fixing a time t_0, knowing only that $Y(0) = 0$, and we consider $X = T'' - T'$, the length of the interval containing t_0 and no roots (i.e. T' = the last root of $Y(t) = 0$ with $t \leqslant t_0$, and T'' = the first root of $Y(t) = 0$ with $t \geqslant t_0$§), then we have the following probability distributions:

(87) for T': $$f(t) = K/\sqrt{[t(t_0 - t)]}, \qquad F(t) = K\sin^{-1}\sqrt{(t/t_0)}, \qquad (0 \leqslant t \leqslant t_0),$$

(88) for T'': $$f(t) = K/\sqrt{[t(t - t_0)]}, \qquad F(t) = K\cos^{-1}\sqrt{(t_0/t)}, \qquad (t \geqslant t_0),$$

(89) for X: $$f(x) = K\int_{\alpha}^{t_0} \mathrm{d}t/\sqrt{[t(t_0 - t)(t + x)(t + x - t_0)]}, \qquad (x \geqslant 0),$$

† This means not only that, in the absence of any contrary hypothesis (the case of an infinite number of roots adherent on the left), there is a probability $= 1$ that $X = 0$ (i.e. X is concentrated at the point $x = 0$), but also that with the single assumption that $x > 0$, all the probability is *adherent* to zero (i.e. however $x_0 > 0$ is chosen, the probability that $X \geqslant x_0$ is zero; this is obvious, because, in any finite interval, however large, there can only be a finite number of intervals containing no roots and of length greater than x_0, whereas there are an infinite number of 'small' intervals containing no roots in every interval of almost all the roots; i.e. excluding the isolated ones).

Note, of course, that the problem would be different if we were talking about an interval containing no roots, and chosen by picking out some point in it. As usual—recall 'sums' at Heads and Tails, 'number' and 'length' of strings, etc.—this procedure would favour the choice of the longest intervals (cf. the comments to follow in the text). The choice must be made by saying, for example, 't' = the starting point of the third interval of length $\geqslant x_0$ (possibly with some additional complications, in order that the restriction to a simple example is seen not to be necessary) after the level $Y(t) = c$ has been reached'. For the case $x_0 \to 0$ (under the assumption $X > 0$), this explicit method of choice does not exist. We can, however, reduce to the previous case by thinking of t' as having been determined in this way by some other person, with x_0 unknown to us, but given by $x_0 = 1/N$, where 'N is an integer chosen at random' (in the sense we discussed in Chapter 3, Section 3.2, and Chapter 4, Section 4.18).

‡ Or $u_n/2 \simeq 0{\cdot}4/\sqrt{n}$: it does not make any difference whether we consider the X of the continuous case as a generalization of the length of a *string*, or of the period spent in the *lead* (i. as L or V of Chapter 8, Section 8.7).

§ If it so happened that $Y(t_0) = 0$ (and this has probability 0), we would have $T' = T'' = t_0$ and $X = 0$.

where $\alpha = 0 \wedge (t_0 - x)$. Similarly, we have the result that, given t' and t'' $(t' < t'')$, the probability of at least one root of $X(t)$ in the interval (t', t'') is equal to $K \cos^{-1}\sqrt{(t'/t'')}$.

The same results hold (by virtue of the usual transformations) if we think, for example, of T' as denoting the abscissa of the maximum (or of the minimum) of $X(t)$ between 0 and t_0 (rather than the last root), and, correspondingly, of (T', T'') as the interval in which the maximum (or minimum) remains constant (i.e. T'' is the last instant up to which $X(t)$ does not exceed the maximum value attained in $(0, t_0)$; and similarly for the minimum), and so on. It is interesting to note—and this ties in with what we drew attention to in Section 9.8 as *seemingly 'paradoxical'*—that these points (maximum, minimum, last root) are more likely to be near the end-points of the interval $(0, t_0)$ than near the centre. More precisely, as a more expressive interpretation, recall that T' is the abscissa of a point 'chosen at random' (i.e. with uniform probability density) on the circumference of a semi-circle having $(0, t_0)$ as diameter (cf. Figure 8.9c).

In the case of Heads and Tails, we saw that, asymptotically, in any interval $(0, t_0)$ with $Y(0) = 0$ (or in any interval (t', t'') with $Y(t') = 0$), the proportion of time during which $Y(t)$ is positive had the arc sine distribution. This property continues to hold, exactly, for the Wiener–Lévy process.

8.9.9. The 'pathological' character of the 'small-scale' behaviour might leave one somewhat puzzled as to the possibility of interpreting the process in a constructive way. For this purpose, Lévy suggests a procedure of definition by successive approximations. It consists of subdividing the interval under consideration ($0 \leqslant t \leqslant 1$, say) into $2, 4, 8, \ldots, 2^k, \ldots$ equal parts, in determining $Y(t)$ at the division points, and in taking as the kth approximation a function $Y_k(t)$, coinciding with $Y(t)$ for those t which are multiples of $1/2^k$, and linear in between them. Given $Y(0) = 0$, and $Y(1)$ determined as a random quantity with a standard normal distribution $(m = 0, \sigma = 1)$, the intermediate points are successively determined by means of the considerations of Section 8.3.2. If t' and t'' are two consecutive multiples of $1/2^k$, and $t = (t' + t'')/2$ is the point at which $Y_{k+1}(t) = Y(t)$ is to be determined, we know that it is sufficient to add to the prevision given by $Y_k(t)\,(= [Y_k(t') + Y_k(t'')]/2 = [Y(t') + Y(t'')]/2)$ a random quantity having a centred normal distribution $(m = 0)$ and standard deviation given by

$$\sigma = \sqrt{[(t - t')(t'' - t)/(t'' - t')]} = 1/2^{(k+2)/2} \qquad \text{(cf. Figure 8.3).}$$

By bounding the probabilities of large values for these successive correction terms, we can conclude (following Lévy) that the $Y_k(t)$ converge almost certainly to a continuous $Y(t)$.

Of course, this means that we are using countable additivity. If one wishes to avoid this, all the difficulties relating to 'small-scale' behaviour could be avoided by imagining, for instance, that the process only appears to take place in continuous time, but, in fact, takes place in discrete time, with time intervals $1/N$ (with N unknown, and having probability zero of being smaller than any arbitrary preassigned integer).†

---

† Note that the same idea can be used in reverse, making any discontinuous process, for example, the Poisson process, *continuous*. It is sufficient to think of the 'jump' +1, at any instant t, as actually a continuous increase taking place in a very short time interval from t to $t + 1/N$ (with N as above; for example, one could take an increment $N\tau$ in $0 \leqslant \tau \leqslant 1/N$, or take $\sin^2(\frac{1}{2}\pi N\tau)$; one could even assume behaviour of the form $1 - e^{-N\tau}$, or $1 - e^{-N\tau} \cos N\tau$, and so on, in $0 \leqslant \tau \leqslant \infty$).

Without countable additivity, there is no unique answer to certain of the more subtle questions. Countable additivity certainly provides unique answers, but this, of course, is no reason to consider the latter as 'well founded' in any special sense.

CHAPTER 9

An Introduction to Other Types of Stochastic Process

9.1 MARKOV PROCESSES

9.1.1. The cases treated so far have been considered at some length, this being a convenient way in which to introduce various of the basic notions and most frequently used techniques. They are, however, nothing more than examples of the simplest and most special form of random process; that is, the linear form, or, more explicitly, the homogeneous process with independent increments. We now give some of the basic properties of other cases of interest, although, given the limits of the present work, the treatment will necessarily be brief.

Processes for which 'given the present, the future is independent of the past', or, alternatively, 'the future depends on the past only through the present' are called *Markov processes*. Processes with independent increments are a special case (they are even independent of the present, and the process depends on the latter only through the fact that the future increment, $Y(t) - Y(t_0)$, is added on to the present value $Y(t_0)$); the Markov property is much less restrictive.

The name derives from the fact that Markov considered this property in a particular discrete situation (involving probabilities of 'linked' events, whence *Markov chains*). To give a simple example, let us consider a function Y_n, taking only a finite number of values, $1, 2, \ldots, r$, say. For a physical interpretation, which may be more expressive, we could think of it as a 'system' which can be in any one of the r 'states', $S_1, S_2, \ldots, S_r$, and which passes from one state into another in a sequence of 'steps' (including the possibility that a 'step' could result in the process remaining in the same state: recall, however, what was said in Chapter 8, 8.2.5 concerning the case $\mu_{ii} \neq 0$).

Such a system is said to be a Markov chain if, given that at time n the system is in state i, the probability that it then occupies state j at time $n + 1$ is given by some value, $p_{ij}(n)$, which is independent of anything one might know about the past.

The simplest case is that of the *homogeneous* chain, for which the *transition probabilities*, p_{ij}, are also independent of the time n. These probabilities form a matrix, $P = \|p_{ij}\|$, and, with the usual definition of matrix product, its square and cube, etc., give the analogous matrices of transition probabilities, $p_{ij}^{(2)}$, $p_{ij}^{(3)}$, etc., for passages from i to j in two steps, three steps, and so on. In fact, we have

$$p_{ij}^{(2)} = \sum_h p_{ih} p_{hj} \tag{1}$$

(the sum, over h, of the probabilities of going from i to j in two steps when the intermediate state is h), and, in general,

$$p_{ij}^{(m+1)} = \sum_h p_{ih}^{(m)} p_{hj}. \tag{2}$$

Of course, the p_{ij} must be non-negative, and, for each i, must have sum $\sum_j p_{ij} = 1$.

If the p_{ij} are all non-zero, or if this is the case for the $p_{ij}^{(m)}$ from some given m onwards, we have the so-called *ergodic* case: as m increases, the $p_{ij}^{(m)}$ tend to limit-probabilities p_j, which are independent of i. In other words, for large n, $\mathbf{P}(Y_n = j) = p_j$, independently of one's knowledge concerning the initial state i. Moreover, as n increases, the proportion of the time in which the system occupies state j during the first n steps tends stochastically to p_j (and $1/p_j$ is, in fact, the prevision of the recurrence time; i.e. the time between two successive passages through j).

If, further, we are unaware of the initial situation Y_0, and if our state of uncertainty causes us to attribute precisely these probabilities p_j as the initial, $\mathbf{P}(Y_0 = j)$, then these will remain our probabilities for the occupation of these states throughout the process, and we have what is called a *stationary* process. In fact, the vectors composed of the p_j are characterized by this property of being a fixed point (i.e. an eigenvector with unit eigenvalue) under the transformation P (and, moreover, under the stated conditions it is the unique admissible such eigenvector; i.e. with non-negative components). The ergodic property ensures that, under these conditions, we approach, asymptotically, this stationary situation. The set-up is often applied to statistical problems (involving, for example, a large number of particles or individuals, etc.), and then the ergodic result has a more concrete interpretation, because it implies the tendency to stationarity of the *statistical distribution*. The reader should compare this situation with those involving illegitimate applications of the ergodic 'principle', outside of the conditions under which the *theorem* holds (cf. Chapter 8, 8.8.4).

A similar set-up can be obtained in continuous time by assuming that the probabilities of passing from S_i to S_j in the time period from t to $t + \mathrm{d}t$ (given that we are at S_i at time t) are given by $\mu_{ij}\,\mathrm{d}t$, where the μ_{ij} may be constant, or perhaps functions of t. This case has already been considered (in Chapter 8, 8.2.5) as background to our discussion of the Poisson process, and we shall not add anything here to our previous discussion. As we remarked at that time, we can, without loss of generality, take $\mu_{ij} = 0$ for $i = j$ (and, except for special cases, it is usually convenient to do so).

9.1.2. Within the framework of the very simple cases which formed the basis of our previous discussion, we outlined several of the main problems and features of interest that arise with these processes. The same kinds of problems and features have been studied for general Markov processes, and, without going into the details, we shall consider a few of these in order to make some appropriate comments.

The kind of relation which we have encountered in the simple form $p_{ij}^{(2)} = \sum_h p_{ih}p_{hj}$ is typical of the Markov set-up (even in continuous time and with continuous state space). Given that we start from some point P_0 at time t_0, the probability of being in a neighbourhood of a point P_1 at time t_1 satisfies a relation involving the sum (or infinite series, or integral, as the case may be) of the probabilities of getting there by passing through the various possible points P at some arbitrary, intermediate time t ($t_0 < t < t_1$). These probabilities are evaluated as the product of the probabilities of the two passages: from P_0 to (a neighbourhood of) P in $[t_0, t]$, and then from P to (a neighbourhood of) P_1 in $[t, t_1]$, the latter probability being independent of P_0. This is the probabilistic version of 'Huyghens' principle', by which, in the deterministic case, one regards the evolution of a system in the period from t_0 to t_1 as being the result of what happens between t_0 and t, followed by what happens from t to t_1, starting from the situation reached at t with no need to recall the past. In our case, the same thing applies, not to the evolution of the system, but to the evolution of the probability distribution on the basis of which we foresee the evolution of the system.

In both cases (Huyghens and Markov), these processes are sometimes referred to as *non-hereditary* (in contrast to *hereditary* phenomena, whose evolution is influenced by the past). Examples are provided by the phenomenon of hysteresis, Volterra integral equations, and so on. One should note, however, that, in these respects, the distinction between 'present' and 'past' is something of a convention. One often believes that a (deterministic) prediction or a (probabilistic) prevision would be determined by the present if only some (unattainable) data or measurements were known. To compensate for their unavailability, one makes use of available data relating to the past (for example, in the case of hysteresis, the characteristics of the present situation are deduced from the history of the magnetic field which has produced them, since it is impossible to explore the state of magnetization at each point of a body). At an even more basic level, it can happen that for some problems 'the present' can be regarded as the position of a particle (or a body, etc.), whereas, for others, we need, in addition, to know the velocity (or the last movement). This is also true in the probabilistic case, and we can consider a *second-order* Markov chain as one in which the probabilities of the possible values for $Y(n+1)$ depend both on the value of $Y(n)$ and of $Y(n-1)$ (but on no others; one could, however, extend the notion and consider chains of arbitrary order). In fact, we could reduce this directly to the first-order case by defining 'the present' at time n to consist of the pair

$$(Y(n-1), Y(n)).$$

In other words, we redefine the 'states' to be the r^2 pairs (i, j), with the obvious restriction that from a state (i, j) one can only move, in one step, to one of the r states of the form (j, h).

9.1.3. Although it may seem a 'natural' condition, we are not claiming that the Markov property holds in all 'non-pathological' cases, nor even for the simplest, standard processes. Simple counterexamples which are of practical interest arise in connection with the Poisson process. For example, let $Y(t)$ be the number of telephone conversations in progress on some telephone system at time t, and let us assume that $N(t)$, the number of conversations which began between 0 and t, has a Poisson distribution, and that the length of any conversation is a random quantity having the same distribution as all the others, and stochastically independent of them. If this distribution is exponential, the process is Markovian (because every conversation in progress then has the same probability, $\lambda\,\mathrm{d}t$, of terminating within an infinitesimal time $\mathrm{d}t$, whatever its duration so far), but, in every other case, knowledge of the duration of the conversation so far modifies the prevision. In other examples of this kind, as here, 'age', or something similar, plays a fundamental rôle. A similar kind of example is that where the 'cumulative effects' have an influence; the prevision at t_0 depends not only on $Y(t_0)$, but also on the sum (or integral, in the continuous case) of the values $Y(t)$ between 0 and t_0. Of course, if the ages or the cumulative values were included as part of the definition of 'the present' (and were observable, or somehow available) then the process, appropriately extended to include these other variables, would turn out to be Markovian.

9.2 STATIONARY PROCESSES

9.2.1. We have already given the basic idea of a *stationary* process. We discussed it in relation to a Markov process, but this is not a necessary condition for stationarity. A sufficient condition is that the probabilities are invariant with respect to a translation along the time axis. For example, it is sufficient that the probabilities of $Y(t)$, $Y(t+t_1), \ldots, Y(t+t_k)$ satisfying the inequalities $y'_i \leqslant Y(t + t_i) \leqslant y''_i$, are independent of t. The above example of a telephone system (along with similar examples) gives a stationary process if we assume either that the system has been in operation for an infinitely long time, or that Y is unknown and that we attribute to the values that it can assume at each instant those probabilities corresponding to the assumption of an infinitely distant beginning. We note that the process is Markovian or non-Markovian according as the distributions of the lengths of conversations are or are not independent exponential distributions.

If $Y(t)$ is a stationary process, then the definition implies, in particular, that the distribution of $Y(t)$ does not depend on t, and so neither does the prevision (if it exists) $\mathbf{P}(Y(t)) = m$, nor the variance $\mathbf{P}(|Y(t)|^2) = \sigma^2$ (we assume, for convenience, and without loss of generality, that $m = 0$ and $\sigma = 1$). The same holds for all other moments and parameters of the distribution. If we consider two distinct times t' and t'', the pair of values $Y(t')$ and $Y(t'')$ has a distribution depending only on the difference $t'' - t' = u$,† and, in particular, all the quantities defined in terms of this distribution depend only on u; above all, this applies to the *correlation*

$$(3) \qquad \phi(u) = r(Y(t'), Y(t'')) = \mathbf{P}(Y(t')Y^*(t'')).‡$$

This correlation—usually referred to as the *autocorrelation* function—characterizes the process so far as second-order properties are concerned (in the sense that it enables one to determine $\mathbf{P}(X)$ for all $X = \sum a_{ij} Y_i Y_j^*$; i.e. for functions of the second degree in the values $Y_i = Y(t_i)$ of $Y(t)$ at any number of arbitrary time points t_i). If we have $m \neq 0$ and $\sigma \neq 1$, we can get back to the original process from the standardized case by noting that the former is equal to $m + \sigma Y(t)$. Similar conclusions hold in the non-stationary case, also, provided that $\mathbf{P}(Y(t))$ is constant, and

$$(4) \qquad \mathbf{P}(Y(t')Y^*(t'')) = \Gamma(t', t'')$$

depends only on the difference between t' and t''. Putting $t' = t''$ gives the second moment, and if this is bounded so is Γ, and the process is called 'second-order stationary'.

9.2.2. In dealing with this topic it is convenient to allow $Y(t)$ to be complex (for the same reasons for which it is convenient to represent harmonic oscillations with e^{it} rather than sines and cosines). The product $Y(t')Y(t'')$ therefore has to be replaced by the Hermitian product; i.e. by $Y(t')Y^*(t'')$

---

† The choice of u is a deliberate attempt to exploit an analogy which makes it convenient to use the same notation, $\phi(u)$, for both the autocorrelation and the characteristic function (cf. the next section).

‡ The asterisk denotes 'complex conjugate'. For the present, it is superfluous, since we are only considering real functions; shortly, however, we shall need the extension to the complex field.

(as we already indicated when we defined ϕ and Γ). This implies that

$$\Gamma(t'', t') = \Gamma^*(t', t''),$$

and, in particular, that $\phi(-u) = \phi^*(u)$. The latter is the more important because it relates directly to the stationary case that we are discussing. Moreover, the real part of $\phi(u)$ is continuous if (and only if) the process is 'mean-square continuous' (a stationary process enjoying this property is known as a *Khintchin process*). This property requires, in the notation of Chapter 6, 6.8.3, that $Y(t) \dashrightarrow Y(t_0)$ as $t \to t_0$, but—and one should be clear about this—it says nothing about the continuity of the function $Y(t)$. We require that the prevision of $[Y(t) - Y(t_0)]^2$ tends to 0. This happens, for example, for a Poisson process, or variants thereof, even for generalized Poisson processes (these only change through discontinuities, which, in the latter case, are everywhere dense), provided the standard deviation is finite (in this case, in fact, $\mathbf{P}(Y(t) - Y(t_0))^2 = K|t - t_0| \to 0$).

Under these conditions, it can be shown that the class of possible correlation functions coincides with the class of characteristic functions (and, of course, in the case of $Y(t)$ with even-valued, and hence real-valued, characteristic functions, we have

$$\phi(-u) = \phi^*(u) = \phi(u),$$

which correspond, as characteristic functions, to symmetric distributions $F(-x) = 1 - F(x)$). In any case, the distribution F has an important significance so far as the process is concerned, not only from a mathematical point of view, but also practically, in all applications, especially to problems in physics, where it has a connection with energy. It gives, in fact, the *spectral function* of the process: i.e. $F(\omega_2) - F(\omega_1)$ is the prevision of the energy corresponding to the frequencies in the interval $\omega_1 \leqslant \omega \leqslant \omega_2$. Expressed in an informal manner, the actual meaning of this in relation to the random function $Y(t)$ defined by the process is the following: let $U(\omega)$ (in general, complex) be the function expressing $Y(t)$ as a mixture of harmonic components (i.e. as a Fourier–Stieltjes transform)

$$(5) \qquad Y(t) = \int_{-\infty}^{\infty} \mathrm{e}^{\mathrm{i}\omega t}\, \mathrm{d}U(\omega).$$

The prevision of the energy corresponding to an individual $\mathrm{d}\omega$ is

$$(6) \qquad \mathrm{d}F(\omega) = \mathbf{P}(|\mathrm{d}U(\omega)|^2),$$

and in terms of F we obtain the correlation function

$$(7) \qquad \phi(u) = \int_{-\infty}^{\infty} \mathrm{e}^{\mathrm{i}\omega u}\, \mathrm{d}F(\omega),$$

which is therefore the characteristic function of the energy distribution.

The spectrum F could contain both concentrated masses (jumps of F)

$$U_k = F(\omega_k + 0) - F(\omega_k - 0),$$

corresponding to 'lines' ω_k, and diffused masses (segments where F is increasing and continuous). To make things as clear as possible, we repeat and extend the previous discussion in the simpler case where we just have concentrated masses U_k, corresponding to a set of particular frequencies ω_k. In this case, we have

$$(5') \qquad Y(t) = \sum_k U_k \,\mathrm{e}^{-\mathrm{i}\omega_k t},$$

and we can deduce that

$$U_k = \lim_{a\to\infty} \frac{1}{2a}\int_{-a}^{a} \mathrm{e}^{-\mathrm{i}\omega_k t} \sum_h U_h \,\mathrm{e}^{\mathrm{i}\omega_h t}\,\mathrm{d}t$$

$$(8) \qquad = \lim_{a\to\infty} \frac{1}{2a}\int_{-a}^{a} \mathrm{e}^{-\mathrm{i}\omega_k t}\, Y(t)\,\mathrm{d}t.$$

The U_k are therefore random quantities which depend on the global behaviour of $Y(t)$; conversely, knowledge of these random quantities determines $Y(t)$ in the way we have indicated. To give the probability distribution for all the U_k is an indirect way of giving all the probabilities of the process leading to $Y(t)$. From the energy viewpoint, we could say that the energy for the frequency ω_k is $|U_k|^2$ with prevision $\mathbf{P}(|U_k|^2)$; the U_k are uncorrelated (i.e. 'orthogonal') in the sense that $\mathbf{P}(U_h U_k^*) = 0$ $(h \neq k)$, and the total energy for frequencies up to and including ω is given by

$$(6') \qquad F(\omega) = \sum_k \mathbf{P}(|U_k|^2)$$

(the sum being taken over all the k for which $\omega_k \leqslant \omega$).

Because we have standardized the process ($\sigma = 1$), the total energy equals 1. The correlation function is given by

$$(9) \qquad \phi(u) = \sum_k \mathbf{P}(|U_k|^2)\,\mathrm{e}^{\mathrm{i}\omega_k u}.$$

Cramèr and Loève have proved that in the case we have considered (with discrete spectrum) the U_k are mutually orthogonal, and that this also holds in the general case (either second-order stationary or Khintchin processes) for the $\mathrm{d}U(\omega)$ for disjoint intervals $\mathrm{d}_1\omega$ and $\mathrm{d}_2\omega$;

$$\mathbf{P}(\mathrm{d}_1 U \,.\, \mathrm{d}_2 U^*) = 0.$$

Conversely, all such processes can be obtained in this way (a result also proved by Cramèr and Loève).

9.2.3. The concepts and techniques which we have discussed are not only applicable to problems in physics—from which we have borrowed the particularly expressive form of terminology—but also to problems in other fields, such as statistics ('time-series' analysis, and so on). However, this was a good place to give an outline of these ideas; it is useful to be able to 'see' the various problems concerning Fourier transforms and their mathematical properties in terms of some concrete framework. The two applications we have encountered are, in some sense, mutual inverses one of the other. In the case of the characteristic function, the concrete datum, or, at any rate, the most immediate, was the distribution, and the transform mainly provided a 'useful image' of it; in the case we have just dealt with, the function $Y(t)$ and the autocorrelation function are more concrete, and the corresponding distributions U and F are the 'images' in a certain wave interpretation.

CHAPTER 10

Problems in Higher Dimensions

10.1 INTRODUCTION

10.1.1. It might be argued that every problem could, or even should, be put in a multi-dimensional framework: indeed, we have seen this over and over again throughout our treatment so far. The subject matter of this chapter is not really new, therefore, and we shall merely emphasize those features and problems which particularly relate to the multi-dimensional nature of certain distributions.

In Chapter 6, 6.9.1, we dealt with the essential points concerning the representation of a distribution over an r-dimensional cartesian space, either by means of the distribution function

$$(1)\qquad F(x_1, x_2, \ldots, x_r) = \mathbf{P}[(X_1 \leqslant x_1)(X_2 \leqslant x_2)\ldots(X_r \leqslant x_r)]$$
$$= \mathbf{P}\left[\prod_i (X_i \leqslant x_i)\right],$$

or, if it exists, by means of the density

$$(2)\qquad f(x_1, x_2, \ldots, x_r) = \partial^r F/\partial x_1 \partial x_2 \ldots \partial x_r.$$

In addition, we can state that a necessary and sufficient condition for a function $F(x_1, x_2, \ldots, x_r)$ to be a distribution function is that f never be non-negative, or, should f not exist, that the expression for which it would be the limit is non-negative. The latter is the probability of the rectangular prism $(x_i' < X_i \leqslant x_i'')(i = 1, 2, \ldots, r)$ given by

$$(3)\qquad \mathbf{P}\left[\prod_i (x_i' < X_i \leqslant x_i'')\right] = \sum \pm F(x_1, x_2, \ldots, x_r),$$

the sum being taken over the 2^r vertices corresponding to all possible assignments of $x_i = x_i'$ or $x_i = x_i''$, with a $+$ or $-$ sign according as these are an even or odd number of x' (for the case $r = 2$, see Figure 6.5 of Chapter 6, 6.9.1 together with the intuitive explanation that accompanied it).

In order to 'see' the meaning of this condition (which is a generalization of the non-decreasing property of the one-dimensional F), it is useful to think of a mass c placed at some given point $P_0 = (x_1^0, x_2^0, \ldots, x_r^0)$ as giving rise to a 'step' of height c in that orthant† of the r-dimensional space of the points whose coordinates are greater than the corresponding x_i^0 (in the plane, this would be the NE quadrant). The function F is given by the superposition of such steps (or as a limit case).

The disadvantage of this is that the function F depends on the coordinate system (often, however, the problem itself has arisen in connection with r given random quantities X_i). A less arbitrary—but less useful—approach would be to assign probabilities over each half-plane (i.e. to assign $F(y)$ for each linear combination $Y = \sum_i a_i X_i$). The justification for this is straightforward, although somewhat indirect, and follows from the fact that this serves to determine the characteristic function, which, in turn, determines the distribution (as we shall see in the next section).

10.1.2. The characteristic function for an r-dimensional distribution of $X_1, X_2, \ldots, X_r$ is a function of r variables, $u_1, u_2, \ldots, u_r$, defined in a completely analogous way to that in the one-dimensional case:

$$(4) \qquad \phi(u_1, u_2, \ldots, u_r) = \mathbf{P}(\mathrm{e}^{\mathrm{i}(u_1 X_1 + u_2 X_2 + \ldots + u_r X_r)}) = \mathbf{P}(\mathrm{e}^{\mathrm{i}\mathbf{u} \times \mathbf{X}}).$$

The vector form is probably the clearer, with vectors $\mathbf{X}$ and $\mathbf{u}$ whose components are the X_i and u_i, respectively ($\mathbf{u}\times$ can, if we so wish, be regarded as a vector in the dual space).

For the cases $r = 2$ and $r = 3$, it is more convenient to avoid the use of subscripts and to write $uX + vY$, $uX + vY + wZ$, respectively (the standard notation for Plückerian coordinates).

The properties of $\phi(u_1, u_2, \ldots, u_r) = \phi(\mathbf{u})$ are (as is fairly obvious) the same as in the one-dimensional case. The inversion formula is also the same: for the case $r = 2$, for example, if the density exists and is bounded it is given by

$$(5) \qquad f(x, y) = \frac{1}{(2\pi)^2} \iint_{-\infty}^{+\infty} \mathrm{e}^{-\mathrm{i}(ux + vy)} \phi(u, v)\, \mathrm{d}u\, \mathrm{d}v.$$

If, in addition, the X_h are independent, we have

$$(6) \qquad F(x_1, x_2, \ldots, x_r) = F_1(x_1)F_2(x_2) \ldots F_r(x_r),$$

$$(7) \qquad \phi(u_1, u_2, \ldots, u_r) = \phi_1(u_1)\phi_2(u_2) \ldots \phi_r(u_r),$$

as well as the converse; i.e. factorization implies stochastic independence.

10.1.3. A number of problems in higher dimensions can be dealt with formally as though they were one-dimensional problems by means of matrix and vector notation. For example, sums of random vectors have the same

† Orthant is the r-dimensional analogue of half-line ($r = 1$) and quadrant ($r = 2$).

properties as sums of random quantities. In particular, if the vector summands (each with prevision zero) all have the same distribution and finite variances, then the sum-vector of n of them, divided by $\sqrt{n}$, has, asymptotically, a normal distribution having the same variances and covariances.

A frequently used and very expressive interpretation is that in terms of a 'random walk' in r-dimensional space, regarded as a random process in discrete time (as an aid to intuition, we shall mainly deal with the cases $r = 2$ and $r = 3$, corresponding to the plane and ordinary space); a step is taken after each unit of time, and at each step we obtain a random vector (always with the same distribution, and stochastically independent). The simplest example is obtained, for example, by simultaneously studying the gain of two (or three) gamblers who bet independently on a sequence of tosses at Heads and Tails (again ± 1 with probabilities $\frac{1}{2}$ and $\frac{1}{2}$ at each toss). This results in a zigzag path (in the plane, each step from (x_n, y_n) to (x_{n+1}, y_{n+1}) is the diagonal of some square in the integer lattice; the same holds in three dimensions with the diagonals of cubes). If (X_n, Y_n) is the 'position after n tosses', then, as n increases, it can be shown that this has, asymptotically a normal distribution with circular symmetry, and standard deviation $\sqrt{n}$ in all directions (and the same holds for the position (X_n, Y_n, Z_n) in three dimensions).

10.1.4. The following is a simple and instructive argument which can be applied to the present case. The probability of a return to the origin after n tosses in the one-dimensional case is given by $u_n \simeq 0{\cdot}8/\sqrt{n}$ for n even, 0 for n odd. In the case of the plane $(X_n = Y_n = 0)$, or ordinary space

$$(X_n = Y_n = Z_n = 0),$$

the respective probabilities are therefore given by $u_n^2 \simeq 0{\cdot}64/n$ and $u_n^3 = 0{\cdot}51/\sqrt{n^3}$: in the general case, we have $u_n^r = K/n^{r/2}$. We observe immediately that, in prevision, the number of returns to the origin is infinite in the plane ($\sum n^{-1}$ diverges) but is finite in three dimensions ($\sum n^{-r/2}$ converges for $r \geqslant 3$). It follows that the return to the origin is practically certain ($p = 1$) for $r = 1$, $r = 2$, but not for $r \geqslant 3$ (where $p = a/(1 + a)$, with $a = \sum n^{-r/2}$; for $r = 3$, for example, $a \simeq 0{\cdot}53$ and $p \simeq 0{\cdot}35$†).

The conclusion concerning the limit distribution (normal, with rotational symmetry and dimensions increasing like $\sqrt{n}$) holds in the general case, also, provided the distribution of every individual step has the same variance in all directions (i.e. equal variances and zero correlation for any two orthogonal directions). Without these conditions, we would have 'ellipsoidal

† If p is the probability of (at least) one return to the origin, $(1 - p)p^h$ is the probability of exactly h returns to the origin, and the prevision of the number of returns is given by

$$a = \sum hp^h(1 - p) = p/(1 - p).$$

contours' instead of spheres (but the latter case can be reduced to the former by making appropriate changes of scale along the axes of the ellipsoids).

10.2 SECOND-ORDER CHARACTERISTICS AND THE NORMAL DISTRIBUTION

10.2.1. To illustrate the use of vector and matrix notation, we shall re-examine certain expressions that we have already encountered in the context of the multivariate normal distribution, pointing out the form that certain properties now take.

The notation we shall introduce will enable us to interpret and understand our formulae in several alternative ways: either in the rather formalistic spirit which derives from algebraic-type theories (vectors and matrices thought of as rows, or columns, or arrays of numbers), or in the geometric, functional analytic spirit.

Vectors will be written boldface: for example, $\mathbf{x}$ (or $\mathbf{X}$, if we are dealing with a random vector). Given r linearly independent vectors $\mathbf{u}_1, \mathbf{u}_2, \ldots, \mathbf{u}_r$ in S_r, $\mathbf{x}$ can be written (in one and only one way) as a linear combination of them; $\mathbf{x} = \sum x_h \mathbf{u}_h$. We may sometimes write $\mathbf{x} = (x_1, x_2, \ldots, x_r)$, but this is simply a convention, and leaves it to be understood (and never forgotten) that the components do not directly relate to the intrinsic meaning of the vector, but only acquire their meaning through the introduction of some arbitrary basis, which can be changed at any time, the choice being simply a matter of convenience (this conflicts somewhat with the algebraic view-point). For a random $\mathbf{X}$, we shall write $\mathbf{X} = \sum X_h \mathbf{u}_h = (X_1, X_2, \ldots, X_r)$. The linear functionals on the vectors of the space S_r themselves form an r-dimensional space, the *dual* space, which we shall denote by S_r^*.

10.2.2. If we introduce a *metric* into the space S_r (i.e. a *scalar product*, which maps each pair of vectors $\mathbf{x}$ and $\mathbf{y}$ to a scalar, $\mathbf{x} \times \mathbf{y} = \mathbf{y} \times \mathbf{x}$, and is linear for each vector, and such that $\mathbf{x} \times \mathbf{x} > 0$ for all $\mathbf{x}$ other than the zero vector), then each dual vector can be expressed as a vector of the original space with the scalar product sign following it. In other words, if $f(\mathbf{x})$ is a scalar depending linearly on $\mathbf{x}$, then there exists a vector $\mathbf{a}$ such that $f(\mathbf{x}) = \mathbf{a} \times \mathbf{x}$, and $f(\cdot)$ can be written as $\mathbf{a}\times$. Given a metric in S_r, it makes sense to define the norm of a vector, $|\mathbf{x}| = \sqrt{(\mathbf{x} \times \mathbf{x})}$, and the orthogonality of two vectors, $\mathbf{x} \times \mathbf{y} = 0$. It then becomes convenient to choose the basis to be an orthogonal set of $\mathbf{u}_h$ with unit norms, in which case we denote them by $\mathbf{i}_h$:

$$(\mathbf{i}_h \times \mathbf{i}_k = (h = k);\ \text{i.e. 1 or 0, according as } h = k \text{ or not}).$$

The scalar product then has a simple representation in terms of the components: $\mathbf{x} \times \mathbf{y} = \sum x_h y_h$ (and $|\mathbf{x}| = \sqrt{(\sum x_h^2)}$). We shall write $\mathbf{a}^*$ instead of

$\mathbf{a}\times$, and $\mathbf{a}^*$ is then interpreted as the 'dual of $\mathbf{a}$' (some authors write $\mathbf{a}^{\mathrm{T}}$, where the superscript denotes 'transpose'; others use $\mathbf{a}_{-1}$, and so on). These alternative notations relate to the interpretation of the vectors in the two spaces as 'column vectors' or 'row vectors', respectively (i.e. matrices with 1 column and r rows, or 1 row and r columns).

From the formal, algebraic point of view, the matrices are also considered simply as arrays of numbers (r rows and s columns). From the geometric or functional analytic point of view, they are linear transformations between some S_r and some S_s. In our particular case, we shall only be considering square matrices.

If A is a matrix (or, better, a *linear transformation*), we have

(8) $\quad \mathbf{y} = A\mathbf{x}, \qquad \text{with} \qquad A(\mathbf{x}_1 + \mathbf{x}_2) = A\mathbf{x}_1 + A\mathbf{x}_2, \qquad A(c\mathbf{x}) = cA\mathbf{x}.$

In terms of components, if $\mathbf{x} = \sum x_h \mathbf{i}_h$, $A\mathbf{x} = \sum x_h A\mathbf{i}_h$, we have

$$A\mathbf{i}_h = a_{h1}\mathbf{i}_1 + a_{h2}\mathbf{i}_2 + \ldots + a_{hr}\mathbf{i}_r, \qquad \text{and}$$

(8′)

$$A\mathbf{x} = \sum_h x_h \sum_k a_{hk}\mathbf{i}_k = \sum_k \left(\sum_h a_{hk}x_h\right)\mathbf{i}_k:$$

in other words, the components of $\mathbf{y} = A\mathbf{x}$ are given by $y_k = \sum_h a_{hk}x_h$. The linear transformation A can therefore be represented (in the given reference system) by means of the r^2 coefficients a_{hk} (which, in the array, corresponds to the hth row, kth column).

10.2.3. We are particularly interested in those linear transformations (or matrices) which, with respect to the metric under consideration, are symmetric and positive; i.e. they correspond to 'positive-definite quadratic forms':

$$A\mathbf{x} \times \mathbf{y} = A\mathbf{y} \times \mathbf{x}, \qquad A\mathbf{x} \times \mathbf{x} > 0 \ \text{ provided } \mathbf{x} \neq 0.$$

If Q denotes such a linear transformation, we shall make the convention that Q will also be used to denote the matrix and the quadratic form. We can write, therefore,

(9) $\quad Q\{\mathbf{x}\} = Q\mathbf{x} \times \mathbf{x} = Q\mathbf{x}^*\mathbf{x} = \mathbf{x}^*Q\mathbf{x} = \mathbf{x}^{\mathrm{T}}Q\mathbf{x} = \sum_{hk} q_{hk}x_h x_k \ (q_{hk} = q_{kh})$

(where the symbols are to be interpreted in an appropriate way). Everything is straightforward, except that, in order to conform with the standard conventions of matrix manipulation, we would need to write $\mathbf{x}A$ instead of $A\mathbf{x}$, $\mathbf{y}\mathbf{x}^{\mathrm{T}}$ instead of $\mathbf{x}^{\mathrm{T}}\mathbf{y}$ (corresponding to $\mathbf{x}^*\mathbf{y}$ or $\mathbf{x} \times \mathbf{y}$), and, therefore, $\mathbf{x}Q\mathbf{x}^{\mathrm{T}}$ instead of $Q\mathbf{x}^*\mathbf{x}$. All vectors are to be understood as row vectors, except when they have a 'transpose' superscript, which transforms them into column vectors (dual vectors; i.e. of the form $\mathbf{a}\times$, but as operators on the right).

Note, therefore, that while $\mathbf{xy}^{\mathrm{T}}$ means $\mathbf{y} \times \mathbf{x}$, $\mathbf{y}^{\mathrm{T}}\mathbf{x}$ means $\mathbf{x} \,.\, \mathbf{y} \times$; i.e. it represents the transformation A which takes every vector $\mathbf{z}$ to $A\mathbf{z} = \mathbf{x} \,.\, (\mathbf{y} \times \mathbf{z})$ (the transformation of rank 1 which transforms all the vectors of S_r into vectors parallel to a particular vector; $\mathbf{x}$ in our case); the entries of the matrix A are given by $a_{hk} = x_k y_h$.† Observe, in particular, that $\mathbf{x} \,.\, \mathbf{x} \times$, or $\mathbf{x}^{\mathrm{T}}\mathbf{x}$ (such that $A\mathbf{z} = \mathbf{x} \,.\, (\mathbf{x} \times \mathbf{z})$), represents the vector which is the projection of $\mathbf{z}$ in the direction of $\mathbf{x}$ (if $\mathbf{x}$ is a unit vector; otherwise, it is multiplied by $\mathbf{x}^2$, which we write instead of $|\mathbf{x}|^2$, i.e. $\mathbf{x} \times \mathbf{x}$).

10.2.4. The covariance matrix—defined in Chapter 4, 4.17.5, for random variables X_h with $\mathbf{P}(X_h) = 0$, by $\sigma_{hk} = \mathbf{P}(X_h X_k)$—can be defined in this set-up as Var($\mathbf{X}$), or simply $V(\mathbf{X})$, by setting, for $\mathbf{X} = (X_1, X_2, \ldots, X_r)$,

$$V(\mathbf{X}) = \mathbf{P}(\mathbf{X} \,.\, \mathbf{X} \times) = \mathbf{P}(\mathbf{X}^{\mathrm{T}}\mathbf{X}):$$

in other words, as the linear transformation which gives, for each vector $\mathbf{u}$,

$$V(\mathbf{X})\mathbf{u} = \mathbf{P}(\mathbf{X} \,.\, (\mathbf{X} \times \mathbf{u})). \tag{10}$$

Since

$$V(\mathbf{X})\mathbf{u} \times \mathbf{v} = \mathbf{P}((\mathbf{X} \times \mathbf{u})(\mathbf{X} \times \mathbf{v})) = V(\mathbf{X})\mathbf{v} \times \mathbf{u},$$

the linear transformation is symmetric, and so we can find an r-tuple of orthogonal directions which are mapped to themselves (i.e. there exist eigenvectors $\mathbf{v}_h$ and eigenvalues λ_h such that $V(\mathbf{X})\mathbf{v}_h = \lambda_h \mathbf{v}_h$); the transformation is also positive ($V(\mathbf{X})\mathbf{u} \times \mathbf{u} = \mathbf{P}(\mathbf{X} \times \mathbf{u})^2$), and so $\lambda_h > 0$.

When we are referring to a fixed $\mathbf{X}$, and there is no danger of ambiguity, we shall simply write V in place of $V(\mathbf{X})$.

We have already seen (in Chapter 4, 4.17.5, and in Chapter 7, 7.6.7) that the normal distribution, in whatever number of dimensions, is characterized by its covariance matrix, and that such a matrix (i.e. symmetric and positive definite) characterizes a unique normal distribution (where throughout we are assuming distributions to be centred at zero). At the point $0 + \mathbf{x}$ the density has the form

$$f(\mathbf{x}) = f(x_1, x_2, \ldots, x_r) = K\,\mathrm{e}^{-\frac{1}{2}Q(\mathbf{x})}, \qquad K = 1/\sqrt{[(2\pi)^r \det Q]}. \tag{11}$$

The relationship of Q and V is given by $V = Q^{-1}$ (and, conversely, $Q = V^{-1}$), by virtue of the fact that the eigenvalues are the variances, σ_h^2, for V, but their inverses, σ_h^{-2}, for Q.

For these reasons, we again get involved with the *ellipsoid of covariance* (or *of inertia*) and the *ellipsoid of concentration*, which we first came across in Chapter 4, 4.17.6, and which we are forced to consider further. What we said in Chapter 7, 7.6.7, concerning the affine properties (for which it is sufficient

† This follows, even without taking into account the geometrical meaning of x_x and y_h, from the fact that the characteristic of the matrix must be 1 (rows and columns are proportional).

to consider the case of spherical symmetry) still holds, whereas any consideration of the ellipsoids only makes sense, and has any use, if it is necessary, or appropriate, to base oneself upon a preassigned metric. (This would be the case, for example, were we dealing with a problem in real, physical space, or if a number of problems, each of which separately would require a different metric for convenience, were considered simultaneously.)

In any case, we lose nothing in the way of generality, and we gain a great deal in terms of simplicity and understanding, if, in order to study this problem, we take the principal axes of inertia as our reference system. In other words, we take as our unit vectors $\mathbf{i}_h$ the eigenvectors of Q and V (necessarily orthogonal)†, whose respective eigenvalues are the variances σ_h^2 of V, and the reciprocals ('weights') σ_h^{-2} of Q.

We obtain, therefore,

(12) $$Q\mathbf{i}_h = \sigma_h^{-2}\mathbf{i}_h, \qquad Q\mathbf{x} = \sum_h x_h\sigma_h^{-2}\mathbf{i}_h,$$

(13) $$Q\{\mathbf{x}\} = Q\mathbf{x}\times\mathbf{x} = \sum_h(\sigma_h^{-2}x_h)x_h = \sum_h\sigma_h^{-2}x_h^2 = \sum_h(x_h/\sigma_h)^2;$$

(14) $$V\mathbf{i}_h = \sigma_h^2\mathbf{i}_h, \qquad V\mathbf{u} = \sum_h u_h\sigma_h^2\mathbf{i}_h,$$

(15) $$V\{\mathbf{u}\} = V\mathbf{u}\times\mathbf{u} = \sum_h(\sigma_h^2u_h)u_h = \sum_h\sigma_h^2u_h^2 = \sum_h(\sigma_hu_h)^2.$$

As we already know, $Q\{\mathbf{x}\}$ is useful when it comes to expressing the density (by means of (11)), which, in the present reference system, becomes (there being no cross-product terms)

(16) $$f(\mathbf{x}) = K\,\mathrm{e}^{-\frac{1}{2}Q\{\mathbf{x}\}} = K\exp\left[-\tfrac{1}{2}\sum_h(x_h/\sigma_h)^2\right] = K\prod_h\exp[-\tfrac{1}{2}(x_h/\sigma_h)^2].$$

This shows (as was obvious anyway) that, in this reference system, the components X_h of $\mathbf{X}$ are stochastically independent (the density is a product of factors each of which is a function of only one x_h). But this implies that the same factorization holds for the characteristic function,

(17) $$\phi(\mathbf{u}) = \prod_h\exp[-\tfrac{1}{2}(\sigma_hu_h)^2] = \exp\left[-\tfrac{1}{2}\sum_h(\sigma_hu_h)^2\right] = \mathrm{e}^{-\frac{1}{2}V\{\mathbf{u}\}},$$

and we therefore see the complementary rôle played by $V = Q^{-1}$ in defining the characteristic function.

The two ellipsoids are given by

$V\{\mathbf{u}\} = 1$ (*covariance* or *inertia*; semi-axes $1/\sigma_h$), and

$Q\{\mathbf{x}\} = 1$ (*concentration*; semi-axes σ_h).

† Apart from irrelevant ambiguities in the case of multiple eigenvalues.

The choice of the different variables **u** and **x** for V and Q is deliberate, and in explaining this choice we will be led to a comparison of the two ellipsoids. The **x** on which Q operates are the actual vectors of the space over which the distribution is defined (the ambit $\mathscr{A}$; e.g. physical space): the **u** on which V operates are essentially the dual vectors (even though, given the introduction of the metric, the two spaces are superposed). This supports the idea that the ellipsoid of concentration is more directly meaningful, as was confirmed, in part, by what we established in Chapter 4, 4.7.6. We must now, as we then promised, consider this further, basing ourselves on the representation in terms of the appropriate normal distribution; i.e. the distribution with the most frequently occurring and stable form having the same previsions and covariances (in mechanical terms, barycentre and kernel of inertia).

As we have seen, the ellipsoids Q = constant are the surfaces on which the density, f, is constant. The special case $Q = 1$ (which gives, therefore, $f = K\,\mathrm{e}^{-\frac{1}{2}}$, which is 0·606 of the maximum at the origin) enjoys a property which justifies one in singling out, and defining as the *body* or *kernel* of the distribution, that part of it contained in $Q \leqslant 1$ (that part corresponding to $Q \geqslant 1$ might be referred to as the *tail*, or *shell*, but no appropriate term seems to exist). The meaning is clearest in one dimension: the kernel is the portion of the distribution with convex density lying between the points of inflexion—cf. Figure 7.6 of Chapter 7, 7.6.6—and the tail consists of the two outside portions with concave density; i.e. tapering away. The same thing applies in the general case, however: *inside* $Q \leqslant 1$ the density is convex (with the same meaning: for each point $\lambda A + (1 - \lambda)B$, $0 < \lambda < 1$, in the segment between A and B, the density has a greater value than the linear interpolation $\lambda f(A) + (1 - \lambda)f(B)$); *outside*, however, i.e. for $Q \geqslant 1$, in the direction of a radius emanating from the origin the behaviour is concave (we are back in the one-dimensional case), and convex in all directions which are conjugate with respect to Q.

10.3 SOME PARTICULAR DISTRIBUTIONS: THE DISCRETE CASE

10.3.1. We shall now look at a few specific problems in more detail, and we begin with those involving discrete distributions.

Many of the problems we have considered for ordinary events can be extended in an obvious manner to the case of multi-events: Instead of a coin, which only has two faces, we could consider a die, which has six faces; instead of an urn with black and white balls, we could have an urn containing balls of r different colours; instead of games which can only result in either victory or defeat, we could consider those in which a draw is also possible, or we could even distinguish a whole range of results (for example, the actual scores, 3–1, 2–2, 0–1, etc., as in football), and so on.

In all these cases, by making various assumptions, there are a whole range of problems that can be considered. In particular, one can try to calculate the probabilities of the r possibilities $1, 2, \ldots, k, \ldots, r$ occurring $n_1, n_2, \ldots, n_k, \ldots, n_r$ times, respectively. This same question can be formulated along different lines, clearly equivalent, but seemingly different at first sight. For example, we might ask how many objects will be given to each of r individuals as the result of some given method of selection (like giving an object to individual k whenever a certain outcome occurs). If the 'objects' are 'particles', and instead of individuals we think of 'physical states', or 'cells' corresponding to them, the different distributions will correspond to different 'macroscopic states'.

10.3.2. The following examples are of this kind, and in order to make them seem more intuitive we shall present them as far as possible in terms of familiar set-ups. They correspond, however, to the fundamental 'statistics' —as they are called by physicists—of Maxwell–Boltzmann (case (*a*)), Fermi–Dirac (case (*b*)) and Bose–Einstein (case (*c*)).

For all these cases, we can think in terms of an urn containing

$$g = g_1 + g_2 + \cdots + g_r$$

balls of r different colours, and then, with respect to different procedures for drawing a total of n balls, we seek the probabilities that the numbers of balls drawn of each of the different colours will be $n_1, n_2, \ldots, n_r$. One should bear in mind, however, that there are many other interpretations which could be considered: for example, how many objects, out of a total of n, will be attributed to individuals (or placed into cells) identified by *colours* $1, 2, \ldots, r$ (i.e. *associated with* balls of these colours). In practice, the individuals could be characterized in any way whatsoever: nationality, sex, marital status, school, etc. (in the case of cells, it might be energy levels). If we stick to colours, this has the advantage of making it clear that, so far as the considerations we are interested in are concerned, the nature of the characteristic on which the classification is based is irrelevant (whereas, of course, this is no longer the case if one wishes to study the particular aspects of some given application).

10.3.3. We now consider the three cases mentioned above. They differ in the form of procedure used in drawing the balls; these correspond to (*a*) with replacement, (*b*) without replacement, (*c*) double replacement, terms which will be made more precise as we go along (equal probabilities being assumed throughout).

(*a*) *With replacement*. We perform n drawings from an urn with replacement. Thinking in terms of our alternative interpretation, we draw n objects in succession, distributing them among the g individuals (or cells) regardless of whether the latter have previously received any or not. This is the obvious extension to higher dimensions of the binomial distribution, and is known as the *multinomial distribution*. At each drawing (independently of the previous outcomes) the probabilities of the various colours are given by $p_k = g_k/g$ (either referring to a drawing of that colour, or in favour of some individual, or cell, identified by that colour). The probability of the various colours appearing $n_1, n_2, \ldots, n_r$ times is therefore given by

$$(18)\qquad \omega^{(n)}_{n_1, n_2, \ldots, n_r} = \frac{n!}{n_1!n_2!\ldots n_r!} p_1^{n_1} p_2^{n_2} \ldots p_r^{n_r} = \frac{n!}{g^n} \prod_k \frac{g_k^{n_k}}{n_k!} = K \prod_k \frac{g_k^{n_k}}{n_k!}.$$

Special case. Taking all the $g_k = 1$ (for example, if all balls, individuals, or cells, are of a different colour; i.e. if we are dealing with the distribution among the g different balls, individuals, or cells, without speaking of colours), we obtain

$$(19)\qquad \frac{n!}{n_1!n_2!\ldots n_r!}\left(\frac{1}{g}\right)^n = \frac{n!}{g^n} \prod_k \frac{1}{n_k!}.$$

10.3.4. (*b*) *Without replacement*. We perform n drawings from an urn without replacement. Thinking in terms of our alternative interpretation, we draw n objects in succession, distributing them only among those of the g individuals who have not yet received any. In this way, we exclude the possibility of an individual (or cell) receiving more than one object (we must therefore assume $n \leqslant g$, and we certainly have $n_k \leqslant g_k, k = 1, 2, \ldots, r$). This is the obvious extension to higher dimensions of the hypergeometric distribution; we obtain

$$(20)\qquad \begin{aligned} \omega^{(n)}_{n_1, n_2, \ldots, n_r} &= \tbinom{g_1}{n_1}\tbinom{g_2}{n_2} \ldots \tbinom{g_r}{n_r} \Big/ \tbinom{g}{n} = K \prod_k \tbinom{g_k}{n_k} \\ &= K \prod_k \frac{g_k(g_k - 1)\ldots(g_k - n_k + 1)}{n_k!}. \end{aligned}$$

In fact, $\binom{g}{n}$ is the number of ways in which n individuals can be chosen out of g (i.e. of distributing n objects among them, not more than one to each). The interpretation of $\binom{g_k}{n_k}$ for colour k is similar, and it gives the number of ways in which a distribution of the given form can take place.

Special case (as above, $g_k = 1$). The possible distributions correspond to the various possible choices of n out of the g balls (or individuals, or cells), and these number $\binom{g}{n}$. They all have the same probability, $1/\binom{g}{n}$, because $\binom{g_k}{n_k}$ is either equal to $\binom{1}{1}$ or $\binom{1}{0}$, and is therefore equal to 1.

10.3.5. (*c*) *Double replacement*. We perform n drawings from an urn, replacing, on each occasion, the ball drawn, together with a further ball of the same colour (so that, after $m = m_1 + m_2 + \ldots + m_r$ drawings of the balls of various colours, the urn contains $g + m$ balls, of which $g_k + m_k$ are of colour k). Thinking in terms of our alternative interpretation, we could imagine that every individual participates at each drawing as though it were a raffle, and, together with his original ticket, has a number of additional tickets, one for each object received so far.†

In this case, we have

$$\omega^{(n)}_{n_1,n_2,\ldots,n_r} = \frac{n!}{n_1!n_2!\ldots n_r!}\frac{\prod_k g_k(g_k+1)(g_k+2)\ldots(g_k+n_k-1)}{g(g+1)(g+2)\ldots(g+n-1)}$$

(21)

$$= \frac{1}{\binom{g+n-1}{n}}\prod_k \binom{g_k+n_k-1}{n_k}.$$

To see this, note that the ratio giving the second factor is precisely the probability of obtaining the required distribution in some preassigned order. In fact, if we write, for example,

$$\text{(22)}\qquad \frac{g_1}{g}\cdot\frac{g_3}{g+1}\cdot\frac{g_3+1}{g+2}\cdot\frac{g_2}{g+3}\cdot\frac{g_2+1}{g+4}\cdot\frac{g_3+2}{g+5}\cdot\frac{g_1+1}{g+6}\cdot\frac{g_2+2}{g+7}\cdot\frac{g_2+3}{g+8},$$

we are expressing, as a product (compound probability), the probability of obtaining, in $n = 9$ drawings, colour 1 twice, colour 2 four times, colour 3 three times, in the order 1–3–3–2–2–3–1–2–2. For a different order, we merely permute the numerator; the denominator does not change. If the order is not taken into account, the required probability is that given above multiplied by the number of permutations (in which the order is preserved *among* $g_k, g_k + 1$, etc.). In the example, the number of permutations is $9!/2!4!3!$; in the general case, we have $n!/n_1!\ldots n_r!$, as in (21).

Pólya's urn scheme (*for 'contagious diseases'*). The process which we have just considered—drawings with double replacement—is known as Pólya's urn scheme (especially in the case $r = 2$, black and white balls), having been introduced by Pólya as a particular model for the spread of 'contagious diseases' (in the sense that the more a colour turns up, the more probable it is to do so again). We observe that, contrary to what one might think initially, results which differ only in the order (permutations!) have the same probability (as we saw in the case of (22)). On the other hand, this also holds in the case of drawings without replacement and in other variants: for example, after each drawing replacing c balls of the colour just drawn and d balls of the other colour. If $d > 0$, we have the possibility of dealing with

† A somewhat more expressive example is the following. The r original individuals act as recruiting officers for companies. New individuals are assigned to companies by randomly selecting someone already present, and then assigning the individual to his company (so that, at any given moment, the largest company has the highest probability of recruiting).

other cases besides the 'contagious' form. If negative values are also permitted for c and d, many conclusions still hold, but the process may—and sometimes certainly does—terminate after a finite number of drawings (it suffices to consider the case $c = -1$ and $d = 0$; the case of drawings without replacement). We could also generalize beyond the model of balls in an urn, and take c and d as non-integer parameters for determining the successive probabilities.

Special case (as above, $g_k = 1$). In this case, we have $g_k + n_k - 1 = n_k$, and hence the product in (21) is equal to 1 (all factors are of the form $\binom{n_k}{n_k} = 1$) and all possible distributions (with any given g and n) have the same probability:

$$1/\binom{g+n-1}{n}. \tag{23}$$

We recall that $\binom{g+n-1}{n}$ is the number of ways of distributing g objects among n individuals, two distributions being considered distinct only if they differ in the *number* of objects (and not in the *particular* objects) attributed to each individual.†

Sometimes, one refers to 'the different distributions that arise when the objects are considered as indistinguishable', and in the interpretation of cases in physics where this turns out to be applicable (experimentally) it is attributed to the fact that the particles in question are 'indistinguishable'. The same interpretation also holds in the general case (g_k arbitrary), and the explanation is practically identical.

However, the interpretation in terms of the Bayes–Laplace scheme (which we shall meet in Chapter 11, 11.4.3) is possibly more satisfactory, and might also be considered.

10.3.6. *Remark.* In the case of the applications in physics to which we have referred, case (b) (drawings without replacement) holds when Pauli's exclusion principle applies, and corresponds to the so-called Fermi–Dirac 'statistics', applicable to electrons, protons, and neutrons (i.e. particles with semi-integer spins). Case (c) (double replacement) holds in all other cases, and corresponds to the so-called Bose–Einstein 'statistics', applicable to photons, mesons, and so on (i.e. particles with integer spin).

Case (a) (drawings with replacement, the Bernoulli scheme) corresponds to classical statistical mechanics (Maxwell–Boltzmann 'statistics'). According to modern theoretical physics this never applies, but it provides, asymptotically, an approximation to both (b) and (c) when the g_k are much larger than the corresponding n_k.

It would be very worthwhile to proceed further with the actual application of these ideas, principally to the problem of determining statistical equili-

† Cf. the Remark in Section 10.4.1.

brium. However, this would take us well beyond our purpose in providing this introductory outline.

10.4 SOME PARTICULAR DISTRIBUTIONS: THE CONTINUOUS CASE

We now turn to the continuous case, where there are a number of interesting problems. We shall only be able to sample a few of them, choosing those which are best suited to illustrating certain useful techniques, and to presenting, in a simple fashion, those distributions most frequently encountered in practice.

10.4.1. *Subdivisions of an interval.* This is a continuous analogue of the problem we have just discussed in the discrete case. Instead of considering the subdivision of some given n objects into r groups, we consider the subdivision of an interval (for convenience, assumed to be of unit length) into r parts. In this way (or as a result of subdividing some other quantity), we end up with a collection of r random quantities $X_1, X_2, \ldots, X_r$, whose sum is equal to one.

There are various ways of performing such a subdivision. Of these, we shall consider one of the most straightforward and 'symmetric', and we shall give it its customary title, referring to it as 'random subdivision'. This has a certain convenience, so long as one does not attempt to read too much into the terminology, thinking of it as endowing this particular method of subdivision with some special significance, rather than being just a matter of convention.

More precisely, when we talk of *random subdivision* of an interval we mean that $r - 1$ division points are chosen independently, each with a uniform distribution. Equivalently, we could say that, after having performed the subdivision into k parts, the kth division point is chosen by first choosing a subinterval—with probability of choice proportional to length—and then choosing a point within this subinterval by means of a uniform distribution over it. This formulation is a little 'artificial' if we are considering subdivision of an interval, but it still makes sense, and has the advantage of also being applicable to the subdivision of an arbitrary quantity (mass, area, sum of money, amount of energy, etc.). The distribution itself has constant density over the range of possible values; i.e. over the $(r - 1)$-dimensional simplex defined by $x_k \geqslant 0$ $(k = 1, 2, \ldots, r)$ and $x_1 + x_2 + \ldots + x_r = 1$. If $r = 3$, for example, it is uniform over the equilateral triangle, as shown in Figure 10.1.

Remark. It is instructive to point out that we are here dealing with the limit case (as we pass from the discrete to the continuous) of the Bose–Einstein 'statistic', as considered above. In that case, in fact, the distributions of the n 'indistinguishable' objects over the g cells correspond to the $\binom{g+n-1}{n}$ ways in which n points (representing the objects) and $g-1$ division bars, together with a bar at either end (which represent the division into cells), can be arranged. For example, the distribution which results in 0, 2, 0, 3, 1 objects in the 1st, 2nd, ..., 5th cells, respectively, would be represented by //**//***/*/. If the number of points n is large in comparison with the number of cells, the bars subdivide the interval in a manner very close to that described above. If we consider the distribution, it is practically uniform over the simplex because the possible points are uniformly distributed over it—the x_k are all multiples of $1/n$—and all have the same probability (i.e. $1/N$, where $N = \binom{g+n-1}{n}$ is the total number of points). Note that the two cases are also analogous notationally (in the 'special case' we have $g = r$; the comparison with the general case led us, however, to prefer to write g rather than r).

10.4.2. *Problems relating to random subdivision* arise quite naturally and frequently in a number of applications. In order to be able to picture the distribution, it will often be useful to consider special cases where r has a small value, and the simplex reduces to an interval ($r = 2$), an equilateral triangle ($r = 3$) or a regular tetrahedron ($r = 4$). We find—for the same reasons, although the purpose is different—that the diagrams we require are the same as those already encountered in Chapter 5 (especially Figure 5.3b), and which represented probabilities p_h with sum equal to one.

In our case, it is the sum of the subintervals x_h which is equal to one. For $r = 3$, subdivision into three corresponds to some point in the triangle $A_1A_2A_3$ (having barycentric coordinates, x_1, x_2 and x_3, with $x_1 + x_2 + x_3 = 1$, where the x_h are the distances from the three sides, and the height of the triangle is taken to be unity). Two simple examples will suffice to illustrate this form of representation, and to show how, with its aid, one can obtain immediately certain conclusions which would involve heavy calculations if arrived at analytically.

In Figure 10.1(a), the areas corresponding to $X_1 \leqslant x_1, X_2 \leqslant x_2, X_3 \leqslant x_3$ (for given x_1, x_2, x_3) are indicated with different forms of shading. The unshaded triangle which remains (with sides $1 - x$, where $x = x_1 + x_2 + x_3$; we clearly have $x \leqslant 1$, and so this does exist) represents the subdivisions in which $X_1 \geqslant x_1, X_2 \geqslant x_2, X_3 \geqslant x_3$. The probability of a subdivision for which this holds is therefore given by $(1-x)^2$ (the ratio of the area of the smaller triangle to the larger), and, by virtue of the homogeneity, one can see immediately that, for arbitrary r, the probability is equal to $(1-x)^{r-1}$.

Figure 10.1(b) ($X_1 > X_2 > X_3$) illustrates the following problem. Suppose that $Z_1, Z_2, \ldots, Z_{r-1}$ are the abscissae of the $r-1$ division points arranged *in increasing order*: what are their probability distributions? We recall that the points are chosen independently and with a uniform distribution over the given interval, but that if we consider them as ordered neither independence nor uniformity continues to hold. It is obvious—but one does not always think of it—that everything changes when the state of information changes (and the latter change may be obscured by the *terminology* used). As an example of this, we note that 'the 1st division point obtained' (in chronological order) may very well be 'the 7th division point obtained' (when they are taken in increasing order—for example, at some given moment when 20 of them have been considered), and that knowing these two facts to coincide changes the state of information, and with it the probability distribution. More concretely, the probability distribution of the chronologically first division point changes after each new division point is obtained if we are informed as to whether the latter is to the right or to the left of the former.

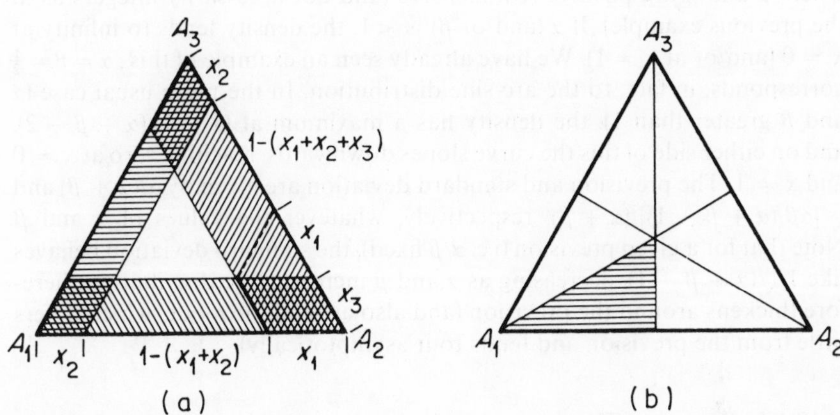

Figure 10.1 (a) The probability that $X_1 \geqslant x_1 . X_2 \geqslant x_2 . X_3 \geqslant x_3$. (b) The probability that $X_1 > X_2 > X_3$

Let us first determine the distribution of the *maximum*, Z_{r-1}. To say that $Z_{r-1} \leqslant x$, amounts to saying that all the $r - 1$ division points are $\leqslant x$; this has probability x^{r-1}, and the distribution we are after is therefore given by

(24) $$F(x) = x^{r-1}, \qquad f(x) = (r-1)x^{r-2} \qquad (0 \leqslant x \leqslant 1).$$

For the *minimum*, Z_1, we have, by symmetry,

(25) $$F(x) = 1 - (1-x)^{r-1}, \qquad f(x) = (r-1)(1-x)^{r-2}.$$

In general, for the kth point, Z_k (taken in increasing order), the density is given by

(26) $$f(x) = (r-1)\binom{r-2}{k-1}x^{k-1}(1-x)^{r-k-1} \qquad (0 \leqslant x \leqslant 1).$$

To see this, note that the probability of one (no matter which) of the $r - 1$ points falling in the interval from x to $x + \mathrm{d}x$ is $(r - 1)\,\mathrm{d}x$, which must then be multiplied by the probability that, of the remaining $r - 2$ points, $k - 1$ fall on the left (with probability x) and $r - k - 1$ on the right (probability $1 - x$).

10.4.3. *The beta distribution.* The distributions which we have just encountered belong, in fact, to the family of beta distributions; a family which finds frequent and important applications. The general form of the density is given by

$$(27)\quad f(x) = Kx^{\alpha-1}(1 - x)^{\beta-1} \quad (K = \Gamma(\alpha + \beta)/\Gamma(\alpha)\Gamma(\beta);\quad \Gamma(n) = (n - 1)!),$$

where α and β are positive *real numbers* (and not necessarily integers as in the previous example). If α (and/or β) is <1, the density tends to infinity at $x = 0$ (and/or at $x = 1$). We have already seen an example of this; $\alpha = \beta = \frac{1}{2}$ corresponds, in fact, to the arc sine distribution. In the more usual case (α and β greater than 1), the density has a maximum at $(\alpha - 1)/(\alpha + \beta - 2)$, and on either side of this the curve slopes downwards, reaching zero at $x = 0$ and $x = 1$. The prevision and standard deviation are given by $\alpha/(\alpha + \beta)$ and $\sqrt{[\alpha\beta/(\alpha + \beta + 1)]}/(\alpha + \beta)$, respectively, whatever the values of α and β. Note that for a given prevision (i.e. α/β fixed), the standard deviation behaves like $1/\sqrt{(\alpha + \beta + 1)}$, decreasing as α and β increase. The distribution therefore thickens around the prevision (and also around the mode, which differs little from the prevision and tends to it asymptotically).

10.4.4. *Extension.* The argument which we gave in the case of the division points extends immediately to the case of any n independent random quantities having the same distribution F, where F can be any distribution at all.†

The distribution of the maximum of $X_1, X_2, \ldots, X_n$ is given by

$$(28)\qquad F_{(n)}(x) = F^n(x), \qquad f_{(n)}(x) = nF^{n-1}(x)f(x),$$

and the density for the kth largest by

$$(29)\qquad f_{(k)}(x) = n\tbinom{n-1}{k-1}F^{k-1}(x)(1 - F(x))^{n-k}f(x).$$

Similar expressions can be obtained under more general conditions.

† Note that n corresponds to the 'number of division points' of the preceding case, where it was denoted by $r - 1$ because there were r subintervals.

If, for the sake of simplicity, we restrict ourselves to the maximum, we obtain, in the general case,

(30) $$F_{(n)}(x) = F(x, x, \ldots, x),$$

where F is the joint distribution function of the n random quantities. If, in particular, the random quantities are independent (but each X_k has a different distribution $F_k(x)$), we have

(31) $$F_{(n)}(x) = F_1(x)F_2(x) \ldots F_n(x).$$

10.4.5. *'Random' subdivision and the Poisson process.* Suppose that, in a Poisson process, n occurrences are known, or are assumed, to have taken place in some given interval: then, in the sense we have defined, they form a 'random subdivision' of the interval. Conversely, if we imagine the subdivision of an interval of length $n + 1$ by means of n points in such a way that each of the $n + 1$ subintervals has expected length 1, then, as n increases, we approach a Poisson process (with intensity $\mu = 1$). The distributions relating to the 1st, 2nd, ..., kth, ... positions now belong to the gamma family instead of the beta as above.

In both these cases, the length of each interval is, in prevision, equal to 1. In general, however, it is important that the method of picking such an interval should be made explicit. If we refer to the 'third interval starting from 0', or 'the first interval after $x = x_0$', then what we have said is true. It is clearly no longer true if we look at 'the shortest', or 'the longest' (in which case, we reduce to the problem considered above, independence holding in the Poisson case, but not for a random subdivision). It is perhaps not quite so obvious that the result no longer holds if we pick out the interval 'containing some given point', but it is clear, on reflection, that this method does favour the longer intervals. In actual fact, the prevision of the length of an interval chosen in this way is *twice* that of the Poisson case, and only a little less than twice that of the case of a random subdivision. In the first case, the prevision of the distance from a given point (division point or not) to the first division point, both on the left and on the right, is equal to 1. In the second case, the point chosen as a reference point plays the rôle of an additional 'point chosen at random'.† This means that the original interval, of length $n + 1$, turns out to be subdivided into $n + 2$ subintervals, each of which has expected length $(n + 1)/(n + 2)$. Two of these subintervals join

† For this to hold exactly, we require the point to be 'chosen at random', and its position to be unknown. In other cases, the result is very little altered except when the point is very close to the end-points (and then one of the two subintervals is necessarily small). This should provide an adequate background to more complicated situations, as well as illustrating how such complications can arise in seemingly harmless formulations of problems if one is not sufficiently careful.

together to form the interval into which the new division point has fallen, and the expected length of this interval is therefore $2(n+1)/(n+2) = 2 - 2/(n+2)$.

10.5 THE CASE OF SPHERICAL SYMMETRY

10.5.1. *Examples with spherical symmetry.* We shall obtain further useful insights by considering—in the plane, in ordinary space, and in an arbitrary number of dimensions—distributions possessing spherical symmetry. In particular, we shall consider the normal distribution. Referring to the three-dimensional case for convenience, this means that the density (provided it exists) is a function of the distance ρ only; i.e. $f(x, y, z) = g(\rho)$, a function of $\rho^2 = x^2 + y^2 + z^2$.

10.5.2. *Distance from the origin.* The distance $(X_1^2 + X_2^2 + \ldots + X_r^2)^{\frac{1}{2}}$ has a probability distribution with density $f(\rho) = Kg(\rho)\rho^{r-1}$. Taking the particular case of a uniform distribution inside the hypersphere (with radius 1), we obtain $f(\rho) = K\rho^{r-1}$, and we note that this is identical to what we obtained for the distribution of the abscissa of the maximum when r points were chosen at random in $[0, 1]$. Observe that, for large r, the volume is concentrated near the surface; i.e., for any given $\varepsilon > 0$, the layer between $1 - \varepsilon$ and 1 includes all the volume apart from a fraction θ which tends to zero as r increases. More precisely, the distance from the surface, as r increases, tends asymptotically to an exponential distribution with prevision $1/r$.

In the case of the normal distribution, the distance is distributed with density

$$f(\rho) = K\rho^{r-1}\,\mathrm{e}^{-\rho^2/2}. \tag{32}$$

For $r = 3$, we note that we obtain *Maxwell's formula* for the distribution of the (absolute values of) velocities in a gas, assuming them to be normally and spherically distributed: $f(v) = Kv^2\,\mathrm{e}^{-v^2/2}$ (where we take the prevision of the square of the velocity to be equal to 3; i.e. equal to 1 for each component).

The distribution given by (32) is widely used in a number of problems. In particular, it occurs in statistics, where one often takes as a basis of comparison the square of some deviation from a 'true' value. In this case, it is known as the χ^2 (*'chi-square'*) distribution. If we take $x = \rho^2$ as the variable rather than ρ, we obtain a gamma distribution

$$f(x) = Kx^{(r-2)/2}\,\mathrm{e}^{-x/2}.$$

In fact, if we temporarily write $f_1(x)$ in order to avoid confusion with $f(\rho)$, we have

$$f_1(x)\,\mathrm{d}x = f(\rho)\,\mathrm{d}\rho\dagger = Kf(x)\,\mathrm{d}x = K\,.\,x^{(r-1)/2}\,\mathrm{e}^{-x/2}x^{-\frac{1}{2}}\,\mathrm{d}x$$

(the constant $\frac{1}{2}$ being included in K).

10.5.3. *Distance from a hyperplane to the origin* (or, alternatively, the coordinate, or projection, onto an arbitrary axis).‡ This has as its distribution the projection of the spatial distribution, and is the same for all axes. In other words, if $f(x)$ gives the density of the distribution of X, then it also gives that of Y and Z, and of any other coordinate $aX + bY + cZ$, where $a^2 + b^2 + c^2 = 1$ (i.e. with the same unit of measurement). Given $g(\rho)$, we have

$$f(x) = K\int_0^\infty g(\sqrt{(x^2 + \lambda^2)})\lambda^2\,\mathrm{d}\lambda, \tag{33}$$

and, for general r,

$$f(x) = K\int_0^\infty g(\sqrt{(x^2 + \lambda^2)})\lambda^{r-1}\,\mathrm{d}\lambda. \tag{34}$$

We have already seen that in the case of the normal distribution (and only in this case) g and f coincide (up to the normalization constant). We have also seen that for a uniform spherical distribution, $g(\rho) = K > 0$ for $\rho \leqslant 1$, $g(\rho) = 0$ for $\rho > 1$, we have $f(x) = K(1 - x^2)^{(r-1)/2}$ (see Section 7.6.8), and we observe that we are dealing with a beta distribution,

$$f(x) = K(1 + x)^{(r-1)/2}(1 - x)^{(r-1)/2},$$

defined over $[-1, 1]$, rather than over $[0, 1]$. Let us take up again the case of a distribution on the surface of a unit sphere; more precisely, we shall consider a spherical layer (points whose distances from the origin lie in the range $1 - \varepsilon$ to 1) whose thickness, $\varepsilon > 0$, we let tend to zero. In this way, we obtain

$$\begin{aligned} f(x) &= K[(1 - x^2)^{(r-1)/2} - ([1 - \varepsilon]^2 - x^2)^{(r-1)/2}] \\ &\simeq K[2(r - 1)\varepsilon(1 - x^2)^{(r-1)/2-1}] = K(1 - x^2)^{(r-3)/2}. \end{aligned} \tag{35}$$

† We take this opportunity of pointing out how a change of variable leads to an altered form of density (obviously: we are dealing with the derivative of a function of a function!). For increasing transformations, this applies directly: for transformations which are not one-to-one, we have to add up the separate contributions. As a practical rule, it is convenient to transform (as we have done) $f(y)\,dy$ into $f_1(x)\,\mathrm{d}x$, rather than writing $f_1(x) = f(y)\,.\,\mathrm{d}y/\mathrm{d}x$.

For transformations of several variables one proceeds in a similar fashion, but multiplying by the Jacobian $\partial(y_1, \ldots, y_r)/\partial(x_1, \ldots, x_r)$ instead of by $\mathrm{d}y/\mathrm{d}x$.

‡ This differs only that, in speaking of *distance*, one needs to take the *absolute value* of the abscissa. Given the symmetry, the density is $2f(x)$ for $x \geqslant 0$, and zero for $x \leqslant 0$, rather than

$$f(x)(-\infty < x < +\infty).$$

We are thus led to the same distribution, but with r reduced by 2. For the particular case $r = 2$, we have $f(x) = K/\sqrt{(1 - x^2)}$; in other words, as was obvious geometrically, we again obtain the *arc sine* distribution. For $r = 3$, we obtain the uniform distribution (as one would expect from the well-known relation between the area of the sphere and of the cylinder). In both of these cases, as in many other cases of this kind, as r increases the projection of the distribution tends to normality.

The distance from a straight line (or plane, or arbitrary Euclidean space with dimension $d < r$) passing through the origin can be shown to lead to a gamma distribution (with parameters $\alpha = d$ and $\beta = r - d$).†

10.5.4. Finally, let us consider the *central projection* of a distribution with spherical symmetry (r-dimensional) onto an arbitrary hyperplane ($(r - 1)$-dimensional); a straight line if $r = 2$, a plane if $r = 3$, and so on. This is clearly the same no matter which hyperplane we take (apart from changes of scale, which can be avoided in any case if we adopt the convention of taking the hyperplane to be unit distance from the origin), and *no matter what distribution one starts with* (in other words, it does not matter what $g(\rho)$ is: in fact, it does not matter how the mass moves along the radii of the projection). We might as well assume, therefore, that the mass is uniformly distributed on the surface of a hypersphere with radius 1 (centred at the origin).

We have just seen, however, that the projection of this distribution onto an axis, $x = \cos\phi$ (see Figure 10.2), has density $K(1 - x^2)^{(r-3)/2}$. A mere change of variable suffices, therefore, to obtain the distribution in terms of either the angle ϕ, or $y = \tan\phi$, or $z = 1/y = \cot\phi$.‡

From $x = \cos\phi$, we obtain

$$(1 - x^2)^{\frac{1}{2}} = \sin\phi, \qquad dx = \sin\phi\, d\phi,$$

$$K(1 - x^2)^{(r-3)/2}\, dx = K\sin^{r-3}\phi\,.\,\sin\phi\, d\phi = K\sin^{r-2}\phi\, d\phi,$$

the distribution for ϕ having density proportional to $\sin^{r-2}\phi$ (i.e., as is well-known from geometry, the area of the ring cut on the hypersphere by cones with semi-angles ϕ and $\phi + d\phi$).

From $y = \tan\phi$, i.e. $\phi = \tan^{-1} y$, we obtain

$$\sin\phi = y(1 + y^2)^{-\frac{1}{2}}, \qquad d\phi = (1 + y^2)^{-1}\, dy,$$

$$K\sin^{r-2}\phi\, d\phi = Ky^{r-2}(1 + y^2)^{-r/2}\, dy.$$

† This problem crops up in connection with problems in theoretical physics: cf. J. von Neumann, *Zeitschr. Phys.*, **57** (1929); A. Loinger, *Rend. S.I.F.*, 1961.

‡ The letters y and z are used here simply for convenience, and not in their usual sense of co-ordinates.

Finally, from $z = 1/y$, i.e. $y = z^{-1}$, we obtain

$$dy = -z^{-2}\,dz,$$

$$Ky^{r-2}(1 + y^2)^{-r/2}\,dy = Kz^{-(r-2)}(1 + z^{-2})^{-r/2}z^{-2}\,dz$$

$$= Kz^{-r}(1 + z^{-2})^{-r/2}\,dz = K(1 + z^2)^{-r/2}\,dz.$$

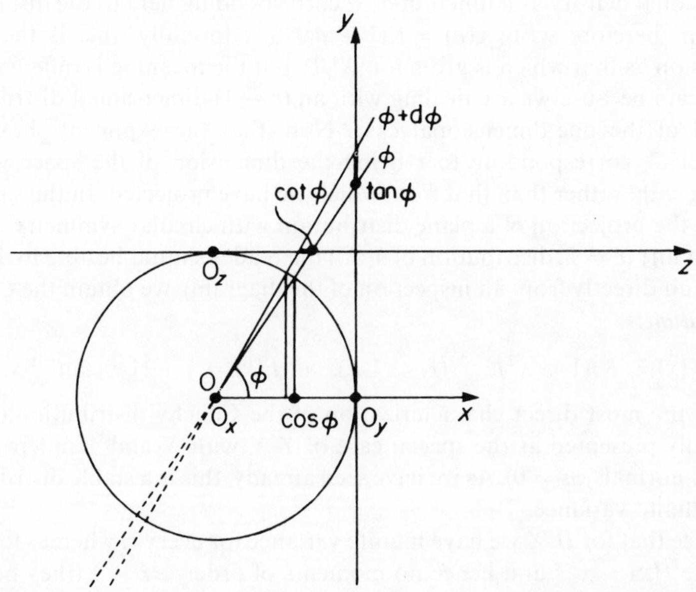

Figure 10.2 The central projection (origin 0) of a spherically symmetric distribution. The functions ϕ, $\cos\phi$, $\tan\phi$, $\cot\phi$, and their derivatives, appear in the various problems which we consider

If $X_1, X_2, \ldots, X_r$ are random quantities whose joint distribution has spherical symmetry, and we set $X = X_1$, $R =$ distance of the point $(X_1, \ldots, X_r)$ from 0 $(R^2 = X_1^2 + X_2^2 + \ldots + X_r^2)$, $D = \sqrt{(R^2 - X^2)} =$ distance of the same point from the x-axis, then the variables previously denoted by x, y, z, and ϕ correspond to X/R, D/X, X/D and $\tan^{-1}(D/X)$, respectively. Their distributions therefore have densities of the form:

(36) X/R $(\cos\phi)$: $f(x) = K(1 - x^2)^{(r-3)/2}$ $(-1 \leqslant x \leqslant 1)$

(37) D/X $(\tan\phi)$: $f(x) = Kx^{r-2}(1 + x^2)^{-r/2}$

(38) X/D $(\cot\phi)$: $f(x) = K(1 + x^2)^{-r/2}$

(39) $\tan^{-1}(D/X)$ (ϕ): $f(x) = K\sin^{r-2}x$ $(-\pi/2 \leqslant x \leqslant \pi/2)$;

where x, as usual, denotes the variable. We note that in the case of D/X with r odd, we would have to include the absolute value sign, or, alternatively, think of K changing its sign as x does. In all cases, the same distributions (if we double K and restrict the range to $x \geqslant 0$) correspond to the absolute values of the random quantities ($|X|/R$, etc.).

The distribution of $D/|X|$ is that of the distance for the distribution projected onto the hyperplane. Dividing by $x^{(r-1)-1}$, we obtain the $(r-1)$-dimensional density as a function of x, corresponding here to the distance ρ. We can therefore write $g(\rho) = K(1 + \rho^2)^{-r/2}$; formally, this is the same expression as that which is given for $|X|/D$, but the meaning is different, and K appears because we are dealing with an $(r-1)$-dimensional distribution instead of the one-dimensional case. Note that the exponent should be $-(r+1)/2$, corresponding to r being the dimension of the space we are dealing with, rather than that from which we have projected. In the simplest case of the projection of a plane distribution with circular symmetry onto a straight line ($r = 2$, distribution of Y/X; the reader should be able to deduce this result directly from an inspection of the diagram), we obtain the *Cauchy distribution*:

$$(40) \quad f(x) = K/(1 + x^2) \qquad (K = 1/\pi), \qquad F(x) = \tfrac{1}{2} + (1/\pi)\tan^{-1} x.$$

This is the most direct characterization of the Cauchy distribution (which is usually presented as the special case of Y/X with X and Y independent, centred normals, $m = 0$). As we have seen already, this is a stable distribution with infinite variance.

Notice that for D/X we have infinite variance for every r, whereas for X/D we have $f(x) \sim x^{-r}$ and hence no moments of order $\geqslant r - 1$ (they become infinite). This latter distribution (X/D, with r arbitrary) is also of great importance in statistics, where it finds wide application as *Student's distribution* (Student being the non-de-plume of W. S. Gosset, who introduced it into statistics; cf. Chapter 12, 12.3.6).

CHAPTER 11

Inductive Reasoning; Statistical Inference

11.1 INTRODUCTION

11.1.1. Within the ambit of the logic of certainty, that is to say ordinary logic, valid arguments are deductive arguments. Conclusions which are *certain* can only be arrived at by establishing that they are implicit in something already known. In other words, we arrive at the particular through the general. In doing so, however, it is clear that we can never enlarge our field of knowledge (except in the sense that certain features of our previously acquired knowledge, of which, perhaps, we were previously unaware, are now made more explicit).

The form of argument leading to conclusions which go beyond what is already known, or what has previously been ascertained, is different; this is the so-called inductive form of argument. We have used 'so-called' because, in fact, we must first of all discuss whether, and in what sense, it is legitimate to refer to it as a form of 'argument' at all (cf. Sections 11.1.3–11.1.4).

The problem of induction arises in every field and at every level: from the examination of arguments for and against various scientific theories, to those concerning the guilt of someone suspected of a crime; from methods for establishing, on the basis of some given data, the conditions for a specific kind of insurance policy, to methods of estimating some quality or other to the required degree of accuracy on the basis of measurements which are inherently imprecise.

It is particularly instructive to consider the process by which new scientific theories are formulated. The first step is an intuitive one, arising out of some particular set of observations, but then various modifications are made as a result of more up-to-date results which suggest that this or that alternative theory provides a better explanation. In essence, it is always a question of analysing the current state of information by means of Bayes's theorem (except that, in this rather open-ended and imprecise context, such applications of the theorem are necessarily qualitative in nature). The most interesting feature of all this is, perhaps, the substantial scope which is left for the personal judgements of individual scientists. In particular, it is interesting to note the prevalent, conservative aversion to any form of novelty; an aversion which some might regard as an alarming symptom of the superstitious faith placed in the 'scientific truths' of the moment.

There are two common fallacies which deserve special mention. One consists in believing that a theory can be disproved merely by discrediting some particular explanation, consequence, or application of it. This is not so: it may well be the case that the particular explanation is not essential, or that the particular application breaks down for some other reason.† The cursory manner in which new ideas are discussed is to be deplored, because such ideas, even though they may turn out to be false trails, usually contain within them the germ of something fruitful. In this respect, the second of the two fallacies is even more dangerous. This fallacy consists in leaving out of consideration certain of the data or observations. In the logic of certainty this is quite legitimate: it is perfectly proper to start from some restricted set of hypotheses and to *deduce* the corresponding restricted set of conclusions (which are, in any case, *correct*). In the logic of probability this is not so (as is obvious—even without a consideration of *likelihood*, Chapter 4, 4.6.1—if one considers the bias that would be introduced if all the evidence against some particular hypothesis were suppressed). It is important to note how easy it is to overlook this fact—albeit inadvertently!

A deeper analysis of the way in which scientific thought evolves, in all its many aspects, would make an extremely interesting study, although one fraught with difficulty. In actual fact, I do not know of any work along these precise lines, nor of any such attempt. It would need to be the history of a continuous series of conceptual U-turns, occasioned by singular minds reflecting upon singular results, and initially greeted with hostility, incomprehension and suspicion, until, finally, the weight of favourable evidence and the resulting improvement in the theoretical formulations renders them acceptable.

We can quote a few examples of this. They may serve to give some idea of how a few of the main aspects of the problem should be tackled in the context of a synthesis which captures the essence of the whole. In the work of Weisskopf (which we have already quoted in the footnote to Chapter 8, 8.8.4), the decisive part played in every field by revolutionary conceptual innovations and changes is stressed and allotted its rightful place within an overall examination of the development of modern

† Here are two examples. Wegener's theory (of 'continental drift') has been rejected on the grounds that the mechanism he suggested by way of explanation is not appropriate. But this in no way excludes the possibility of the theory being correct (with the explanatory detail revised, under the assumption of some other mechanism). Velikovsky's theory (concerning certain aspects of the planetary system) was considered absurd because, among other things, it implied that the temperature of Venus could be ridiculously high .... Such temperatures were, in fact, confirmed by Mariner II, and by other observations. This in no way proves the correctness of the theory (which is rather speculative–at least, in terms of current views), but it is sufficient to discredit the claims of those scientists who believe they have the right to make their superficial judgements, without even bothering to examine the numerous, careful arguments put forward.

scientific conceptions. The intuitive basis of an idea and the way in which it develops as one searches for evidence supporting it is vividly described by James D. Watson in *The Double Helix*, Weidenfeld and Nicholson, London (1968), an autobiographical description of the events leading up to the discovery of the structure of DNA (the substance of which the genetic code, etc., is made up). A critical analysis of the academic establishment's attitude to 'disturbing' theories can be found in the article 'The scientific reception system' by Alfred de Grazia, in the volume *The Velikovsky Affair: The Warfare of Science and Scientism*, University Books, New York (1966), which he edited. Various considerations closely bound up with the themes of this book are developed in a paper of mine, 'Remore e freni sul cammino della scienza', appearing in *Civiltà delle macchine* (1964). Everyone is familiar with the background to the struggle for the establishment of new ideas; from relativity theory to quantum physics, from the theory of evolution to that of Mendelian heredity. It would be instructive to obtain a critical compilation of all this material in order to ascertain whether, and to what extent, the situation (mutatis mutandis) has improved since the time of Galileo.

Another aspect of the problem, and one more strictly in line with the subject matter of this work, and, in particular, of this present chapter, is that of examining these processes of discovery and acceptance in the context of the probabilistic basis of the inductive argument. As we have remarked already, we are dealing with rather imprecise situations, so that any estimation of the probabilities involved could only be attempted by experts attempting to put themselves in the place of the scientists of the period in question, assuming only the knowledge available to them. It might be possible to do something worthwhile in this connection.† There is a vast literature in this area, but, in my opinion, it is inspired by considerations which are too abstract and formalistic (in particular, this applies to the work of R. Carnap and K. Popper).‡ In contrast, the critical comments of H. Jeffreys in 'Logic and scientific inference' (which forms Chapter I of his book *Scientific Inference*, cited in the footnotes to Chapter 7, 7.5.5) are beautifully penetrating and witty. By means of a brilliant, imaginary dialogue between a logician and a botanist, he attempts to establish the inevitably uncertain and tentative nature of all scientific 'truths' or 'laws', and, for this reason, the necessity of making probabilistic logic the basis of every argument.§ However, the treatment stops short of actually using this idea, or examining it more deeply; neither does it mention the possibility of applying it to the grander problem of synthesis; i.e., to the problem of choice among competing theories.

11.1.2. The range of problems for which the inductive argument can be carried out specifically as an application of the calculus of probability in a technical sense is more modest, and concerns rather special problems arising in the context of an already accepted approach. We shall confine

† Research carried out by K. Pearson and G. M. Morant, in which they classify and evaluate all the ingredients which could be put forward in any discussion concerning the authenticity of Cromwell's skull (and the conclusion reached is that, to all intents and purposes, it is authentic), might, in some sense, serve as an example which could possibly be extended to more interesting problems. Cf. quotations and comments in B. de Finetti and L. J. Savage, 'Sul mondo di scegliere le probabilità iniziali', in *Bibl. del Metron*, C, Vol. I, Rome (1962), pp. 130–131 and 153.

‡ A comparative and critical study of the various ideas put forward in this area can be found in a useful paper by Imre Lakatos: 'Changes in the problem of inductive logic', in *The problem of Inductive Logic*, North-Holland Publ. Co., Amsterdam (1968).

§ Cf. similar considerations put forward in the Appendix (especially Section 13).

ourselves to these kinds of problems, and, in particular, to the most basic and straightforward cases. We begin by identifying, rather crudely, three possible meanings, or forms, of induction; the second form is the one where we encounter the standard applications, and we shall concentrate on this one.

First form. Here we obtain conclusions of a (more or less strictly) *deterministic* kind. From the realization that in some given number of more or less similar cases some given event has always occurred in precisely the same way, we are often led to expect that the same thing will continue to happen in the future, or even to believe that it must necessarily happen on account of there being some 'law'. This is the most extreme case.

Second form. From the realization that some given event has occurred in some given way *almost always*, or *with some given frequency* (eg., 37·2%), we are often led to expect that in the future it will continue to happen almost always, or with that given frequency, as the case may be. This is the most typical example of *statistical inference* as it is commonly understood.

Third form. From the knowledge of the behaviour of some given event in a collection, or sequence, of more or less similar cases in the past, we are often led to make some kind of forecast of the future. For example: that an apparent tendency for a frequency to decrease will continue; or that this will apply to the tendency of successes to group together in runs, or to alternate with failures; and so on. This can be regarded as the most general case.

We cannot claim that there is any really clear-cut distinction among the three cases (especially between the last two), nor that the distinction between past and future has any real importance. The inductive argument can equally well be used to make conjectures about behaviour at or before the time for which observations are available. The three categories should therefore simply be treated as a convenient way of concentrating attention on certain aspects of the problem, and for reflecting upon the questions raised.

If one were interested in the difficult and complex questions raised by considering the notion of indeterminism (*marginal*, or *static*—with or without experimentation—or *dynamic*, using stochastic models), one might adopt a different classification (for example, that suggested by J. Neyman).† We shall not deal with these problems, however.

11.1.3. If by 'argument' we mean something based upon logic—the logic of certainty, ordinary logic—then it is clear that the 'inductive argument' is not an 'argument'. Even in the first form of induction, where the conclusion

† Jerzy Neyman, 'Indeterminism in science', etc., in *Jour. Amer. Math. Ass.* (1960); cf. comments in B. de Finetti and F. Emanuelli, *Economia delle assicurazioni*, Vol. XVI of *Tratto italiano di economia* (Edited by C. Arena and G. Del Vecchio), Utet, Torino (1967), pp. 17–19.

(whether valid or not) has the logical meaning of a statement, it is clear that logic cannot provide any proof of its validity. Knowing that an event has never occurred in the past in no way excludes the possibility of its occurring in the future (even if we admit, in order to pre-empt any hair-splitting objections, that an event which has never been observed has, in fact, never occurred). So far as the other forms are concerned, there is always the objection that from the knowledge of the past (or, at least, of that which has already been observed or ascertained) nothing can be logically concluded concerning the future (or, in general, concerning that which is as yet unobserved or unknown, and which could be anything at all, even the unimaginable). Moreover, in these cases, the 'conclusions' themselves do not even have the logical status of precise statements (leaving aside the question of their validity).†

Within what 'logical' ambit, then, might it be admissible to assert that the 'inductive argument' is an 'argument'? From our standpoint the answer is straightforward. It is admissible within the ambit of probabilistic logic; i.e., that of (subjective) probability theory, which, for us, is the only form of logic required over and above that of ordinary logic. In fact, in what follows (in this and the final chapter) we shall illustrate all the questions which arise in the context of induction by presenting them within the framework of the subjectivistic–probabilistic interpretation. The vital element in the inductive process, and the key to every constructive activity of the human mind, is then seen to be Bayes's theorem.

11.1.4. Those who do not accept this point of view (and they are, unfortunately, in the majority) come up against a dead-end, and are in a different situation. If one accepts, in its totality, the subjectivistic interpretation, probability theory constitutes the logic of uncertainty; this complements the logic of certainty, and the two together form a unified and complete framework within which to conduct any argument. Those who reject this point of view find themselves without any coherent foundation on which to build. Between the logic of certainty and probability theory—reduced now to a fragmentary collection of those aspects which can be provided with an objectivistic disguise—there is a void; any attempt to fill this must be without foundation, and consists, in the final analysis, of empty phrases. A useless attempt is made to enlarge and extend the rôle of the calculus of probability (and the applications thereof—referred to nowadays by the

† In the 'third form', we could also encounter statements of a deterministic kind, expressing, for example, a tendency to decrease or fluctuate according to some precisely stated law (like the exponential, or the sine curve, etc.) suggested by extrapolation. In this case, the second objection (that of imprecision) no longer holds; a third objection arises, however (or one might say that the first objection becomes more serious), because of the large degree of arbitrariness which attaches, in general, to the choice of an extrapolation formula.

Anglo-Saxon title of 'statistics') in a manner which cannot be justified within the terms of the objectivistic assumptions, and which, in any case, falls far short of the required generality, a condition met only by the subjectivistic interpretation. As is evidenced by the ever increasing proliferation of *ad hoc* methods for special cases and subcases (Adhockeries!), and the disputes to which these give rise within the ranks of the supporters of objectivistic conceptions, all such efforts fall short of being either satisfactory or sufficient. The gap remains.

In order to be able to provide 'conclusions'—but without being able to state that they are *certain*, because they are undoubtedly not so, and not wanting to say that they are *probable*, because this would involve admitting subjective probability—a search is first made for words which appear to be expressing something meaningful, it is then made clear that they do not, in fact, mean what they say, and then, finally, a strenuous attempt is made to get people to believe that it is wise to act as if the words did, in fact, have some meaning (though what it is heaven only knows!).

As examples of such *words*, we are said to '*accept*' or '*reject*' an '*hypothesis*', and to give an '*estimate*' of a quantity which is not known precisely.

In order to be dealing with a *concept* rather than a mere *word*, we should require that an *estimate* be some value arising in the context of the probability distribution attributed to the unknown quantity: for example, the prevision, the median, or whatever, especially if selected in a manner appropriate to some specific decision. If probabilities and probability distributions are not mentioned, any reference to an 'estimate' is a nonsense.

Similarly, if we use 'accept' and 'reject' to mean that the probability attributed to some given hypothesis is large enough (or small enough) for us to behave, in certain respects, as if the hypothesis were true (or false), then again we would be dealing with *concepts* and not mere words.

It is convenient at this point to enter a further reservation, this time in connection with the use of the word 'hypothesis'. What we have said above only makes sense if we are referring to an 'hypothesis' for which it is possible to verify directly whether or not it is true. If, instead, the 'hypothesis' is somewhat of an abstraction, used solely as an interpretative device, suitable only for summarizing certain features of the problem, and depending on certain given facts, the latter neither requiring it nor capable of ruling it out, then it would be illusory, or, at least, suspect (as is the case when we ask whether 'light is a wave or particle phenomenon', or whether 'a particular individual is intelligent'). Strictly speaking, one would need to replace such statements with a precise list of the verifiable, factual circumstances which one would accept as a substitute. If the 'hypothesis' expresses an opinion about probabilities (either implicitly or explicitly) then matters are even worse. As examples, we could take the following: 'this coin is perfect, in the sense that $p = \frac{1}{2}$'; 'sunspots influence economic life (in the sense of there being a probabilistic correlation)'; 'the fact that having been the cause of a car accident increases the risk of one's being involved in other accidents in the future'. In these cases, it would be necessary to substitute formulations—if any such exist—which could be expressed in terms of (subjective) probabilities referring exclusively to facts and circumstances which are directly verifiable, and of a completely objective, concrete and restricted nature.

11.1.5. The final sentence above re-emphasizes something we have pointed out on numerous occasions. The objectivistic conception of probability and statistics, by misguidedly attempting to make everything objective (including things which cannot be so), in fact has the opposite effect: instead of objectivity being granted its rightful, important place, it is discredited by being claimed in contexts where it is inappropriate. The same thing would happen if someone tried to raise the status of the property of 'rigidity' by referring to all solid bodies as 'rigid bodies' (including those which are elastic or plastic). The effect would be to deprive the notion of 'rigidity' of any meaning or applicability, even in those situations for which it was originally introduced and served a useful purpose, and where it needs to be free of any distortions of meaning, ambiguities or artificial interpretations.

In the philosophical arena, the problem of induction, its meaning, use and justification, has given rise to endless controversy, which, in the absence of an appropriate probabilistic framework, has inevitably been fruitless, leaving the major issues unresolved. It seems to me that the question was correctly formulated by Hume (if I interpret him correctly—others may disagree) and the pragmatists (of whom I particularly admire the work of Giovanni Vailati†). However, the forces of reaction are always poised, armed with religious zeal, to defend holy obtuseness against the possibility of intelligent clarification. No sooner had Hume begun to prise apart the traditional edifice, then along came poor Kant in a desperate attempt to paper over the cracks and contain the inductive argument—like its deductive counterpart—firmly within the narrow confines of the logic of certainty.

The remainder of this work can be seen as an attempt to do away with such nonsense once and for all. In both the general philosophical context, and in the more technical mathematical–statistical sense, we shall try to show that these questions, which are, in themselves, perfectly clear and straightforward, can be formulated in a perfectly clear and straightforward

† Cf. G. Vailati, *Scritti* (Edited by Seeber), Florence (1911). Giovanni Vailati, a mathematician of the Peano school, was an original, profound and committed supporter of pragmatism in Italy (which had several features—which I, in fact, approve of—distinguishing it from the American version of Peirce, James, etc.). The beginnings of a work on pragmatism (which was to be with Mario Calderoni, but was unfinished because of Vailati's death) are published in two articles (CCX and CCXI) to be found in the above volume (pp. 420–432 and 933–941): 'Le origini e l'idea fondamentale del pragmatismo' and 'Il pragmatismo e i vari modi di non dir niente'. Cf., also (CLIX, pp. 684–694), 'Pragmatismo e logica matematica'.

manner. All that is required is that we abandon the traditional pursuit of creating for ourselves pretentious and misleading malformations.

11.2 THE BASIC FORMULATION AND PRELIMINARY CLARIFICATIONS

11.2.1. In our formulation, the problem of induction is, in fact, no longer a problem: we have, in effect, solved it without mentioning it explicitly. Everything reduces to the notion of conditional probability (introduced in Chapter 4), and to the considerations which were developed there (particularly in Chapter 4, Section 4.14, albeit rather concisely) concerning 'stochastic dependence through an increase in information'.

What is required now is a more systematic study of this topic, oriented specifically towards the questions which presently concern us. These questions only differ from the general case—that of the 'effect of an increase in information'—insofar as the information in our case may well be obtained by design, by means of observations, or even through appropriate experimentation. However, this distinction is of no real importance.

11.2.2. By virtue of having observed, or having obtained the information, that some given complex A of events have occurred, what *are* we entitled to say about some future event E? (Or about some collection of future events? Or about events for which 'future' is replaced by 'not yet known'?) The answer is . . . nothing! Nothing 'certain', that is, because nothing justifies our making any *prediction* about a future event E unless it is assumed to fall within the ambit of some never-failing 'laws' (this might apply, for example, to an eclipse of the sun; even in this case, however, if one wished to be rigorous it would be necessary to add 'assuming no violent changes to the planetary system, outside of what previous observations could have led us to expect', and, as a blanket qualification, 'unless the above-mentioned laws are disproved'). And nothing can be said, in any objective sense, concerning probability or *prevision*. This means that no restrictions can be made: every prevision—i.e., every evaluation of probability—can be made freely, and is entirely a matter for the subjective judgement of the individual.

It is only *within* this process of subjective judgement that certain restrictions occur. These are the restrictions imposed by coherence, from which derives all that can legitimately be said concerning 'inductive reasoning', and which essentially reduces to the theorem of compound probabilities (or to its corollary, Bayes's theorem; the latter often being the more expressive form).

Suppose that $\mathbf{P}(E)$ represents the probability evaluated on the basis of our assumed information; i.e. that of knowing of the occurrence of the complex of events A (perhaps by means of certain observations). If H_0 denotes the entire complex of initial information (cf. Chapter 4 Section 4.1), and $\mathbf{P}^0(E) = \mathbf{P}(E|H_0)$ denotes the corresponding probability, then, in fact, $\mathbf{P}(E) = \mathbf{P}(E|H_0A)$, the probability corresponding to the original information plus that provided by the knowledge of A: Bayes's version of the theorem of compound probabilities implies in this case that we must have

$$\mathbf{P}(E) = \mathbf{P}^0(EA)/\mathbf{P}^0(A) = \mathbf{P}^0(E)\mathbf{P}^0(A|E)/\mathbf{P}^0(A). \tag{1}$$

Remarks. There is a delicate point here which requires some attention. When we defined conditional probability (Chapter 4. Section 4.1). we stated that the H appearing in $\mathbf{P}(E|H)$ means that this is the probability You attribute to E if 'in addition to your present information, i.e. the H_0 which we understand implicitly, *it will become known to You that H is true* (*and nothing else*)'. It would be wrong, therefore, to state, or to think, in a superficial manner, without at least making sure that these explanations are implicit, that $\mathbf{P}(E|H)$ is the probability of E once H is known. In general, by the time we learn that H has occurred, we will already have learnt of other circumstances which might also influence our judgement. In any case, the evidence which establishes that H has occurred will itself contain, explicitly or implicitly, a wealth of further detail, which will modify our final state of information, and, most likely, our probabilistic judgement.

In the Appendix (Section 16), we shall present some further critical comments relating to this topic. In any case, it should always be borne in mind when dealing with a problem of inductive reasoning (and, were it not for fear of annoying the reader, we should certainly stress this more frequently).

11.2.3. This, then, is what 'inductive reasoning' is all about. It is often said to reveal how it is that one 'learns from experience', and this is true, up to a point. It must be made clear, however, that experience can never create an opinion out of nothing. It simply provides the key to modifying an already existing opinion in the light of the new situation. The complex A (the experience) *by itself* determines nothing, nor does it provide bounds: to reach a conclusion—i.e. to determine a new ('posterior') opinion $\mathbf{P}$—we require the conjunction of A with $\mathbf{P}^0$ (the initial, or 'prior', opinion). This should not be interpreted as the experience (represented by A) disproving $\mathbf{P}^0$, or forcing one to discard it in favour of $\mathbf{P}$. On the contrary, the adoption of $\mathbf{P}$ in the new state of information is the only way of remaining consistent with what was adopted as the initial opinion in the initial state of information.† So far as the terms 'prior' and 'posterior' are concerned, they simply signify 'before' and 'after' the acquisition of the information A. One should avoid giving too much weight to this, lest the impression is given that 'prior' refers to some mysterious circumstance of being 'prior to any experience', or to a state of 'absolute ignorance', or 'total indifference', and so on, or even that we are referring to different kinds of probability (as was the case with the old terminology relating to *a priori* and *a posteriori* probabilities).

† Recall—or, preferably, re-read—the discussion given in Chapter 4, 4.5.3, and in Chapter 5, Section 5.9; see also Section 11.3.1 of this chapter.

Better still, remembering that there are two sides to every relationship, we could say that (1) merely reveals the possibility of evaluating $\mathbf{P}(E)$ in two different ways: directly, having in mind the final state of information, H_0 plus A, or by evaluating $\mathbf{P}^0(E)$ and $\mathbf{P}^0(AE)$, thinking only in terms of our previous state of information H_0. Coherence requires that the two answers be the same. If one tries it both ways and finds a difference, then the evaluations should be reconsidered, their reliability checked by each method, and adjustments made on this basis until they coincide. This is not a question of deduction (albeit within the ambit of evaluations of subjective probabilities) so much as an invitation to reflect on one's own opinions in order to make them compatible with the requirements of coherence. We point this out explicitly, largely because—for obvious reasons of simplicity—our own exposition will always follow the same path; first evaluating $\mathbf{P}^0$, subsequently observing A, and, finally, arriving at the conclusion $\mathbf{P}$.†

11.2.4. One fact to note is that the explanation which we have given has only one form, and is suitable for every application of inductive reasoning, with no exceptions. This will seem natural to those who have entered into the spirit of the subjectivistic conception of probability, and would scarcely be worth mentioning at all, were it not that certain other approaches consider *statistical induction*—usually referred to as *statistical inference*—as a case apart, and, indeed, as the only case in which probability theory finds any legitimate application.

According to these other approaches, statistical inference is the special form of reasoning to be applied when a large quantity of related data is available. For example, when the frequency of some given phenomenon in a large number of trials is known, or when we know the percentages of people in a given population who possess certain characteristics, and so on. The conclusions which are put forward on this basis derive their overall justification from the fact of there being a large quantity of data. They are valid, therefore, insofar as the quantity of data is sufficient for them to be regarded as such, and not otherwise.

† This last comment might help to reduce the impact of one rather obvious objection that springs to mind: if from $\mathbf{P}$ we require to trace back to $\mathbf{P}^0$, should we not trace back from $\mathbf{P}^0$ to some $\mathbf{P}^{00}$, and so on, *ad infinitum*? Where, then, would the very first evaluation have come from? The question is rather sophistical, since the procedure which we have given loses its force when carried out in situations which are too far removed from reality. On the other hand, it is well-known—and we shall see examples of this—that even very vague prior evaluations are often sufficient to yield conclusions of more than adequate practical precision (and this holds, *a fortiori*, if we retrace from one 'beginning' to another). Eventually, perhaps, we should need to have recourse to an explanation based on 'instinct', or to experience in the form of genetic inheritance, or something of that kind (I do not want to insist too seriously on these suggestions relating to fields in which I am not expert). For a more detailed discussion, see B. de Finetti and L. J. Savage (1962).

To use a classical form of terminology, we would be dealing with a property connected with the existence of an 'aggregate'. So long as we are dealing with just a few objects, they do not form an aggregate and no conclusions can be reached. If, however, we have a large number of objects, then we do have an aggregate, and then, and only then, does the argument go through. If we add in objects one at a time, nothing can be said until the number of objects becomes sufficient to be considered an aggregate; then the conclusion appears (just like that? in passing from 99 to 100? or from 999 to 1000? . . .), as that which is not yet an aggregate at last becomes one. Now it will be objected that this version is a travesty: there is no sharp break of this kind, but rather a gentle transition. The non-aggregate passes through a to-be-or-not-to-be-an-aggregate phase, inclining first one way and then the other, and only subsequently does it gradually transform itself into a real and genuine aggregate. But this does not answer the original objection raised against the distinction, here put forward as being of fundamental conceptual importance, between the 'aggregate effect' and the 'effect of individual elements'. To recognize that a clear-cut separation cannot exist, even though this admission may perhaps resolve certain of the apparent paradoxes, does not get to the real root of the problem, and, indeed, serves to underline the weakness and the contradictory nature of the whole approach.

The problem is only resolved by acknowledging that distinctions of this kind have no significance. The conclusions one arrives at on the basis of a large quantity of data are not the consequence of some aggregate effect, but simply the cumulative effects of the contributions of the individual pieces of information. The modification of a prior opinion into a posterior opinion through knowing the outcome of some given set of trials is precisely the same as that obtained by considering each item of data separately, and effecting the appropriate modifications (in general, minor) one at a time. This is so no matter whether the number of trials is large or small, and is an important fact to bear in mind if serious misunderstandings are to be avoided. We are aware that we have, perhaps, given undue emphasis to this point, but the fact remains that the germs of such misunderstandings seem to permeate the very air we breathe.

11.2.5. In what follows, problems of the 'statistical type' will receive their due emphasis; they are undoubtedly interesting from a theoretical point of view, and certainly important so far as practical applications are concerned. They will, however, simply be a special case—or, more precisely, a collection of rather ill-defined special cases—having in common the following general characteristics: that past experience consists of a number of observations of 'more or less similar facts' (and often the case of interest is that of a large number of such observations). Of course, analogy *per se* is a rather marginal and irrelevant factor, but it often leads one to considerations of some kind of symmetry in the evaluations of probability, and this is what really concerns us (even though, from a descriptive point of view, it is often useful to mention the analogy in question).

As a more expressive statement of the way in which such an analogy is translated into probabilistic terms, we could say that the analogy leads us to make the conclusions depend only, or, at least, mainly, on *how many* 'analogous' events occur out of some given number, and not on *which* of them occur. This is intended to give a broad view of what we mean, however, and should not be interpreted in any literal sense.

It is within this kind of framework that we consider the problem of evaluating probabilities on the basis of observed frequencies (and although we have touched on this topic before—cf. Chapter 5, 5.8–5.9 and Chapter 7, 7.5.5–7.5.6—we have not done so in a systematic fashion).

We shall soon see, however, that even in this case, the simplest, there is no unique answer. In fact, one is permitted—and, indeed, obliged—to choose any initial opinion from among those possible (and the latter will turn out to correspond to the set of functions which increase from 0 to 1 over the interval [0, 1]). Conversely, we shall also see that, starting from an examination of the same series of results, it may be natural to express opinions which involve extending the whole approach, although the qualitative features originally advanced as being characteristic of problems of the statistical type remain unchanged (or are changed in minor respects only).

11.2.6. In order to give a more concrete presentation of the various possible attitudes to the way in which a given set of results should influence us, we shall examine a particular example. Let us suppose that we have observed 50 events, $E_1, E_2, \ldots, E_{50}$, and that the results are the following:

1111001111 0111000111 1100001110 0000111011 1000000010,

where 1 denotes a success, and 0 a failure.

What can we say on the basis of these results? What probability should we attribute to some other event, E_{51}, or E_{312}? Or to the proposition that there will be k successes ($k = 0, 1, 2, \ldots, 100$) out of some other collection of 100 events (either a particular, preassigned collection, or just some collection chosen, in some specified sense, 'at random')?

It does not make sense to pose these questions in this abstract fashion. We have got to know what kind of events we are dealing with, and what information we have concerning them, no matter how limited it may be. To say there is 'no available information' is too glib: were this actually the case, we would not even know what kinds of events we were dealing with (they would simply exist as $E_i, i = 1, 2, \ldots, 50$). In such a case, there could be no possibility of considering their probabilities, nor any interest in doing so. Even in this case, however, in trying to figure out why the author has presented such an example, the reader would form some opinion, albeit tentative, and it would be this opinion that was relevant, rather than the so-called state of 'no available information'.

11.2.7. In real-life examples, one will have some idea of which features or attendant circumstances might lead to different probabilities of success (in the sense that one feels inclined to treat as meaningful, and to take into future account, any significant departures from the norm in the frequencies for events possessing these features, or being dependent on these circumstances). If, for example, in considering the deaths resulting from an epidemic one finds significant differences for individuals with different blood pressures, or for those born at different times of the year, or on different days of the week, there will be a tendency to regard the differences as meaningful in the first case, but not in the others.

Another circumstance which may or may not appear as meaningful is that of order. In our example, we assumed the events to be numbered from 1 to 50: in many cases, such a numbering is just a matter of convention and is completely irrelevant (registration numbers, passport numbers, etc.), but in others it will correspond to chronological order, and then may well be meaningful, in that it could reveal a tendency for the frequency of successes to increase or decrease with time, or to oscillate. Moreover, in cases where the fact of two trials being consecutive is meaningful, study of the order may reveal a difference in frequencies for trials depending on whether they follow a success or failure (and one could easily consider other variants of this idea).

In any actual example, there are innumerable different factors which could be considered in this way, the vast majority of which would certainly be meaningless. But there may be other examples in which these same factors in various combinations, will appear, to some extent at least, meaningful. In any case, this rather general summary finds its genuine expression only in the evaluation of the probabilities for all the constituents constructed on the basis of the events under consideration. Alternatively, if one prefers to look at it in this way (the two approaches are equivalent), we consider the probabilities conditional on every possible combination of results out of any group of observed events, these being taken over every possible combination of other events. Within this framework, everything can be expressed in a complete form, and this is true for all possible cases.

11.2.8. It is clear, however, that it would be very difficult to consider simultaneously all possible factors, and, in any case, this would only cause confusion. In studying this topic from a theoretical viewpoint, therefore, one restricts oneself to considering certain relatively simple cases in which only a small number of factors (and sometimes only one) are considered. In any practical application, of course, one must not lose sight of the fact that simplified schemes of this kind are likely to be inadequate in certain respects.

We should also make it clear that the various *schemes* to which we shall make reference (those of Bernoulli, Poisson and Markov, together with the exchangeable and partially exchangeable cases, contagion models, and so on) should not be interpreted as fixed slots into which real applications are to be fitted. Still less should they be viewed primarily as mathematical inventions, whose complications are merely evidence of mathematical playfulness, and which are devoid of interest so far as applications are concerned. They should be seen rather as simplified schemes serving as possible representations of the one and only realistic 'scheme'—that which includes all possible distinctions in all possible combinations. The schemes we shall deal with are useful in practice, but, again we note, only as simplified representations of more complicated situations which themselves cannot be represented straightforwardly.

11.3 THE CASE OF INDEPENDENCE AND THE CASE OF DEPENDENCE

11.3.1. *Independence.* We first consider the case in which the possibility of observed results influencing subsequent evaluations is specifically excluded. This is the case of independence, with which we are already familiar, and *for which the problem of inference does not exist.* The case would not concern us, therefore, were it not for the following two considerations, the first of which is of a technical nature. It turns out that the most convenient way of attacking problems of interdependence is to reduce them, if possible, to appropriate combinations of independent schemes (as we shall see in Section 11.3.5, and subsequently). The second consideration is of a critical nature: the statement that the problem of inference does not exist in the case of independence, although obvious, often gives rise to misunderstandings (more precisely, it is misguidedly dismissed by those who have not properly understood what it actually says). What it says is that any possibility of 'learning through experience' is *excluded*—'ruled out by the principle of contradiction'—if the original opinion is based on independence, because the latter, by definition, requires that the original opinion will not be modified on the basis of any observation of results.

Let us consider, as an example, the case of Heads and Tails, with the assumption that the two probabilities are equal (i.e. $\frac{1}{2}$), and that trials are independent. The evaluations of probabilities for successive trials remain unchanged, no matter what results are observed (like, for instance, those considered above in Section 11.2.6, supposing them to be the results of the first 50 trials).

The same could be said in the case of a die if, for example, we considered the face '1' as a success, and the others as failures, and assumed throws to be independent (we merely have $p = \frac{1}{6}$ in place of $p = \frac{1}{2}$). In the case of independence—i.e. if the original opinion is based on an assumption of independence—every possibility of 'learning through experience' is ruled out (it would not be consistent with the original opinion).

Someone might, perhaps, argue as follows (in the context of the example of Section 11.2.6). If the die gives me face '1' 26 times out of 50 (instead of about 8 times), I am inclined to believe that it is 'loaded' (i.e. that it favours '1', and perhaps there is a weight in the opposite face): it also happens that 18 times out of 26 '1' is followed by '1', and 16 times out of 23 '0' is followed by '0'; I suspect that the way the die is thrown favours 'repeats', and this leads me to revise my original assumption of independence and to drop it. Moreover, noticing that the number of times '1' occurs in the five blocks of ten decreases from 8, 6, 5, 5, to 2, I am led to think that the loading which originally favoured '1' was temporary, and subsequently ceased to operate, so that the die is now perfect. Perhaps, if I continue, I shall notice a number of other things!

Now it may be that arguments of this kind are acceptable in themselves (this is a matter of opinion), but it is necessary that they be formulated correctly, so as to avoid any possibility of misunderstanding. Insofar as they seem to be in conflict with the previous assertion concerning the contradiction involved in changing one's mind having assumed independence, we can deduce that either the form in which they are expressed, or the manner in which they are interpreted, is mistaken.

The mistake, in fact, is in referring to stochastic independence as if it were an 'hypothesis' which the facts can 'dispute', enabling us, and possibly obliging us, to change our minds. If we are to be able to 'change our mind', the original opinion must be expressed in a form which is compatible with such a possibility of revision. Such an opinion could, at most, be a 'first approximation' to the case of independence, in that it might, for example, consist of a mixture of evaluations, most of which correspond to the case of independence (with some preassigned p), but some of which, although having little weight, correspond to various *alternatives* (like those mentioned for the above example).

It is only by admitting such alternatives that a 'revision' can take place; and, indeed, not simply by admitting their possible existence, but rather through their actual presence in the original opinion, *which, therefore, can no longer possess the property of independence.* The so-called revision—i.e. the passage from the original opinion to a different subsequent opinion—takes place, in fact, as a result of outcomes which give rise to a strong likelihood for such an alternative: in other words, roughly speaking, if we suspect that the occurrence of a certain event should be attributed to an alternative explanation under which it would have a higher probability.

When discussing this topic previously (cf. Chapter 7, the first footnote to Section 7.5.8), we emphasized that one should speak of *suspicious* cases, rather than calling them 'strange' or 'unlikely', as is often done. The reason for this is the one we have just given, but it will be useful to provide further illustration, in order to clarify the contrast between our terminology and the terminology which we reject (not simply on the grounds that it is inappropriate, but also because it leads to the construction and application of methods which have no proper foundation). We note that from a conceptual viewpoint the considerations which we have put forward hold completely generally: our detailed concentration on the Bernoulli scheme—in particular the special case of Heads and Tails—is purely for the purpose of fixing ideas.

Those who think in terms of a 'revision'—or even a 'disproof'—of the original opinion, without having in mind, or referring to, any alternatives, could not regard what has occurred as 'suspicious', since the word is meaningless unless alternative explanations are admitted. Instead, it would be described, having in mind the original opinion, as "strange', 'unlikely', 'exceptional', 'very improbable', or 'very unexpected'.

More specifically (and, for convenience, we deal with the simplest cases), the circumstances which would characterize the cases 'disproving' the original opinion would be one or other of the following:

the *distance* from the prevision; this ties up with *hypothesis testing*, for example whether or not something belongs to a 'confidence interval' (with some given 'tail area' probability), or to the interval $m \pm 3\sigma$, or something similar (cf. Chapter 12, 12.6.4);

the *small probability* of the case which has occurred;

some observed *peculiarity*; for example, that all the 0s come before all the 1s, or that they alternate, or—should one happen to spot the fact!—that the binary sequence is the coding of some celebrated historical date.

These are circumstances which may turn out to be useful in practice; not in themselves, however, but rather if, and insofar as, they serve to strengthen more usual forms of 'suspicion' (like those regarding 'cheating', 'malfunctioning', etc.). With regard to 'small probabilities', one should say immediately that the whole thing is rather ambiguous. Is it to be taken as referring to the probability of the particular sequence of 50 1s and 0s, or to the probability of the frequency; i.e. of all those sequences in which a 1 occurs 26 times?†

To speak in terms of objective circumstances, rather than suspicions relating to other alternatives (and to make use of criteria based upon such objective circumstances), means, as usual, that one is attempting to draw conclusions on the basis of a single possibility, neglecting the necessary comparative possibilities.

Everyone is free to choose his prior opinion in whatever way he likes. The choice can only be made once, however. If I choose to base my prior opinion

† Not to mention the fact that in more complicated cases, or if one takes other circumstances into account, the 'observed result' (whatever it may be) always has an arbitrarily small probability if one describes it in a sufficiently precise way.

upon the assumption of independence, it means that I exclude, once and for all, any circumstance that might in future be pointed out to me as rather 'strange'. I refuse to consider it as a possibility; i.e. as something capable of modifying my opinion. If the future occurrence of this 'strange' circumstance would, in fact, lead me to suspect 'cheating' (or whatever), then I should make it clear from the very beginning that my opinion is not based on the assumption of independence, but that it accepts the dependence deriving from admitting the possible suspicion (which, although negligible at the outset, could, under certain circumstances, come to the fore). If I omit to say this, then, at best, I have expressed myself rather superficially (this might be excused, however, if I was aware what I was doing).

There would be no excuse, on the other hand, if a change of opinion was explained in a distorted fashion, by attributing it to the fact of experience having disproved the original opinion, dictating its replacement by another. Nothing can oblige one to replace one's initial opinion, nor can there be any justification for such a substitution. From a logical point of view—and, it might even be argued from the 'moral' point of view—one would be adopting the same contradictory posture (or indulging in the same unfair subterfuge) as a person who regards himself as released from a promise to help a friend if a certain event occurs, given that the event in question has already occurred.

In order to retain the right of being influenced by experience, it will therefore be necessary to express an initial opinion differing from that of independence.

11.3.2. *Exchangeability*. Having abandoned independence, the simplest choice open to us is to continue to regard the order as irrelevant. Given n events, the probabilities $\omega_h^{(n)}$ that h of them occur ($h = 0, 1, 2, \ldots, n$) are now arbitrary† (and are no longer necessarily those of the binomial distribution, as in the case of independence); however, the combinations of h 1s and $n - h$ 0s all have the same probability $\omega_h^{(n)}/\binom{n}{h}$. This is equivalent to simply saying that all products of n events have the same probability $\omega_h^{(n)}$.‡

In this case, the events are called *exchangeable* (the reasons for the terminology being contained in what we have already said): knowledge of the n results can only have an influence through the reporting of n and the frequency (i.e. n and h). whereas any other aspect connected with order will be ignored. We shall return to this topic shortly (in Section 11.4).

It is intuitively obvious that drawings from an urn with unknown composition are exchangeable (eg. an urn containing an unknown number of black

† If, however, the events are at least potentially infinite in number, then there may be restrictions (cf. Section 11.4).

‡ In fact, it suffices to observe from (11) of Chapter 3, 3.8.4 that we have

$$\omega_h^{(n)} = \binom{n}{h} \sum_{r=0}^{n-h} (-1)^r \binom{n-h}{r} \omega_{h+r}^{(h+r)} = \binom{n}{h} \Delta^{n-h} \omega_h^{(h)}. \tag{2}$$

and white balls, with the standard method of drawing with replacement). The same applies to tosses of a possibly asymmetric coin, and, more generally, to all those cases which are commonly referred to as 'repeated trials with a constant but unknown probability of success'.† It is less obvious, but none the less true, as we shall see, that we still have exchangeability in the case of drawings without replacement, or with double replacement (the 'contagion' model; cf. Chapter 10, 10.3.5).

In the example we have been considering, the only significant fact is that we had $h = 26$ successes out of $n = 50$ tosses. This can also be expressed by saying that the two numbers $n = 50$ and $h = 26$ are 'sufficient statistics' (i.e. they constitute an exhaustive summary of the data). In other words, so far as 'learning from experience' is concerned, it does not matter whether we observe the complete sequence, or whether we simply observe that $n = 50$ and $h = 26$; this is a consequence of the assumption of exchangeability.

11.3.3. *Partial exchangeability.* We obtain a somewhat less restrictive condition (although at the expense of some additional complication) by thinking of the events under consideration as divided into various classes (in order to fix ideas, we shall consider two classes), and of exchangeability as holding within both classes. In other words, the probability that out of $n = n' + n''$ events (n' of the first class, n'' of the second) $h = h' + h''$ occur (h' of the first class, h'' of the second) is the same, no matter how the n' and n'' events are chosen, and no matter which of them are among the h' and h'' successes. The probability of obtaining a total of h' and h'' successes is $\omega_{h',h''}^{(n',n'')}$, and the probability of them occurring for a particular, preassigned sequence of events is that just given, divided by $\binom{n'}{h'}\binom{n''}{h''}$. Obvious, trivial examples are that of exchangeability *per se* (for which ω depends only on $n' + n''$ and $h' + h''$), and that of independence between the classes (within each of which there is exchangeability; ω is then the product of the $\bar{\omega}_{h'}^{(n')}$ and $\bar{\bar{\omega}}_{h''}^{(n'')}$ for the two cases). Actual cases of partial exchangeability fall into an intermediate category: put in a rather imprecise form, but one which conveys the general idea, we have interdependence between all the events, but a rather stricter one among those in the same class. This would be true, for example, of drug trials carried out on patients of both sexes.

In the particular case of events occurring in chronological order, the division into classes may depend on the result of the previous trial; in such a case we have the Markov form of partial exchangeability.

If one suspects that an outcome is influenced by the preceding result, then one would not initially regard all sequences having the same numbers of successes and failures as equally likely (in the example, this would be those

† The terminology is incorrect (cf. Chapter 4, 4.8.3–4.8.4), but is expressive (and the meaning it suggests is essentially correct).

with 26 1s, and 24 0s). Instead, the judgement of equal probability would apply to all those sequences having the same number of successes and failures following the occurrence of a 1 (18 and 8, respectively), and the same number (7 and 16, respectively) following the occurrence of a 0. We might expect some similarity with the case of a Markov chain with probability 18/(18 + 8) (about 70%) of a success following a success, and 7/(7 + 16) (about 30%) of a success following a failure ... (however, there are various reservations, as in the previous case, and these become more serious the more complicated the situation becomes). In any case, with the above assumption one requires n', n'', h' and h'' for an exhaustive summary (n and h alone no longer suffice).

11.3.4. *Other cases.* Similar conclusions hold in the other case we mentioned in Section 11.2.7; that in which one suspects a progressive increase in one of the two probabilities at the expense of the other (right from the beginning; recall that no suspicion can arise if it is not present initially). For example, we might suspect that under certain circumstances (for instance, a black ball being drawn) white balls may turn into black balls during a series of drawings with replacement from an urn of unknown composition. A study of the outcomes provides information concerning the composition of the urn if we consider the tendency for the frequency of white balls to decrease with time (this frequency being the only thing we can get hold of). It also provides a basis for making conjectures about the past history—and hence about the future—of the unobservable process by which white balls are gradually turned into black balls.

The general case follows along the same lines as all the examples which we have considered. Given an arbitrary prior probability distribution $\mathbf{P}^0$, which attributes probability to each $A = E_1'E_2' \dots E_n'$ (where E_i' stands for either E_i or $\tilde{E}_i$, and n is arbitrary), the problem is solved by simply stating that, knowing A, the (posterior) probabilities are given by $\mathbf{P}$, where

$$\mathbf{P}(E) = \mathbf{P}^0(EA)/\mathbf{P}^0(A).$$

The extreme simplicity of this mathematical statement is, however, misleading. In general, straightforward application of the method is precluded by the requirement that one provide the $\mathbf{P}^0(A)$ directly for all A. This is only really feasible if the situation can be represented in terms of simple formulae.

11.3.5. *Mixtures of distributions which assume independence.* The straightforward case of independence is itself uninteresting; we have, simply, $\mathbf{P}^0(EA) = \mathbf{P}^0(E)\mathbf{P}^0(A)$, and hence $\mathbf{P}(E) = \mathbf{P}^0(E)$ for all E which are defined in terms of 'future' trials (or, at least, do not depend on the observations A). As we mentioned in Section 11.3.1, however, it turns out that in a number of cases it is extremely useful to consider the possibility of expressing $\mathbf{P}^0$ as a *mixture* of such distributions $\mathbf{P}_i$: in other words, we take linear combinations

$$\mathbf{P}^0 = c_1^0\mathbf{P}_1 + c_2^0\mathbf{P}_2 + \ldots + c_m^0\mathbf{P}_m = \sum_{i=1}^{m} c_i^0\mathbf{P}_i,$$

with non-negative coefficients c_i^0 having sum equal to 1 (or limit cases thereof). The latter may take the form of infinite series, or of integrals; in either case, it is sufficient to describe it as a $\mathbf{P}^*$ for which, given any arbitrary $\varepsilon > 0$, there exist **P**s having the form of finite linear combinations such that $\sup_E|\mathbf{P}^*(E) - \mathbf{P}(E)| < \varepsilon$.

It is easily seen that if $\mathbf{P}^0$ is a mixture, then so is any **P** to which it leads as a result of (arbitrary) observations A (the coefficients varying for different experiences, A). In fact, we have

$$\begin{aligned} \mathbf{P}(E) = \frac{\mathbf{P}^0(EA)}{\mathbf{P}^0(A)} &= \frac{c_1^0\mathbf{P}_1(A)\mathbf{P}_1(E) + \ldots + c_m^0\mathbf{P}_m(A)\mathbf{P}_m(E)}{c_1^0\mathbf{P}_1(A) + \ldots + c_m^0\mathbf{P}_m(A)} \\ &= c_1\mathbf{P}_1(E) + \ldots + c_m\mathbf{P}_m(E), \end{aligned} \tag{3}$$

where $c_i = Kc_i^0\mathbf{P}_i(A)$ (K = the normalization factor $= 1/\sum c_i^0\mathbf{P}_i(A)$).

The expression in mixture form may correspond to an actual mixture, in which case there exist events $H_1, H_2, \ldots, H_m$ (exclusive and exhaustive) such that the $\mathbf{P}_i$ represent probability distributions conditional on the H_i: $\mathbf{P}_i(E) = \mathbf{P}(E|H_i)$. In other cases, where this does not apply, it may, nevertheless, turn out to be useful to proceed, formally, *as if* such events existed.

11.3.6. In the *exchangeable* case, if we think in terms of an urn of unknown composition, the H_i represent the events (or 'hypotheses') that the proportion of white balls is θ_i (and under such an hypothesis we consider the drawings to be independent and of constant probability, $p_i = \theta_i$). If we think in terms of a biased coin, or of the Pólya urn scheme, objective circumstances of this kind (i.e. observable in principle, even though we cannot actually observe them) do not exist. However, we shall soon see (in Section 11.4.2) that, in the exchangeable case, $\mathbf{P}^0$ always has the form of a mixture. This will then permit us to argue as if the coefficients c_i^0 and c_i were the probabilities of events H_i, conditional on which we have independence and probability of success equal to p_i.

11.3.7. In the *Markov* case (dependence on the preceding result), we can still reduce to mixtures by considering distributions $\mathbf{P}_i$ under which the trials are independent with probability p_i' or p_i'', depending on the outcome of the previous trial.

In the third example (that of decreasing probabilities), it may or may not be possible to reduce to a mixture, depending on the way the initial opionion is stated. The statement considered previously does not permit us to do this. On the other hand, we do have a mixture of distributions if the latter are taken to be of the form

$$\mathbf{P}_i(E_h) = f_i(h),$$

where the E_h are independent according to the $\mathbf{P}_i$, and the $f_i(h)$ are arbitrary (eg., $e^{-\lambda_i h}$, if one requires them to be decreasing).

11.4 EXCHANGEABILITY

11.4.1. We shall now consider the general notion of exchangeability, and, in particular, exchangeable events and exchangeable random quantities. What we are considering, in fact, is the most fundamental and widely used form of statistical inference.

The definition of exchangeability in the case of events has already been given, but we shall re-express it in such a way as to include, also, the case of exchangeable random quantities. The definition is the following: for arbitrary n, the distribution function, $F(\cdot, \cdot, \ldots, \cdot)$, of $X_{h_1}, X_{h_2}, \ldots, X_{h_n}$ is the same, no matter how the X_{h_i} are chosen (in particular, F must be symmetric, because the X_{h_i} could simply be permuted). More generally, every condition concerning n of the X_h has the same probability, no matter how the X_h are chosen or labelled.

We shall come across applications of exchangeable (and partially exchangeable) random quantities in Chapter 12. For the time being, we shall restrict ourselves to establishing a particular property which we shall make use of in the special case of exchangeable events (a topic to which we shall return shortly).

Let us consider exchangeable X_h having *finite variances*, and, in particular, we shall look at two large groups of such quantities. What we shall prove, roughly speaking, is that their arithmetic means, Y'/n' and Y''/n'', are almost certainly equal (where Y' and Y'' are the sums of the n' quantities in the first group and the n'' in the second, respectively). More precisely, we shall show that the square of their difference tends to zero in prevision as n' and n'' increase. This gives us Cauchy convergence in mean-square, and hence weak convergence, and a limit distribution F for the mean Y_n/n of a large number of terms: $F_n \to F$, where $F_n(x) = \mathbf{P}(Y_n/n \leqslant x)$.

The proof is as follows (and is given under conditions which are less restrictive than those of exchangeability). We assume that the X_h have the same, finite, previsions and variances, m and σ^2, and the same pairwise correlation coefficient, r.†

Expanding the square of $n''Y' - n'Y''$, we obtain $n'n''(n' + n'')$ terms of the form X_hX_k with $h = k$, and the same number‡ (but with the opposite sign) having $h \neq k$. The previsions are $m^2 + \sigma^2$ and $-(m^2 + r\sigma^2)$, respectively, so that

$$(4)\quad \mathbf{P}\left(\frac{Y'}{n'} - \frac{Y''}{n''}\right)^2 = \frac{n' + n''}{n'n''}[(m^2 + \sigma^2) - (m^2 + r\sigma^2)] = \left(\frac{1}{n'} + \frac{1}{n''}\right)\sigma^2(1 - r).$$

We could have set $m = 0$ at the very outset (it disappears in the formulation of the problem), but it is sometimes useful to have the formula available for $m \neq 0$. This is particularly so in the case of exchangeable events, because $\mathbf{P}(E_h^2) = \mathbf{P}(E_h) = \omega_1$ and $\mathbf{P}(E_hE_k) = \omega_2 (h \neq k)$, and hence

$$(4')\quad \mathbf{P}\left(\frac{Y'}{n'} - \frac{Y''}{n''}\right) = \left(\frac{1}{n'} + \frac{1}{n''}\right)(\omega_1 - \omega_2).$$

Returning now to the case of exchangeable events, and thinking of them as a sequence (but one whose ordering is arbitrary and irrelevant), we can characterize them as a stochastic process with the same representation as we used for Heads and Tails: all paths leading from the origin to some given point are equally probable. In this case, we shall speak of an *exchangeable process*.

For such a property to hold, it is sufficient that the probabilities $p_h^{(n)}$ and $\tilde{p}_h^{(n)}$ of steps of $+1$ or -1, respectively, when leaving a given point $[n, h]$,§ depend on the vertex in question, but not on the path travelled in order to reach it, and that the probability of successive steps of $+1$ and -1 remains unchanged if they are reversed: i.e. $p_h^{(n)}\tilde{p}_{h+1}^{(n+1)} = \tilde{p}_h^{(n)}p_{h+1}^{(n+1)}$. This condition lends itself to an elegant geometrical interpretation. If, at each vertex, the probabilities of the next step are expressed as a vector $(1, p - \tilde{p})$ (where $p = p_h^{(n)}$) pointing at the barycentre (or prevision) of the possible points of arrival, then the condition may be stated as follows: for any vertex $[n, h]$ and the two following, $[n + 1, h]$ and $[n + 1, h + 1]$, the three corresponding vectors *meet at a point* (cf. Figure 7.2 of Chapter 7, 7.3.3). By induction, this condition is itself sufficient to ensure exchangeability (provided it holds at all vertices of the lattice).

As a function of the ω, we have

$$(5)\quad p_h^{(n)} = \frac{h + 1}{n + 1}(\omega_{h+1}^{(n+1)}/\omega_h^{(n)}).$$

† We remind the reader that if (at least in principle) the X_h are infinite in number then $r \geqslant 0$ (cf. Chapter 4, 4.17.5). So far as we are concerned, the X_h must be infinite in number—or, at least, very numerous—and so r will always be positive—or, at worst, negative but very small (and $1 - r$ will be $\leqslant 1$, or just greater than 1).

‡ The 'number of terms' is to be understood in an algebraic sense (i.e. counted as -1 if it has a sign opposite to that being understood).

Note that the two groups are assumed to be disjoint; if they had c terms in common, we have a tighter bound (as one might have expected). The factor $n' + n''$ becomes $n' + n'' - 2c$ (i.e. $2c/n'n''$ is subtracted from $(1/n') + (1/n'')$).

§ Cf. Figure 7.1 in Chapter 7, 7.3.2.

Note that each path from the origin to $[n+1, h+1]$ has probability $\omega_{h+1}^{(n+1)}/\binom{n+1}{h+1}$, whereas those paths coming from $[n, h]$ have probability $[\omega_h^{(n)}/\binom{n}{h}] \cdot p_h^{(n)}$. Equation (5) follows on comparing the two probabilities.

11.4.2. Some processes necessarily come to an end in a finite number of steps (for example, drawings without replacement from an urn containing N balls; if H are white, $0 < H < N$, we have $\omega_H^{(N)} = 1$, and so we cannot continue with the $\omega_h^{(n)}$ for $n > N$): others can be considered as if they could be continued indefinitely.

All exchangeable processes which end after N steps are mixtures of the *hypergeometric* process. The mixtures over the possible cases $H = 0, 1, \ldots, N$, with probabilities $c_0, c_1, \ldots, c_N$ (H unknown, and perhaps chosen at random in some way or other), coincide, within the N steps, with every process which has, for $t = N$, the given distribution for Y_N: i.e.

$$\mathbf{P}(Y_N = 2h - N) = \mathbf{P}(S_N = h) = \omega_h^{(N)} = c_h \quad (h = 0, 1, \ldots, N).$$

The idea is obvious: if the probabilities of passing through the various vertices of the vertical line $t = N$ are made to coincide (by balancing the drawing), then, for any two process, all the probabilities relating to occurrences before time t (displayed on the left in the figure) also coincide. These latter probabilities are all well determined, since all the paths ending at a given point have equal probabilities.

More important, however, is the case of exchangeable processes which can be continued indefinitely. Clearly, those obtained by mixtures of Bernoulli processes, i.e. such that

$$(6) \qquad \omega_h^{(n)} = \int_0^1 \binom{n}{h}\theta^h(1-\theta)^{n-h}\,\mathrm{d}F(\theta),$$

are of this type. In the discrete case, or if a density exists, we have

$$(6') \qquad \omega_h^{(n)} = \sum_{i=1}^{m} c_i\binom{n}{h}\theta_i^h(1-\theta_i)^{n-h}$$

and

$$(6'') \qquad \omega_h^{(n)} = \int_0^1 \binom{n}{h}\theta^h(1-\theta)^{n-h}f(\theta)\,\mathrm{d}\theta.$$

Conversely, it can be shown that every exchangeable process which can be continued indefinitely is a mixture of this form. In order to prove this, it is sufficient to refer to the previous case. For any given N, we know how to construct an exchangeable process coinciding (in $0 \leqslant t \leqslant N$) with our given process. As we have seen, this can be achieved as a mixture of hypergeometric processes with N steps: it suffices that the urn having composition H/N (H white balls out of N) be chosen with the same probability as is attributed to the frequency H/N ($=S_N/N$) in the given process. If

$$F_N(\theta) = \mathbf{P}(S_N/N \leqslant \theta)$$

denotes the distribution function of the frequency in N trials, and Ber(N, θ) and hyp(N, θ) are used to denote, symbolically,† the Bernoulli and hypergeometric processes that result from drawing *with* and *without* replacement, respectively, from an urn containing N balls, $H = N\theta$ of which are white, then our process is given by the mixture

(7) $$\int_0^1 \text{hyp}(N, \theta)\, \mathrm{d}F_N(\theta).$$

But we know that, as $N \to \infty$, hyp$(N, \theta) \to$ Ber(N, θ), and $F_N \to F$ (cf. Section 11.4.1), so that, in the limit, the form given in (7) tends to that of (6). There would be no difficulty in providing a rigorous treatment, but it seems more instructive to emphasize the basic idea, and to give an intuitive understanding of the general validity of the mixture form (and in doing so, we have opened up the way for a rigorous proof).‡

This representation in mixture form enables us to obtain, in the way we have indicated, the modified distribution resulting from the knowledge of some given number of trials, yielding r successes and s failures, say. We find that $F(\xi)$ must be replaced by $\bar{F}(\xi)$, where

(8) $$\mathrm{d}\bar{F}(\xi) = K\xi^r(1 - \xi)^s\, \mathrm{d}F(\xi).$$

We still have a process consisting of exchangeable events, but now with a probability distribution modified in proportion to the likelihood, $\xi^r(1 - \xi)^s$. In other words, proportional to the ξ and $(1 - \xi)$ deriving from the effect of each success and failure, respectively.

In particular, the probability for each individual trial, which is given initially by $\omega_1^{(1)} = \int \xi\, \mathrm{d}F(\xi)$ (i.e. by the abscissa of the barycentre of the distribution F), becomes, similarly, after r successes and s failures, the barycentre of $\bar{F}$: that is to say,

(9) $$p_r^{(r+s)} = \int \xi \,.\, \xi^r(1 - \xi)^s\, \mathrm{d}F(\xi).$$

† We are not dealing with an abstraction, but rather with a convention of notation for indicating that in place of hyp(N, θ) we could put $\omega_h^{(n)}$, or any other probability (or prevision), whose value in our process will be given as a mixture by (7).

‡ It can happen that by following through a sequence of logical steps one is forced willy-nilly to concede the truth of something without ever seeing what Federigo Enriques used to call the *wherefore*. I happen to believe that the wherefore is all important (a point I have repeatedly emphasized, and do not wish to dwell upon here). See, e.g., B. de Finetti, 'Sulla suddivisione casuale di un intervallo: spunti per riflessioni', in '*Rend. Sem. Mat. e Fis.*', **XXXVII**, Milan (1967) (especially numbers 1, 2, 5 and 6).

11.4.3. We shall give the details for a very simple special case: that for which the initial distribution is uniform ($f(\xi) = 1$ $(0 \leqslant \xi \leqslant 1)$, $F(\xi) = \xi$). This is the classical Bayes–Laplace version, which corresponds to the idea that 'knowing nothing about the probability' obliges one to assume the uniform distribution as the 'probability of the unknown probability'. We do not regard the uniform distribution as having any special status, and still less do we subscribe to these kinds of underlying assumptions; indeed, we regard them as meaningless and metaphysical in character. On the other hand, there is some value in considering a simple, clear example; especially one which provides us with an opportunity to make some useful points. We have, in fact, already mentioned this case (in Chapter 10, 10.3.5 and 10.4.1) in relation to the problem of subdividing an interval, and in connection with Pólya's urn scheme for contagion models.

In the subdivision of the interval [0, 1], the division point chosen first, P_0, has, like any other division point, a uniform distribution. Knowing its position ξ, the event that any particular one of the other division points $P_1, P_2, \ldots, P_n, \ldots$ falls to the left of P_0 will have probability ξ, independently of the others. If ξ is not known, the probability that h out of some other n division points fall to the left of P_0 is $1/(n + 1)$ for every h (i.e. all the frequencies are equally probable), because P_0 is equally likely to be any one of the $n + 1$ ordered division points 'chosen at random'. If we assume that we know there to be r out of n points to the left of P_0, then the probability of a success with the next division point (i.e. that P_{n+1} falls to the left of P_0) is given by $(r + 1)/(n + 2)$, because the $n + 1$ points divide the interval into $n + 2$ pieces, $r + 1$ of which are to the left of P_0 (and all have exactly the same probability of containing the new division point—assuming that nothing is known about their lengths, etc.). The probability distribution of P_0, which is initially uniform, is no longer such if we know that out of a further n 'random' subdivision points r have fallen to the left of P_0. It is, instead, the beta distribution $f(\xi) = K\xi^r(1 - \xi)^{n-r}$, because P_0 is then the $(r + 1)$st point from the left out of $n + 1$ 'random' points.

In this way, we have again displayed the likelihood factors, the equal probabilities of the frequencies, and also the value of the probability after observing r successes out of n trials. In other words, we have found the barycentre of the beta distribution without evaluating the integral (which, in any case, would give the same result):

$$(10) \qquad K\int_0^1 \xi \,.\, \xi^r(1 - \xi)^{n-r}\,\mathrm{d}\xi = (r + 1)/(n + 2)$$

$$\left(K = \left[\int_0^1 \xi^r(1 - \xi)^{n-r}\,\mathrm{d}\xi\right]^{-1}\right).$$

This result can be expressed more appealingly by saying that, in the Bayes–Laplace case, the probability for any future trial is given by the observed frequency, modified by adding in two fictitious observations, one a success, the other a failure. This is Laplace's celebrated 'rule of succession'.

11.4.4. The same rule reveals, on the other hand, the identity of the Bayes–Laplace scheme and that of Pólya's contagion model. In the latter, in fact, one adds to two initial balls, one white and the other black, as many white and black balls as there have been drawing from the urn (the result of double replacement). After n drawings, r of which resulted in the drawing of a white ball, we shall have $n + 2$ balls in the urn, $r + 1$ of which are white. The probability of drawing a white ball is then $(r + 1)/(n + 2)$.

This establishes that the probabilities are all identical to those of the previous case: in particular, the drawings are exchangeable events, the frequencies (out of a given number of drawings) are equally probable, and so on. It follows that not only is $7/10(=(6 + 1)/(8 + 2))$ the probability of drawing a white ball at the 9th drawing after 6 of the previous 8 have resulted in white (in which case, we know that at this point the urn contains 10 balls, 7 of which are white), but it is also the probability of drawing a white ball on any occasion for which we do not know the outcome, provided that, out of 8 observed drawings, 6 (for instance, the 3rd, 8th, 19th, 52nd, 53rd, 100th) resulted in white balls and 2 in black (the 1st and 92nd, say). And this will hold for the 2nd drawing (even though it is certain that at that moment there were three balls in the urn, 1 of which was white), the 4th (even though there were then 5 balls in the urn, either 2 or 3 of which were white), the 20th, 50th, 200th or 1000th, or any other (although in determining the proportion of white and black balls the need for information becomes less and less). At the second drawing, if I only knew the outcome of the first (black), I would attribute a probability of $\frac{1}{3}$ to white, there being certainly one white and two black balls in the urn. Although this is clear-cut, the knowledge of the subsequent outcomes leads me to attribute a probability of $\frac{7}{10}$ to obtaining a white ball in that same drawing. This is because the subsequent prominence of white balls leads me to assume that their percentage increased due to a number of drawings of white balls—including, perhaps, on the second drawing.

The resolution of what appeared at first sight to be a paradox is instructive, because it makes one aware of the traps that one can so easily fall into. In this way, one's attention is drawn to the kinds of misunderstanding which may persist (due, in part, to an inability to rid oneself of past habits), even without one noticing, and even if one has thought carefully about what we have said so far, and has made an effort to adjust to our perspective and terminology. We have stated repeatedly that probability can only mean probability as evaluated by someone on the basis of available information. In this sense, the Bayes–Laplace and Pólya schemes are identical, because anyone who adopts a given prior probability distribution and has the same information (concerning the outcomes of certain events in the scheme) must evaluate the probabilities in the same way. There may, however, be a temptation to regard these probabilities of ours as less concrete or less valid than other things which might more justifiably be called true probabilities: for example, the actual and unchanging composition of the urn in the first case, or any of the momentary compositions in the ever-changing Pólya scheme. On the contrary, these other things are either irrelevant, or even illusory. The composition of the urn (in the Bayes–Laplace sense) does make sense if we are actually dealing with drawings from an urn, and is connected with the idea of probability conditional on the knowledge of such a composition. But this is irrelevant, because it is assumed that we do not have a knowledge of the composition of the urn. Nevertheless, it may serve to highlight the interpretation of the distribution as a mixture.

On the other hand, *to posit an imaginary urn for the purpose of giving a more concrete interpretation to the expression in mixture form, and to the symbols in that expression in mixture form, and to the symbols in that expression which replace the probability*, and then, in this context, to refer to the 'true, unknown probabilities', is a distortion which leads to an immediate confusion of the issues. It would be equally illusory, and just as much a distortion, to imagine that behind every set of exchangeable events with an initial distribution judged to be uniform, there exists, or can be assumed to exist, a Pólya scheme whose probabilities, from drawing to drawing, are to be interpreted as a composition obtained as a result of drawing with double replacement. We have seen that changing the order changes everything, even when the above scheme actually exists.

The *one genuine and real factor* is the *probability* (albeit subjective, and relative to the person making the evaluation—and, indeed, precisely because of this) that one evaluates in the actual situation obtaining (and in future situations, with respect to certain hypothetical and as yet unavailable information, which will subsequently be obtained). If we step out of this ambit, we not only find ourselves unable to reach out to something more concrete, but we tumble into an abyss, an illusory and metaphysical kingdom, peopled by Platonic shadows.

11.4.5. The considerations we have put forward in the preceding sections should be carefully studied. Not only do they provide the necessary basis for a valid conceptual approach, but they also serve to give one a clear practical awareness of how, under conditions like those which characterize the case of exchangeable events, one can justify evaluating probabilities on the basis of observed frequencies for events which are, in some sense, 'similar'. The safest and most down-to-earth approach consists, as always, in confining attention to just those particular events which are of interest to us, and, within this framework, considering the smallest number possible (without positing any infinite sequences, or any imaginary, fictitious underlying schemes). For example, if we have observed r successes and $n - r$ failures, then, in the exchangeable case, the probability which we attribute to a success on any other trial is given by

$$(11)\quad p_r^{(n)} = \left[\frac{\omega_{r+1}^{(n+1)}}{\binom{n+1}{r+1}}\right]\Big/\left[\frac{\omega_r^{(n)}}{\binom{n}{r}}\right] = \frac{r+1}{n+2}\Big/\left[1 + \left(1 - \frac{r+1}{n+2}\right)\left(\frac{\omega_r^{(n+1)}}{\omega_{r+1}^{(n+1)}} - 1\right)\right]$$

(as is easily verified). This shows that, provided the probabilities attributed initially to the two frequencies $r/(n + 1)$ and $(r + 1)/(n + 1)$ out of $n + 1$ trials do not differ greatly, this probability itself differs little from the frequency (or the modified frequency, as in the Bayes–Laplace scheme).

11.4.6. If one uses the properties of the likelihood and the mixture form, a stronger conclusion can be obtained, although somewhat indirectly. After n trials, r of which are successes, the function $\xi^r(1 - \xi)^{n-r}$, which represents the likelihood (and, in the Bayes–Laplace case, the density), increases in the range 0 to $\xi = r/n$, where it attains its maximum, and then decreases as we move from r/n to 1. It vanishes at the end-points 0 and 1 (provided, of course, that $0 < r < n$), and, if r and $n - r$ are large, it is practically 0 everywhere except in the immediate neighbourhood of the maximum. We can see this clearly by observing that, as n increases with $r/n = \bar{\xi}$ held fixed, one obtains, in the limit, the density function of the *normal* distribution, centred at the frequency, $\bar{\xi} = r/n$, and having standard deviation

$$\sqrt{[\bar{\xi}(1 - \bar{\xi})/n]}$$

(i.e. the same standard deviation that we have for the difference between the frequency and the probability for n events having constant probability $\bar{\xi}$).

In fact, setting $x = (\xi - \bar{\xi})/\sqrt{[\bar{\xi}(1 - \bar{\xi})/n]}$, we have, asymptotically,

$$\begin{aligned} K\xi^h(1 - \xi)^{n-h} &= K[\xi^{\bar{\xi}}(1 - \xi)^{1-\bar{\xi}}]^n \\ &= K\left(1 + \frac{x}{\bar{\xi}}\sqrt{\{\bar{\xi}(1 - \bar{\xi})/n\}}\right)^{n\bar{\xi}} \\ &\quad \times \left(1 - \frac{x}{1 - \bar{\xi}}\sqrt{\{\bar{\xi}(1 - \bar{\xi})/n\}}\right)^{n(1-\bar{\xi})} \rightarrow K\,\mathrm{e}^{-x^2/2}. \end{aligned}$$

To prove this, all we need to do is to take the logarithm of the penultimate expression.† Omitting the constant K, we obtain

$$\begin{aligned} & n\bar{\xi}\log\left(1 + \frac{x}{\bar{\xi}}\sqrt{\{\bar{\xi}(1 - \bar{\xi})/n\}}\right) + n(1 - \bar{\xi})\log\left(1 - \frac{x}{1 - \bar{\xi}}\sqrt{\{\bar{\xi}(1 - \bar{\xi})/n\}}\right) \\ &= n\bar{\xi}\left[\frac{x}{\bar{\xi}}\sqrt{\{\bar{\xi}(1 - \bar{\xi})/n\}} - \frac{1}{2}\frac{x^2}{\bar{\xi}^2}\bar{\xi}(1 - \bar{\xi})/n + O(n^{-\frac{3}{2}})\right] \\ &\quad + n(1 - \bar{\xi})\left[-\frac{x}{1 - \bar{\xi}}\sqrt{\{\bar{\xi}(1 - \bar{\xi})/n\}} - \frac{1}{2}\frac{x^2}{(1 - \bar{\xi})^2}\bar{\xi}(1 - \bar{\xi})/n + O(n^{-\frac{3}{2}})\right] \\ &= -\tfrac{1}{2}x^2[(1 - \bar{\xi}) + \bar{\xi}] + O(n^{-\frac{1}{2}}) \rightarrow -\tfrac{1}{2}x^2. \end{aligned}$$

This establishes directly that, in the Bayes–Laplace case, the posterior distribution (which is a beta distribution, with observed frequency $\bar{\xi}$, and very large n) is asymptotically normal. This conclusion holds more generally, provided that the limit distribution $F(x)$ obeys certain qualitative conditions. More precisely, it is sufficient that a density exists and is 'practically constant'

† There is no mystery in the disappearance of the factor in square brackets: it does not involve x, which is the only thing we are interested in, and we have subsumed it in K.

in the neighbourhood of $\xi = \bar{\xi}$, and that it is not too small in comparison with distant masses, which, if they were very large, would otherwise give an appreciable contribution to the product, even though multiplied by the likelihood factor, which itself would be quite small. Such a condition—which we prefer to express in this rather vague form, because we are interested in ensuring a good approximation for large n, rather than for the asymptotic case of $n \to \infty$—can be summarized (following L. J. Savage) by saying that the distribution F must be *diffuse* (in the neighbourhood of the point of interest).

11.4.7. The same argument also applies in cases where the scope is much wider (like those involving exchangeable random quantities rather than events), and it can be proved that the normal distribution arises quite naturally, and under relatively weak conditions even in these cases. The cases we are discussing are, of course, very different from those involving the limiting normal distribution that we discussed previously. There we were dealing with the distribution of a quantity defined as a function of a large number of other independent quantities (in particular, as the sum, but also in other ways); here, we are dealing with the form of the posterior distribution after a large number of items of information have been acquired.

From a conceptual point of view, the reason for the appearance of the normal distribution is clear (albeit in outline form) if one thinks of the genesis of the beta function in the example we have considered. It arose as the product of a number of terms ξ and $1 - \xi$, each of which was the likelihood factor corresponding to an observation (the outcome of an event). In the case of observations of random quantities, also (e.g. performing a measurement, which is affected by error, of the quantity in which we are interested, or of others of which it is a function, etc.), under similar conditions the likelihood factor for the totality of such observations will be the product of the factors corresponding to individual observations:

$$\nu(\xi) = \nu_1(\xi)\nu_2(\xi) \ldots \nu_n(\xi).$$

Let $\bar{\xi}$ denote the 'maximum likelihood' point: i.e. the point at which $\nu(\xi)$ has an absolute maximum (which we shall assume to be unique; and we further assume that $\nu(\xi)$ is much less than $\nu(\bar{\xi})$, except in a neighbourhood of $\bar{\xi}$ small enough for whatever purposes we have in mind). If, in the neighbourhood of $\bar{\xi}$, we replace $\nu_i(\xi)$ by the linear approximation

$$\nu_i(\bar{\xi}) + \nu_i'(\bar{\xi})(\xi - \bar{\xi}) = K[1 + a_i(\xi - \bar{\xi})]$$

(with $\sum_i a_i = 0$, in order that $\nu'(\bar{\xi}) = 0$), the product can be replaced by a polynomial for which we can repeat essentially the same form of argument as we used for the case of the beta.† By including with the $v_i(\xi)$ the factor $f(\xi)$ = prior density, the product becomes the posterior density, and the conclusions can be applied to it. If the contribution of this factor is irrelevant when compared with the others, then $v(\xi)$ alone already provides an approximation to the posterior density. Both the latter and the likelihood have, asymptotically, the form of the normal density.

11.4.8. Similar considerations can be made in cases where something less restrictive than exchangeability is assumed (as in those cases which we pointed out for the sake of giving examples in Section 11.3.3). In some cases the conclusions are rather similar; in others they are markedly different. The starting point and the basic ideas remain the same, however, and are always clear and straightforward. We have merely to apply to these various cases the theorem of compound probabilities, or, more directly, Bayes's theorem, which can be expressed simply in the form:

$$\text{posterior probability} = \text{constant} \times \text{prior probability} \times \text{likelihood}.$$

We should point out that many of the methods used in statistics for purposes similar to those we have been considering do not follow the lines we have indicated. They are based upon a very different set of underlying concepts, and we shall not make use of them, nor shall we advocate their use. They will, however, be mentioned in Chapter 12, which is devoted specifically to statistical applications, and where it will obviously be necessary to examine and compare a number of different viewpoints and the methods they give rise to, especially those which are widely used in practice. Above all, it will be important to discuss the question of whether, and to what extent, those methods which have been introduced and justified on the basis of approaches which we consider invalid (i.e. non-Bayesian) can, in fact, be seen as legitimate (i.e. Bayesian) by suitably reinterpreting their underlying assumptions.

† In order to obtain the exact value of $v''(\bar{\xi})$, one must take into account $v''(\xi)$, by writing

$$v_i(\xi) \simeq K[1 + a_i(\xi - \bar{\xi}) + \tfrac{1}{2}b_i(\xi - \bar{\xi})^2].$$

This then gives the approximation

$$v(\xi) = K\left[1 - \tfrac{1}{2}(\xi - \bar{\xi})^2 \sum_i (a_i^2 - b_i)\right].$$

(Note that $\sum_{i \neq j} a_i a_j = \sum_i a_i \sum_j a_j - \sum a_i^2$.)

CHAPTER 12

Mathematical Statistics

12.1 THE SCOPE AND LIMITS OF THE TREATMENT

12.1.1. A brief account of mathematical statistics within the confines of the final chapter of this book must necessarily offer a limited perspective. Nevertheless, its inclusion serves a very definite purpose.

In Chapter 11, we have already encountered certain of the problems which fall within the purview of the subject matter of mathematical statistics. Specifically, we examined applications of inductive reasoning based on statistical data; that is, data involving a number of observations (possibly a large number) which are, in a certain sense, similar to one another. We have also explained the Bayesian approach to such problems (an approach which constitutes an integral part of the subjectivistic conception), noting that a unified coherent structure cannot be maintained if this is abandoned in favour of other approaches involving a variety of more or less empirical 'ad hoc' methods.

In this chapter, we shall attempt to give a more explicit account of the problems with which mathematical statistics is concerned, and of the implications, both for these problems and more generally, which flow from the adoption of one or other of the competing points of view.

12.1.2. In addition to the strictly probabilistic aspects, with which the previous considerations are concerned, we shall have occasion to examine other topics relating to decision theory. There are two basic reasons for this, and they correspond to two different questions that can be posed within the same framework.

The first of these concerns applications where the entire enterprise is more or less explicitly geared to arriving at a decision. Obvious examples of this are batch testing, or quality control, performed on a sample of a certain product in order to decide what to do with the rest of the stock (whether or not to reject it), or how to produce it in an optimal fashion (whether or not process parameters need adjusting), and so on. Although one does not always have such immediate actions in mind, it could be said that there is always some practical purpose for which one is somehow seeking guidance.

The second reason may or may not be relevant, depending on the particular application: more precisely, it depends on whether or not one is able to experiment. If this is possible—i.e. if one has certain choices regarding the way in which observations are to be obtained—then there is an additional dimension to the problem, because this choice is itself a decision, and must be made in the most appropriate way. What is appropriate will in this case, of course, depend on the final decision, itself dependent on the outcome of the experiment. More precisely, the whole question of what is appropriate needs to be set in the context of the theory of compound decisions; a framework including both decisions concerning the experiment to be performed, and the final decision to be taken after the results of the experiment are known.

12.1.3. General issues will be dealt with in a somewhat summary fashion in what follows, and most of the discussion will centre around specific examples. In this way, we hope to provide a straightforward account which will make best use of the limited space available to use. The examples will, in fact, involve some of the most important and commonly occurring cases, and so, in a sense, they do provide a general perspective.

12.2 SOME PRELIMINARY REMARKS

12.2.1. The cases which are usually considered in mathematical statistics are those which can, in various ways, and in a somewhat loose sense, be regarded as generalizations of the case of exchangeable events. This is in the nature of an aside, however, and simply provides a convenient reference to, and reminder of, the contents of the previous chapter: in fact, in the standard terminology of mathematical statistics one never comes across any reference to exchangeability. In order not to introduce a further difficulty into the task of comparing the various viewpoints, we shall conform to standard usage in this regard. Before doing so, however, we offer the following preliminary clarification of the approach we shall adopt.

We have seen (in Chapter 11, Sections 11.3 and 11.4) that the notions of exchangeability and partial exchangeability can be reduced to that of conditional independence (albeit sometimes in a merely formal sense). More precisely, we have seen that the probability distribution in such cases is always a mixture (i.e. a convex linear combination) of distributions representing independence. If to each case of independence there corresponds an objectively defined 'hypothesis'—like, for example, the proportion of white balls in an urn of unknown composition—then the mixture has an objectively meaningful interpretation. Where this is not the case, the representation is merely formal (as, for example, in the case of a biased coin). There is, however, no difficulty—apart from that of a conceptual nature—in dealing with such 'hypotheses' in these cases as if they were objectively meaningful: for example, one might refer, quite improperly, to 'the hypothesis that the unknown probability of obtaining Heads with the bent coin has the value p'. On the other hand, this pseudo-interpretation could always be treated as an asymptotic interpretation of a property which can be defined in a finitistic way by referring to 'frequency in a large number of trials' instead of to 'unknown probability'.†

We shall therefore adopt, in line with our previous remarks, the standard practice of talking in terms of 'hypotheses', irrespective of whether these exist objectively, or merely formally (in which case, they might be interpreted, if at all, in the asymptotic sense given above). Within this framework, the Bayesian approach consists in considering an initial distribution of probability among these hypotheses, this distribution being modified as new information becomes available.

12.2.2. The enormous range of possible applications might lead one to expect a large number of different theoretical models. On the other hand, if one thinks in terms of the basic simple forms of representation, the possibilities are more limited. (Indeed, one might argue that from a Bayesian point of view there is only one form of the problem since everything reduces, in the final analysis, to an application of Bayes's theorem.) In fact, those cases which form the bulk of mathematical statistics can probably be reduced to one or other of the two forms already mentioned: exchangeability or partial exchangeability. There is a meaningful distinction to be drawn between these two cases, and, indeed, partial exchangeability embraces a wide range of possible deviations from the exchangeable case. Moreover, within the two categories there are a number of problems of detail, and, depending on the field of events under consideration, these may present various levels of difficulty (without there necessarily being any great conceptual difficulties).

† We observe that although the present approach may, at first sight, appear to be very similar (or even equivalent) to that based on 'limit-frequency', there is, in fact, a great deal of difference. We in no way assume the existence of any limit to which the frequency Y_n/n must tend (either with certainty, or in some probabilistic sense—like weak, mean-square, or strong convergence, and so on). Nor do we utilize any probabilistic form of Cauchy convergence (Chapter 6, 6.8.7), even though this holds (cf. Chapter 11, 11.4.2) under the assumption of exchangeability (it does not define a 'limit random quantity', and, in any case, requires an infinite number of 'trials'). We base ourselves solely on the frequency Y_n/n of successes in the trials actually considered, whatever they may be, and however many there are, and on the fact that their distribution F_n (the probability distribution of Y_n/n according to the evaluation made at the beginning of the trials) provides an approximation (which improves as n increases) to the limit distribution F, whose existence is therefore established (and this is the only thing we need!).

Exchangeability and partial exchangeability can also arise in the context of multi-events in general (as we mentioned in Chapter 11), as well as for vectors (r-tuples of random quantities), functions, . . . , and random elements of any space whatsoever.

What is important in these cases is not so much their actual form, or that of the space to which they belong, but rather the kinds of 'hypotheses' that are assumed; i.e. the corresponding distributions. There are three main distinctions worth making in this respect: the *discrete* case (involving a finite or countable set of hypotheses); the *parametric* case (involving a set of hypotheses which can each be represented in terms of a fairly restricted set of parameters; i.e. by a vector in a parameter space whose dimension is not too large); *the non-parametric case* (where either the individual hypothesis cannot be represented in terms of a finite number of parameters, or, alternatively, the number of parameters involved is prohibitively large).

Similar distinctions can be made in the case of forms of representation required for *partial* exchangeability (and we shall shortly give a more precise account of these).

12.2.3. In presenting the mathematical development of these ideas, we shall normally deal with problems involving random quantities in the parametric case; in particular, with just one parameter. This is the most straightforward and meaningful case, and hence the most convenient for illustrating the mathematics. It will be immediately obvious, however, that our treatment is completely general, provided the expressions given, and the comments made, are interpreted in an appropriate manner.

Specifically—to use what we regard as the correct terminology—we shall be dealing with a collection of exchangeable random quantities X_h (cf. Chapter 11, Sections 11.3 and 11.4). These can be represented in precisely the same way as we saw earlier in the case of exchangeable events: in other words, they can, in a formal sense at least, be thought of as 'independent conditional on some given set of hypotheses'. More precisely, each 'hypothesis' indexes a distribution, and we assume that, conditionally, 'all the X_h have this same distribution, and are stochastically independent of one another'.

This implies that their joint distribution is a mixture of products of individual factors (corresponding to the case of independence). As we remarked earlier on, we shall take this representation as our starting point; the interpretation in terms of exchangeability then becomes merely a preliminary clarification.

We shall follow the usual practice in mathematical statistics and work in terms of probability densities (for a justification of this, see the remarks in Section 12.4.3). Expressed mathematically, our basic assumptions then become the following:

there exists a set of 'hypotheses', a general element of this set being denoted by θ (a point in the hypothesis space); in particular, we shall first consider the case where each hypothesis can be represented by a single real-valued parameter θ;

conditional on each hypothesis θ (i.e. on the value of θ for the case in question), all the X_h have exactly the same distribution, i.e. the same density $f(x|\theta)$, and are stochastically independent; this implies that the joint density $p^m(x^1, \ldots, x^m|\theta)$† for m of the X_h (no matter how they are chosen or labelled) is given by the product of the densities

$$p^m(x^1, x^2, \ldots, x^m|\theta) = f(x^1|\theta)f(x^2|\theta)\ldots f(x^m|\theta);$$

over the set of hypotheses we have prior probability with density $\pi_0(\theta)$; in the case we are considering, we have a non-negative function of θ such that

$$\int_{-\infty}^{\infty} \pi_0(\theta)\,\mathrm{d}\theta = 1.$$

We note that this latter assumption is the hallmark of the Bayesian approach, whereas other approaches attempt to do without it. We shall develop our treatment within the Bayesian framework, but, as we proceed, we shall discuss the techniques which are used by those who eschew the use of 'prior probabilities'.

It follows immediately from these assumptions that the marginal (prior) distributions for any individual X_h, or for m of them, are, expressed as densities, given by

$$f_0(x) = \int f(x|\theta)\pi_0(\theta)\,\mathrm{d}\theta, \tag{1}$$

$$\begin{aligned} p_0^m(x^1, x^2, \ldots, x^m) &= \int p^m(x^1, x^2, \ldots, x^m|\theta)\pi_0(\theta)\,\mathrm{d}\theta \\ &= \int f(x^1|\theta)f(x^2|\theta)\ldots f(x^m|\theta)\pi_0(\theta)\,\mathrm{d}\theta. \end{aligned} \tag{2}$$

Here, and elsewhere, it is to be understood that the integrals are to be taken over the entire range of the distribution (and there is no harm in thinking of this as the whole real line, since any range where the density is zero will give a zero contribution). Note that, if we interpret the quantities involved in an appropriate manner, these expressions apply equally well to

† The reason for using superscripts will become clear in Section 12.2.5. Note that f is a special case of p^m for $m = 1$ ($f = p^1$, and, in what follows, $f_n = p_n^1$, etc.). Usually, however, we shall use f in preference to p^1 in order to make the case $m = 1$ more immediately distinguishable, and also to avoid the superscript.

any abstract spaces (and, in particular, to the case of several parameters, where θ represents a vector).

From the point of view of interpretation, note that f_0 and p_0^m give the previsions of f and p^m if the latter are considered as functions of the random quantity θ. Also note that p_0^m, like p^m, is a symmetric function of the x^h (in line with our original assumption of exchangeability).

12.2.4. It is equally straightforward to derive expressions, similar to the above, for the evaluations conditional on knowledge of the values of any of the X_h, or of any n of them. We shall denote these random quantities by X_1, or $X_1, X_2, \ldots, X_n$, partly for convenience, and partly to fix ideas for the case in which we observe them in chronological order (and although this might be useful, the reader should remember that it is not an essential part of the argument). The choice of which particular X_h (or set of them) we are interested in calculating the conditional evaluation for is equally irrelevant, and the reader should again realize that we denote these by

$$X_{n+1}, X_{n+2}, \ldots, X_{n+m}$$

purely for convenience.

We shall see, in fact, that the evaluations conditional on the knowledge of the values of the first n of the X_h (which we shall denote by f_n and p_n^m) can be expressed in essentially the same form as the f_0 and p_0^m above (the special cases corresponding to $n = 0$; i.e. the initial evaluations, prior to any knowledge of the X_h). In fact, it turns out to be sufficient to determine the distribution $\pi_n(\theta|x_1, x_2, \ldots, x_n)$† for the parameter θ, conditional on the values $X_1 = x_1$, $X_2 = x_2, \ldots, X_n = x_n$, and to substitute this in place of $\pi_0(\theta)$ in the expressions for f_0 and p_0^m. Let us first see this for the case $n = 1$.

After having observed the value of any one of the X_h, $X_1 = x_1$, say, the probability distribution of the parameter (or, more accurately,‡ the distribution 'conditional on the hypothesis $X_1 = x_1$') becomes

$$\pi_1(\theta|x_1) = K\pi_0(\theta)f(x_1|\theta) \tag{3}$$

$$\left(\frac{1}{K} = \int f(x_1|\theta)\pi_0(\theta)\,\mathrm{d}\theta = f_0(x_1)\right).$$

This is a straightforward application of Bayes's theorem, or, if one prefers, it is sufficient to observe that the joint density for (θ, X_1), is given both by $\pi_0(\theta)f(x_1|\theta)$ and by $f_0(x_1)\pi_1(\theta|x_1)$.

---

† N.B. For the sake of brevity, we shall sometimes write this as $\pi_n(\theta)$ omitting any explicit mention of $x_1, x_2, \ldots, x_n$ (which must of course be understood).

‡ 'More accurately' by virtue of what we said in Chapter 11, 11.2.2 (and what we shall say in Section 6 of the Appendix).

It follows immediately that

(4) $$f_1(x|x_1) = \int f(x|\theta)\pi_1(\theta|x_1)\,d\theta,$$

(5) $$p_1^m(x^2, x^3, \ldots, x^{m+1}|x_1) = \int p^m(x^2, x^3, \ldots, x^{m+1}|\theta)\pi_1(\theta|x_1)\,d\theta.$$

12.2.5. Before we go on with our development, we need to make a comment about notation. The use of superscripts for the x (x^h rather than x_h) is necessary in order to distinguish the use of the values x^h of X_h as 'names of coordinates' for the distribution of X_h (as yet unknown, or considered as unknown), from the use of the x_h as observed values (or values assumed to be known). The practical effect of this can be observed in the formulae, where superscripts precede the vertical bar and subscripts follow it (except when rôles are reversed, as happens during the application of Bayes's theorem; cf. (11) and (11′)). In the case of something like $f_1(x|x_1)$, it is clear that it would be superfluous to write $f_1(x^2|x_1)$, because the superscripts are only useful for distinguishing between the x^h when there is more than one of them. For a single (generic) coordinate, it is sufficient to denote it by x.

12.2.6. The expressions for f_1 and p_1^m (and we recall that the former is a special case of the latter; $f_1 = p_1^1$) can be rewritten in a different form, so as to show up certain interesting features more clearly:

(6) $$f_1(x|x_1) = K\int f(x|\theta)[f(x_1|\theta)\pi_0(\theta)\,d\theta]$$

$$\left(\frac{1}{K} = \int [\ldots] = f_0(x_1)\right);$$

(7) $$p_1^m(x^2, x^3, \ldots, x^{m+1}|x_1) = K\int p^m(x^2, x^3, \ldots, x^{m+1}|\theta)[f(x_1|\theta)\pi_0(\theta)\,d\theta]$$
$$= K\int \prod_{i=2}^{m+1} f(x^i|\theta)[f(x_1|\theta)\pi_0(\theta)\,d\theta].$$

In this way, we emphasize the fact that f_1 and p_1^m are mixtures of f and p^m, with 'weights' as given in square brackets. Alternatively, we could remove the separation between factors in x_1 and those in x^i ($i = 2, 3, \ldots, m+1$) and write instead

$$f(x|x_1) = K\int \{f(x|\theta)f(x_1|\theta)\}\pi_0(\theta)\,d\theta$$

(8) $$= Kp_0^2(x_1, x) = \frac{p^2(x_1, x)}{f_0(x_1)},$$

or even

$$f_1(x|x_1) = f_0(x)\frac{f_1(x_1|x)}{f_0(x_1)}. \tag{9}$$

In this way, we directly emphasize the interpretation in terms of the theorem of compound probabilities and Bayes's theorem.

Proceeding in a similar fashion, we can derive the more general result

$$p_1^m(x^2, x^3, \ldots, x^{m+1}|x_1) = \frac{p_0^{m+1}(x_1, x^2, x^3, \ldots, x^{m+1})}{f_0(x_1)}. \tag{10}$$

In order to derive a form analogous to that of (9), i.e.

$$p_1^m(x^2, x^3, \ldots, x^{m+1}|x_1) = \frac{p_0^m(x^2, x^3, \ldots, x^{m+1}) f_m(x_1|x^2, x^3, \ldots, x^{m+1})}{f_0(x_1)}, \tag{11}$$

we must introduce the f_m for $m > 1$; the result is then immediate.

12.2.7. It would have been perfectly straightforward, and more in line with the approaches more commonly adopted in statistics, to have considered right away the distributions conditional on n values,

$$X_1 = x_1, X_2 = x_2, \ldots, X_n = x_n,$$

instead of on just one. Our main consideration in starting off with the case $n = 1$ is that it enables one to bring out the fact that the effect of n observations is simply the combined effect of considering them one at a time, rather than some magical consequence of there being enough of them to be 'statistically' relevant.

The probability distribution of the parameter θ, given the observed values $x_1, x_2, \ldots, x_n$, is given by

$$\begin{aligned} \pi_n(\theta|x_1, x_2, \ldots, x_n) &= K\pi_0(\theta)p^n(x_1, x_2, \ldots, x_n|\theta) \\ &= K\pi_0(\theta)f(x_1|\theta)f(x_2|\theta)\ldots f(x_n|\theta), \end{aligned} \tag{3'}$$

where

$$\frac{1}{K} = \int p^n(x_1, x_2, \ldots, x_n|\theta)\pi_0(\theta)\, d\theta = p_0^n(x_1, x_2, \ldots, x_n).$$

As we remarked earlier, it is sufficient to replace π_0 by π_n in order to obtain the distributions for one of the X_h, or for m of them:

$$(4') \quad f_n(x|x_1, x_2, \ldots, x_n) = \int f(x|\theta)\pi_n(\theta|x_1, x_2, \ldots, x_n)\,\mathrm{d}\theta$$

$$(6') \quad = K \int f(x|\theta)\left[\prod_{h=1}^{n} f(x_h|\theta) \,.\, \pi_0(\theta)\,\mathrm{d}\theta\right]$$

$$(8') \quad = p_0^{n+1}(x_1, x_2, \ldots, x_n, x)/p_0^n(x_1, x_2, \ldots, x_n)$$

$$(9') \quad = f_0(x) \,.\, p_1^n(x_1, x_2, \ldots, x_n|x)/p_0^n(x_1, x_2, \ldots, x_n);$$

$$(5') \quad p_n^m(x^{n+1}, x^{n+2}, \ldots, x^{n+m}|x_1, x_2, \ldots, x_n)$$

$$= \int p^m(x^{n+1}, \ldots, x^{n+m}|\theta)\pi_n(\theta|x_1, \ldots, x_n)\,\mathrm{d}\theta$$

$$(7') \quad = K \int \prod_{i=n+1}^{n+m} f(x^i|\theta) \,.\, \prod_{h=1}^{n} f(x_h|\theta)\pi_0(\theta)\,\mathrm{d}\theta$$

$$(10') \quad = p_0^{n+m}(x_1, \ldots, x_n, x^{n+1}, \ldots, x^{n+m})/p_0^n(x_1, \ldots, x_n)$$

$$(11') \quad = \frac{p_0^m(x^{n+1}, \ldots, x^{n+m})p_m^n(x_1, \ldots, x_n|x^{n+1}, \ldots, x^{n+m})}{p_0^n(x_1, \ldots, x_n)}.$$

The last four expressions include all the others as special cases: more precisely,

(5′), (7′), (10′), (11′) give for general n and m, what

(5), (7), (10), (11) give for $n = 1$ (with m arbitrary), and

(4′), (6′), (8′), (9′) give for $m = 1$ (with n arbitrary) and

(4), (6), (8), (9) give for $n = m = 1$.

The interpretations are identical to those which we gave in the simplest case (i.e. that of $n = m = 1$), and, when contemplating extensions to cases which are more complicated (insofar as the formulae are concerned, anyway), it may be useful to bear this case in mind.

Given $x_1, x_2, \ldots, x_n$, the likelihoods for θ and x are

$$\text{(a)} \quad \prod_h f(x_h|\theta) \qquad \text{(as a function of } \theta\text{)},$$

$$\text{(b)} \quad \int f(x|\theta) \prod_h f(x_h|\theta)\pi_0(\theta)\,\mathrm{d}\theta \qquad \text{(as a function of } x\text{)},$$

respectively. In fact, any function differing from (a) by a factor independent of θ, or from (b) by a factor independent of x, could be taken as the respective likelihood.

12.2.8. We now turn to the case of 'partial exchangeability'. The account we shall give will be even shorter than the above, and we shall rely on the examples to clarify our interpretation and approach to the problem.

In terms of our formulation, this case differs from the previous one in that, conditional on each of the 'hypotheses' characterized by θ, the X_h are still stochastically independent, but now may have different distributions. The latter depend not only on the parameter θ, but also on certain observable quantities y_h which relate to the X_h. Like θ, y may be real-valued, or a vector, or whatever (irrespective of the form of θ).† In order to keep the presentation on a simple level, we shall take y to be real-valued (the general case presents nothing new from a conceptual viewpoint).

Formally, instead of starting from $f(x|\theta)$ we consider $f(x|\theta, y)$. So far as the prior distribution for θ is concerned, nothing changes; we begin, as before, with some $\pi_0(\theta)$. The distribution $p_0^m(x^1, x^2, \ldots, x^m)$ (knowledge of which enables one to derive everything else) will, however, also depend on the values that y takes for each of the $X_1, X_2, \ldots, X_m$, and has the form

$$(12) \quad p_0^m(x^1, x^2, \ldots, x^m) = \int f(x^1|\theta, y_1) f(x^2|\theta, y_2) \ldots f(x^m|\theta, y_m)\pi_0(\theta)\, \mathrm{d}\theta,$$

where y_h corresponds to X_h. If we wish this to be made explicit, we must write the left-hand side as

$$p_0^m(x^1, \ldots, x^m|y_1, \ldots, y_m).$$

On the other hand, if we do make systematic use of this explicit form the expressions become rather cumbersome—particularly those which are already complicated, even without this additional detail.

The following are intended as examples of the kinds of y_h that might be observed‡ and considered as possibly influencing the distribution of X_h: the temperature at the time at which the experiment yielding X_h took place; the age of an individual whose reaction to some given drug is measured by X_h; the precision of the instrument which performs the measurement giving X_h; and so on. We shall shortly give an example involving the latter possibility.

12.3 EXAMPLES INVOLVING THE NORMAL DISTRIBUTION

12.3.1. Given that the normal distribution is widely used (and somewhat abused) in statistics, it is natural that the most familiar problems of inference are those which involve this distribution. The prevision m and the variance σ^2

† In other words, one could be a vector and the other a real number, etc.
‡ See the remark at the end of Section 12.3.3.

suffice to characterize the distribution, which is usually denoted by $N(m, \sigma^2)$. The density, as we already know, is given by

$$f(x) = \frac{1}{\sqrt{(2\pi)}\sigma} \exp\{-\tfrac{1}{2}(x - m)^2/\sigma^2\}.$$

By far the most important case is that in which m corresponds to the unknown parameter θ (while σ^2 is known), but we shall also deal with the opposite case ($\theta = 1/\sigma^2$, m known), and with the case in which both parameters are unknown (θ is the 'vector' (θ_1, θ_2), $\theta_1 = m$ and $\theta_2 = 1/\sigma^2$).† On the basis of these examples, all under the assumption of complete exchangeability, we can discuss variants corresponding to partial exchangeability. The simplest such variants are obtained by replacing the assumption 'σ^2 known' (or 'm known') by ' $= y$', known for each X_h, but possibly different for different h'.

12.3.2. *The case where m is unknown.* This is, above all, the case considered in the theory of errors (experimental or observational) as applied in astronomy geodesy, physics, and so on. What is unknown is the *true* value of the quantity that is being measured; i.e. $\theta = m$. The accuracy of the instrument (as represented by σ^2) is assumed known, and the distribution of the observed value is assumed to be $N(m, \sigma^2)$; i.e. a normal distribution centred at the *true* value and having the given precision.

We have, therefore,

(13) $$f(x|\theta) = K \exp\{-\tfrac{1}{2}(x - \theta)^2/\sigma^2\},$$

and (apart from the constant factor K) this is the likelihood for θ given by an observation x. The likelihood given by n observations $x_1, x_2, \ldots, x_n$ is

$$\prod_h f(x_h|\theta) = \exp\left\{-\frac{1}{2\sigma^2}\sum_h (x_h - \theta)^2\right\}.$$

Noting that

$$\begin{aligned}\sum_h (x_h - \theta)^2 &= \sum_h (x_h^2 - 2x_h\theta + \theta^2) = \text{const.} - 2\theta\sum_h x_h + n\theta^2 \\ &= n\left(\text{const.} - 2\theta\frac{1}{n}\sum_h x_h + \theta^2\right) \\ &= n[\text{const.} + (\bar{x} - \theta)^2],\end{aligned}$$

† In the terminology used in the theory of errors, the reciprocal $1/\sigma$ of the standard deviation is called the *precision*, and the reciprocal $1/\sigma^2$ of the variance is called the *weight* (although sometimes a different unit of measure is used; e.g. precision $1/\sqrt{2}\sigma$, weight σ_0^2/σ^2, where σ_0^2 is chosen as appropriate for the problem under consideration). It might seem rather unnecessary to have four terms available, but this is not entirely the case.

We shall take the *weight* σ^{-2} as the parameter θ, instead of the more customary variance, σ^2, since this turns out to simplify the formulae.

where $\bar{x} = 1/n \sum_h x_h$ is the mean of the x_h, we see that the likelihood can be rewritten in the form

$$\exp\left\{-\frac{n}{2\sigma^2}(\bar{x} - \theta)^2\right\}. \tag{14}$$

In other words, it has the same form as the likelihood of a single observation equal to the mean $\bar{x}$, and with *standard deviation* $\sigma/\sqrt{n}$ (i.e. reduced in the ratio 1 to $1/\sqrt{n}$, which is equivalent to *precision* increased in the ratio 1 to $\sqrt{n}$, *variance* reduced in the ratio 1 to $1/n$, *weight* increased in the ratio 1 to n).

The posterior distribution for θ is therefore given by

$$\pi_n(\theta) = K\pi_0(\theta)\exp\left\{-\frac{n}{2\sigma^2}(\bar{x} - \theta)^2\right\}. \tag{15}$$

Since the likelihood is maximized for $\theta = \bar{x}$, and decreases as we move away from this value (the decrease being sharper for larger n), the posterior distribution concentrates around $\bar{x}$. In particular, if the prior distribution is taken to be normal, $N(m_0, \sigma_0^2)$, say, then the posterior distribution is also normal. More precisely, the posterior (or final†) distribution is $N(m_f, \sigma_f^2)$, where

$$m_f = \frac{m_0\sigma_0^{-2} + n\bar{x}\sigma^{-2}}{\sigma_0^{-2} + \sigma^{-2}}, \qquad \sigma_f^{-2} = \sigma_0^{-2} + n\sigma^{-2}. \tag{16}$$

In words: the posterior *weight* (1/*variance*) is the sum of the weights from the prior and the likelihood; the posterior mean is the weighted mean of the prior mean and the mean from the likelihood (m_0, and n times $\bar{x}$; i.e. a function of m_0 and $x_1, x_2, \ldots, x_n$), the weights being the respective *weights* (thus revealing the aptness of the terminology).

12.3.3. The extension to the case of 'observations made with different precisions' is immediate. Let us assume, for instance, that we know that the n observations are performed with different measuring instruments, the errors of which have standard deviations $\sigma_1, \sigma_2, \ldots, \sigma_n$. It is clear that (by an argument similar to that used above) these observations are equivalent to a single observation whose value is given by the weighted mean of the x_h (with weights σ_h^{-2}), and having weight equal to the sum of the weights. If the prior distribution is $N(m_0, \sigma_0^2)$, the posterior distribution is given by $N(m_f, \sigma_f^2)$, where m_f and σ_f^2 are determined by the weighting process just described, except that we now also include m_0 with weight σ_0^{-2}.

† *Translators' note.* The terms *prior* and *posterior* seem firmly established in English publications relating to applications of Bayes's theorem, and we have used them in preference to the terms *initial* and *final*. The Italian version uses the latter, and the notation m_f and σ_f^2 derives from this usage.

This is an example of 'partial exchangeability' with $y_h = \sigma_h^2$ (or we could take $y_h = \sigma_h^{-2}$), and

$$f(x|\theta, y) = K \exp\{-\tfrac{1}{2}(x-\theta)^2/y\}.$$

We should draw attention to the fact that the y_h must actually be known and observed for each X_h under consideration. In our example, we must know with what precision each measurement has been performed. One should be careful not to think of it as being sufficient to know that each measurement has been performed using instruments of various precisions (for example, by choosing each time at random from among some given collection of measuring instruments, but without registering which were actually used and how often). Under this latter assumption, one would have a case of exchangeability with

$$f(x|\theta) = \sum_k c_k f_k(x|\theta), \qquad f_k(x|\theta) = K \exp\{-\tfrac{1}{2}(x-\theta)^2/y_k\}, \qquad c_k \geqslant 0, \qquad \sum_k c_k = 1$$

(i.e. no longer a normal distribution, but a mixture of normals). In the same way, in the other examples it would be necessary to have actually noted the temperature, age, etc., in each case.

12.3.4. *Comments*. The choice of the normal form for the prior distribution in the case just considered is convenient in that the posterior distribution is then always a member of this same family. We shall see that in other cases, too, we can find distributions for which this property holds.

On the other hand, this convenience does not justify our making such a choice if it is not compatible with our actual prior opinion; neither does it provide any *a priori* justification for regarding such distributions as in any way playing a special role. A reasonable approach involves adopting 'convenient' distributions if and insofar as they provide a sufficiently accurate representation of one's actual opinions (and this is especially useful in those problems where the precise form chosen has little influence on the final outcome).

If the influence on the final outcome is going to be practically negligible, one might even 'omit' the factor $\pi_0(\theta)$ altogether; i.e., to be more precise, one might consider the limit case of a 'constant density'. This improper distribution could be interpreted, for example, as the limit of the normal distribution $N(0, \sigma^2)$ as $\sigma^2 \to \infty$, or of the uniform distribution

$$\pi_0(\theta) = \tfrac{1}{2}a(-a \leqslant \theta \leqslant a) \text{ as } a \to \infty.$$

As we shall see, other forms of improper prior distribution may be more appropriate, depending on the form of the problem.

Sometimes the use of the improper, uniform prior distribution is interpreted as representing 'total ignorance'. This is nonsense: every distribution reflects some sort of opinion, and none of these have any special status—not even in the negative sense of representing no opinion at all. Moreover, one should note that the uniform distribution is not invariant under changes in the parametrization (e.g., θ into $\log\theta$, or e^θ, etc.). A number of useful observations of this kind can be found in Lindley, Vol. II (with particular reference to this topic, see p. 145).

Remarks. The above considerations are all dependent on a certain mathematical point which should be clearly understood, because it serves to clarify the particular practical consequences of the above.

Rigorously speaking, a *density is not just a point function but rather a function of the point and of the measure* assumed over the space under consideration. For instance, it is well known that in the case of measures defined in terms of coordinate systems a change of coordinates alters the density by multiplying it by the Jacobian (and the same thing holds more generally). It follows, for instance, that we could always arrange to have a constant density (it suffices to take the distribution corresponding to such a density as the underlying measure).

The *likelihood*, on the other hand, actually is a point function, and 'equating' it to a density is a meaningless idea. We can always achieve what we want, however, by an appropriate choice of the measure (which is never significant from a theoretical viewpoint), taking it over the most convenient reference system (or one which is sufficiently convenient reference system (or one which is sufficiently convenient) in order to make calculations as straightforward as possible.

We shall therefore find it useful (and a number of examples of this will be given) to choose a family of prior distributions with density 'equal' to the likelihood. In the terminology introduced by Raiffa and Schlaifer, these constitute the *conjugate* family for the problem. One should note, however, that this notion has no absolute meaning, but can be useful relative to some given standard formulation of a problem.

12.3.5. *The case where σ^2 is unknown.* We again consider the normal distribution, $N(m, \sigma^2)$, but now with the variance σ^2 unknown (and it is convenient to set $\theta = 1/\sigma^2$ = 'weight') and the mean m known. This case arises, for example, if one wishes to calibrate a new measuring instrument (i.e. to determine its precision as measured by $\theta = 1/\sigma^2$) by making repeated measurements of a given known quantity m.

In this case we have

$$f(x|\theta) = K\theta^{\frac{1}{2}}\exp\{-\tfrac{1}{2}\theta(x-m)^2\}. \tag{17}$$

As a function of θ (and leaving aside factors independent of θ), this is the likelihood for θ given by an observation x.

The likelihood for θ given by n observations $x_1, x_2, \ldots, x_n$ is, therefore,

$$\prod_h f(x_h|\theta) = \theta^{\frac{1}{2}n}\exp\left\{-\tfrac{1}{2}\theta\sum_h (x_h-m)^2\right\} = \theta^{\frac{1}{2}n}\,e^{-\frac{1}{2}\theta S^2}, \tag{18}$$

where $S^2 = \sum_h (x_h - m)^2$ (the constant K^n having been omitted).

Since the form of this expression is that of the density of a gamma distribution, any choice of prior from within the gamma family will ensure that the posterior distribution belongs to the same family (and the comments of Section 12.3.4 should be understood in this case too). Taking

$$\pi_0(\theta) = K\theta^{\alpha-1}\,e^{-\lambda\theta},$$

we obtain

(19)
$$\pi_n(\theta) = K\theta^{\alpha-1+\frac{1}{2}n}\,\mathrm{e}^{-(\lambda+\frac{1}{2}S^2)\theta}.$$

12.3.6. *The case where both m and σ^2 are unknown.* This arises in the context of errors of observation, as in Section 12.3.2, except that we also assume the precision of the measuring instrument to be unknown. It is also the most frequently studied case in statistics, where we have a population (of individuals, objects, experiments, etc.) in which some given quantity (X_h for the hth individual) is known (or assumed) to be normally distributed, but with neither the mean (central) value, nor the variance, known.

The example involves two parameters—those encountered separately in the previous two cases. We put $\theta_1 = m$, $\theta_2 = 1/\sigma^2$, and hence we obtain

(20)
$$f(x|\theta_1, \theta_2) = K\theta_2^{\frac{1}{2}} \exp\{-\tfrac{1}{2}\theta_2(x-\theta_1)^2\}.$$

The likelihood for θ_1 and θ_2 after having observed $x_1, x_2, \ldots, x_n$ is given by

(21)
$$\theta_2^{n/2} \exp\left\{-\tfrac{1}{2}\theta_2 \sum_h (x_h - \theta_1)^2\right\} = \theta_2^{n/2} \exp\{-\tfrac{1}{2}\theta_2[\nu s^2 + n(\bar{x}-\theta_1)^2]\},$$

where

$$\nu = n-1, \qquad s^2 = \sum_h (x_h - \bar{x})^2/\nu, \qquad \bar{x} = \sum_h x_h/n$$

(the steps are the same as those given in Section 12.3.2, except that the constant νs^2 can now no longer be omitted because it is multiplied by the parameter θ_2).

The standard assumption for the prior distribution is, in this case, the improper uniform distribution for both θ_1 and $\log\theta_2$: this results in an improper 'density proportional to $1/\theta_2$'

(over the half-plane $-\infty < \theta_1 < +\infty, 0 < \theta_2 < +\infty$).

Strictly speaking, this assumption is only made by Bayesians (and even then only by those who have no objections to improper distributions), but there is some justification for referring to it as the standard assumption because it leads to the same conclusions as those arrived at by non-Bayesian statisticians using other methods of approach.

With these assumptions, i.e. supposing that we are prepared to express the prior density in the form

(22)
$$\pi_0(\theta_1, \theta_2) = K/\theta_2,$$

we obtain

(23)
$$\pi_n(\theta_1, \theta_2) = K\theta_2^{(n-2)/2} \exp\{-\tfrac{1}{2}\theta_2[\nu s^2 + n(\bar{x}-\theta_1)^2]\}.$$

The marginal posterior densities for θ_1 and θ_2 (obtained by integrating out the other variable) are†

$$(24) \qquad \pi_n^{(1)}(\theta_1) = K\{1 + n(\bar{x} - \theta_1)^2/\nu s^2\}^{-\frac{1}{2}n} = K(1 + t^2/\nu)^{-\frac{1}{2}(\nu+1)}$$

$$\left(t = \sqrt{n}\frac{\bar{x} - \theta_1}{s}\right),$$

$$(25) \qquad \pi_n^{(2)}(\theta_2) = K\theta_2^{\nu/2} \exp\{-\tfrac{1}{2}\nu s^2\theta_2\}.$$

For θ_2, we still have a gamma distribution, just as in the case m known (Section 12.3.5). For θ_1, on the other hand, the normal distribution, which we obtained in the case σ^2 known (Section 12.3.2), has been replaced by Student's distribution (cf. the very end of Chapter 10). The effect of not knowing m and σ^2 is that they are replaced by $\bar{x}$ and s^2 (which are 'reasonable estimates' of them), and that, in the case where we are ignorant of σ^2, the normal is replaced by the Student distribution which has much fatter tails (although it tends to the normal as $n \to \infty$). For large n, therefore, the difference is practically negligible.

12.4 THE LIKELIHOOD PRINCIPLE AND SUFFICIENT STATISTICS

12.4.1. Given that we started out by adopting the Bayesian approach, and that we have adhered to it coherently throughout, the 'likelihood principle' inevitably appears to be rather obvious, and certainly not worth getting excited about. It simply states that the information available from any set of observations is entirely contained in the corresponding likelihood function. Since this is, in fact, the factor which transforms the prior opinion into the posterior, this is all we require and, indeed, all we can ask for.

If, however, one proposes *ad hoc* methods—more or less on a trial and error basis—it might well happen that they conflict with this 'principle'. For this reason, non-Bayesians have debated among themselves as to whether this principle (or a variant thereof) should be rejected, or whether, on the contrary, one should reject methods which do not comply with it (or whether such methods could be considered valid as approximations).

From the Bayesian standpoint there are two possible reasons for wishing to mention the principle: firstly, to warn against superficial interpretations of it; secondly, as a starting point for developing the topic of 'sufficient statistics'.

The warnings are the obvious ones (but, on the other hand, mistakes are often the result of overlooking the obvious), and concern a too literal interpretation of the following statement of the principle:

† For the details of the calculations, see, for example, Lindley, 5.3 and 5.4, but note that he takes $\theta_2 = \sigma^2$ (whereas by setting $\theta_2 = 1/\sigma^2$, we have obtained a certain amount of simplification in the formulae and calculations).

The information contained in a set of observations is completely summarized by the likelihood function, and, provided it can be combined with similar results, etc., it is quite sufficient to quote this.

All is well, provided we also enter the following reservation: '*so long as the basic assumptions remain unchanged*'. If, for instance, one starts off by assuming that certain errors 'are normally distributed', and then begins to wonder whether they, in fact, have some other distribution, the data expressed in the form of the 'likelihood function' can no longer provide all relevant information.

The discussion about '*sufficient statistics*' could begin by our pointing out that the likelihood function is itself a sufficient statistic (that is to say, it provides an exhaustive summary of the information contained in the data). It follows, therefore, that the summaries of the data which characterize the likelihood, when considered altogether, themselves form a sufficient statistic. In the cases which we have examined, for example, we have the following:

Section	Case	Sufficient statistic	
12.3.2	(m unknown, σ^2 known and constant)	the pair	$n, \bar{x}$;
12.3.3	(m unknown, σ^2 known but varying)	the pair	$\sum y_h, \sum y_h x_h$;
12.3.5	(σ^2 unknown, m known and constant)	the pair	n, S^2;
12.3.6	(m and σ^2 both unknown)	the triple	$n, \bar{x}, s^2$.

12.4.2. For the sake of completeness, we now give a few of the basic notions relating to the concept of a sufficient statistic.

In what follows, it will be convenient to denote the *data* (which in general consist of n observations, x_h; $h = 1, 2, \ldots, n$) simply by x; the *parameters* (no matter whether there is just one, θ, or several, $\theta_i, i = 1, 2, \ldots, s$) by θ; and the *sufficient statistic* (consisting either of a single real-valued function of the data, $t = t(x)$, or of several; $t_j = t_j(x)$; $j = 1, 2, \ldots, r$) by t: in the same way $p(x|\theta)$ denotes something of the form $p^n(x_1, x_2, \ldots, x_n|\theta)$, and so on. From a conceptual point of view, the argument is precisely the same, whether $x, \theta, t, \ldots$ are real numbers, or vectors, or whatever.

Expressing our comments about sufficient statistics in the form of a definition, we have the following (also known as the 'sufficiency principle'): $t(x)$ is a sufficient statistic for the family $p(x|\theta)$ if and only if, for any prior $\pi_0(\theta)$, the posterior distribution is the same no matter whether we condition on x or on $t(x)$; i.e. $\pi(\theta|x) = \pi(\theta|t(x))$.

A necessary and sufficient condition for $t(x)$ to be a sufficient statistic for $p(x|\theta)$ is that the latter can be written as

$$p(x|\theta) = f(t(x), \theta) \,.\, g(x), \tag{26}$$

where f and g are arbitrary functions. The necessity of the condition is obvious. From the definition, it follows that

$$p(x|\theta) = p(t(x)|\theta)p(x|t(x), \theta),$$

and this is then equal to

$$p(t(x)|\theta)p(x|t(x)).$$

If we now take the first factor as f and the second as g this is in the required form. The sufficiency part is known as *Neyman's factorization theorem.*

What we have stated above is true in the general case (i.e. x does not necessarily have to consist of the results of 'independent observations from the same distribution'). If we go back to the special case (complete exchangeability) we can pose some further problems. In this case, we may be interested in knowing, for example, whether, as n varies, we can always obtain a sufficient statistic of fixed dimension (for example, having r components: $t = (t_1(x), t_2(x), \ldots, t_r(x))$).

Ignoring the finer points, the condition for this to happen is that the family of distributions $f(x|\theta)$ is a member of the *exponential family*: i.e. that it is of the form

$$f(x|\theta) = F(x)G(\theta) . \exp\left\{\sum_{j=1}^{r} u_j(x)\Phi_j(\theta)\right\}, \tag{27}$$

where F, G, u_j, Φ_j, are arbitrary functions. In this case, a sufficient statistic, given any n observations $x = (x_1, x_2, \ldots, x_n)$, is provided by the r functions

$$t_j(x) = \sum_{i=1}^{n} u_j(x_i) \qquad (j = 1, 2, \ldots, r), \tag{28}$$

together with n (although the latter is sometimes left to be understood).

In this case, the likelihood function (for θ, on the basis of the given x) is

$$p(x|\theta) = K . G(\theta)^n \exp\left\{\sum_{j=1}^{r} t_j(x)\Phi_j(\theta)\right\}. \tag{29}$$

If the prior distribution $\pi_0(\theta)$ is proportional to the form

$$G(\theta)^a . \exp\left\{\sum_{j=1}^{r} b_j\Phi_j(\theta)\right\},$$

then the posterior distribution will also have this same form:

$$\pi(\theta|x) = K\pi_0(\theta)p(x|\theta) = KG(\theta)^{a+n} . \exp\left\{\sum_{j=1}^{r} [b_j + t_j(x)]\Phi_j(\theta)\right\}. \tag{30}$$

This explains how it is always possible to define *conjugate families* of distributions whenever we are dealing with a member of the exponential family (and the same advantages are obtained as we saw previously for the normal and gamma distributions).

Recall, however, that the concept lacks any genuine substantial foundation, as we explained in the final paragraph of Section 12.3.4.

12.4.3. The time has now come for us to explain—in line with what we said in Section 12.2.3—why we have restricted ourselves to cases in which a probability density exists.

Firstly, of course, it is quite natural to restrict oneself to the most straightforward and meaningful practical cases: these are the ones we have mentioned; either the discrete cases or those where a density exists.†

Over and above this, however, it is necessary to point out a far more essential reason; one which I do not think I have heard put forward before, nor have had occasion to mention myself. In order for inferences to be valid independently of the indeterminism that arises on account of 'probabilities conditional on events of zero probability', together with related questions concerning 'non-conglomerability' (Chapter 4, Sections 4.18 and 4.19, and Chapter 6, 6.9.5), it is necessary to confine oneself to problems which can be dealt with by using only probabilities conditional on hypotheses having non-zero probability. This happened trivially in the case of 'concentrated masses', and it happens directly in cases where a density exists, provided one assumes—as, fortunately, seems to be 'inevitable' from an empirical point of view—that knowledge of observed values x_h is not 'exact', but that, at best, it involves 'belonging to a neighbourhood of $(x_1, x_2, \ldots, x_n)$' small enough to make it possible to argue in terms of a density, but not in terms of the point itself.

12.5 A BAYESIAN APPROACH TO 'ESTIMATION' AND 'HYPOTHESIS TESTING'

12.5.1. The natural way to present the solution of any problem of statistical inference is to give the relevant probability or probability distribution. In the cases we have considered, this involved the posterior distribution given the observed data. Unfortunately, however, such a solution cannot be regarded as 'natural', insofar as it is not 'familiar' to most people. It is for this reason, perhaps, that attempts have been made to replace the posterior distribution with some sort of crude summary conveying a more immediate message.

† Or even mixed cases; a distribution admitting a density, plus a few 'concentrated masses' at particular values of interest: for example, the percentage of some given compound in an alloy when the value zero (the absence of the compound) can occur with non-zero probability.

A typical example is the problem of King Hiero's crown (to which the episode of Archimedes' 'Eureka!' refers). Was there silver in the crown? (Cf. L. J. Savage *et al.*, *The Foundations of Statistical Inference*, London, Methuen (1961).) Another example is given by the correlation between two genes (0 corresponds to their being on separate chromosomes).

Two such crude approaches to summarizing the distribution of a random quantity X are widely used: the first consists in providing a unique value $\hat{x}$, around which the distribution is concentrated; the second in providing an interval $[x', x'']$, enclosing a large proportion of the distribution. These descriptions are rather vague, but they can be made more precise in various ways and, in so doing, we obtain the various methods of *estimation*. More specifically, in the first case we refer to $\hat{x}$ as a *point estimate*, while in the second we call $[x', x'']$ an *interval estimate*. Similar considerations apply in higher dimensions (where the form of the 'interval' may be much more general).

There are other cases in which one poses the inferential question in a different way, but where one requires solutions formally similar to those given above. It may be that there is a value x_*, or an interval $[x'_*, x''_*]$, for which we wish to know whether or not X is equal to x_* (either exactly or approximately), or whether or not it lies between x'_* and x''_*. In such cases, one refers to *tests of hypotheses*, because an answer of either YES or NO is required in relation to the so-called '*null hypothesis*'

$$(X = x_*, \text{ or } x'_* \leqslant X \leqslant x''_*).$$

The contrary hypothesis, or the various hypotheses into which the complement may be divided, are known as '*alternative hypotheses*'.

12.5.2. The traditional approach to these problems, and still the most popular, is based on *ad hoc* methods, which, in contrast to Bayesian methods (based on a systematic and coherent theory), are largely rule-of-thumb.

In the present context, we wish to examine the extent to which they can be modified to fit into the Bayesian framework. In other words, we shall consider them not as separate and distinct methods leading to an alternative set of techniques, but rather as useful summaries of certain aspects of the actual, complete solution—i.e. the description of the posterior distribution.

In certain respects, it is clear that the solution will depend on some value relating to the (posterior) distribution; for example, the prevision (for a fair bet), or some other mean, or the median, etc. Such a value might well be referred to as the (point) estimate for the problem; i.e. the appropriate *mean* in the Chisini sense.

In many cases, it is clear that giving an interval in which the random quantity of interest might plausibly be thought to lie is more informative than any attempt at actually pin-pointing it. From the Bayesian standpoint, we would give an interval having some stated probability of containing X (usually a high probability; e.g. 95%, 99%: in general, $100\beta\%$). In such a case, following Lindley, we could refer to this interval $[x', x'']$ as a $100\%\beta$ (*Bayesian*) *confidence interval for* X. The qualification 'Bayesian' will be implicit in what follows, and the reader should note that a '$100\beta\%$ confidence interval' is a very different concept in a non-Bayesian context (as we shall see), and that it is important to distinguish between the two.

In general, there are infinitely many such intervals for any given level. The standard procedure is to choose the shortest one (in a certain sense, it is the most informative). In many cases—for instance, those for which the density has a unique maximum and decreases on either side of it—this interval is characterized by the fact that at each point inside it the density is greater than at every point outside it. One should note, however (in order that this criterion should not appear more 'natural' than it actu ılly is), that both the length and the density change, in general, if X is transformed into some function of itself. For example, if $[x', x'']$ is a 95% confidence interval for X, the interval $[e^{x'}, e^{x''}]$ remains such for e^X, but if the former is the interval of shortest length, the latter, in general, is not.†

12.6 OTHER APPROACHES TO 'ESTIMATION' AND 'HYPOTHESIS TESTING'

12.6.1. Those who reject the Bayesian approach cannot base their inferences on the posterior distribution even if they wished to—it does not make any sense so far as they are concerned. As a result, they are forced to have recourse to *ad hoc* criteria, and hence to open the floodgates to arbitrariness. This has led to an enormous proliferation of such techniques. For the sake of completeness, and to provide a basis for certain critical comparisons, we shall give a short account of the most important and best known of these.

The basic reason why non-Bayesians are unable to refer to the posterior distribution lies in their rejection of the use of a prior distribution.‡ The best they can then do is to base themselves on the likelihood function; failing that, they simply resort to playing with formulae that are without any real foundation.

The situation can be summarized as follows.

A method for obtaining a point estimate $\hat{x}$ given the data x_h $(h = 1, 2, \ldots, n)$, reduces in the final analysis, to providing a formula which expresses $\hat{x}$ as a function of the x_h: $\hat{x} = \phi_n(x_1, x_2, \ldots, x_n)$. (The same thing applies to finding the end-points x' and x'' for an interval estimate.) At the very beginning the choice of the criterion consists in defining a random quantity

$$\hat{X} = \phi_n(X_1, X_2, \ldots, X_n),$$

a function of the X_h, whose value $\hat{X} = \hat{x}$ is to be taken as the estimate.

† This is an obvious consequence of what we have seen more generally (cf. *Remark*, Section 12.3.4).

‡ The paper by B. de Finetti and L. J. Savage, 'Sul modo di scegliere le probabilità iniziali', which we have already quoted several times (cf. Chapter 11, footnote to Section 11.1.1), and my talk at the Saltzburg conference in 1968, published as B. de Finetti, 'Initial probabilities: A prerequisite for any valid Induction', *Synthese*, **XX**, 1 (1969), are devoted to a refutation of this, and to the clarification of various problems connected with it.

Related topics were mentioned at the conference by Vetter, Hintikka, von Kutschera and Frey; in particular, see the paper by I. J. Good, 'Discussion of Bruno de Finetti's paper', which reveals the differences in attitudes existing within the subjectivistic conception.

The problem must always be interpreted as follows (and we express it in a form which should be sufficiently vague to be acceptable to everyone): the X_h are either approximate measurements of some 'true value' x_0, which we would like to know, or they are the values of some given quantity as observed in a sample, and the value x_0 which we wish to know is some typical value (the mean, median, mode, ...) of the distribution of that quantity in the population. We seek to 'estimate' x_0 by $\hat{x}$.

12.6.2. There are, essentially, three different levels at which this problem can be formulated and dealt with.

At the very lowest level one simply ignores the probabilistic nature of the problem (or, at least, it is not taken into account in the formulation). At this level, we can only examine the formal properties of the proposed function, and judge on empirical grounds the extent to which these are appropriate. It is rare to find this approach adhered to in any systematic way, but considerations of this kind do crop up incidentally now and again (and there have been attempts to put forward abstract theories of 'methods of measurement' at this level).

The methods proposed by objectivistic statisticians are at an intermediate level. The probabilistic framework is accepted for that which takes place conditional on certain given hypotheses, but any reference to a probability distribution for the hypotheses themselves is rejected. To relate this to our previous considerations, the 'hypotheses' are the various values of the parameter θ, and what is rejected is the prior distribution $\pi_0(\theta)$ (and hence the posterior $\pi(\theta|x)$).† All that one is permitted to work with is the assumption that the X_h are stochastically independent with the same distribution, $f(x|\theta)$, conditional on each value of θ.‡

The implication of this for problems of estimation (and similarly for 'tests of hypotheses') is that the function ϕ can only be made to depend on the $f(x|\theta)$, whereas in the unrestricted (i.e. Bayesian) formulation one must also make it depend on $\pi(\theta)$.

12.6.3. One way of avoiding the difficulty is to use the Bayesian approach (either consciously or unconsciously), but omitting $\pi(\theta)$: in other words, by implicitly adopting the (possibly improper) prior $\pi(\theta)$ = constant. In this way, the conclusions obtained are necessarily valid, although it should be noted that indiscriminate use of this prior may result in its adoption in situations where neither the individual using it, nor the majority of other people, find it reasonable. Worst of all, actual contradictions can arise if

† There is no objection, however, in problems where objectivists adjudge there to be an 'objective' prior. In such cases, the approach will be the same as a Bayesian would adopt.

‡ There seems little point in complicating this brief account by extending it to include the more general cases (e.g. those with $f(x|\theta, y)$, and so on).

the approach is used independently in related problems (as a trivial example, taking first θ to have a uniform prior and then, later in the same problem, taking $1/\theta$ to be uniform).

One way of running head on into the difficulty—whilst claiming at the same time to have solved the problem—is to assert that nothing can be said concerning the probability of the statement of interest being true (e.g., that x_0 is 'close to $\hat{x}$', or that it lies between x' and x''). Having decided against overcoming this problem by the use of prior probabilities, it suffices ... to pretend that the solution we require is, in fact, a different one, and concerns the probability of the statement of interest being true conditional on the (false!) hypothesis that x_0 is known. In fact, the statements would be similar in appearance only. We could gloss over it by saying, in either version, that '*in any case*, it is almost certain that x_0 and $\hat{x}$, the true and estimated values respectively, are close to one another', as if the phrase 'in any case' had some abstract and absolute meaning, both when it refers to 'whatever the true value might be' and when it refers to 'whatever the estimated value might be'.

The fallacy in confusing the two cases is obvious. In fact, under the usual assumptions, it is almost certain that the mean of the measurements obtained from n observations will turn out to be near the true value (whatever it may be), since we are assuming the error distribution to be the same no matter what the true value is. When we consider the mean resulting from a set of given observed values, however, we are by no means entitled to conclude that the true value is almost certainly near this mean. It may well be that, finding the latter conclusion hard to believe, one considers it as much more plausible (even though *a priori* quite improbable) that, by chance, the observations have turned out to be affected by large errors acting in that particular direction.†

---

† If we call X the true value, Y the estimated value and $Z = Y - X$ the error, it is clear that the distribution of Z given $X = x_0$ is not the same thing as the distribution of Z given $Y = y_0$. If $f(x, y)$ represents the joint density for (X, Y), then, in the two cases, the distributions of Z are given by $Kf(x_0, x_0 + z)$ and $Kf(y_0 - z, y_0)$, respectively. These can only coincide if X and Y both have improper uniform densities, and Z is independent (i.e. $f(x, y) = Kg(y - x)$ with $K = 0$, in the usual sense). In the case of Section 12.3.2 (X and Z independent and normally distributed), $f(z|x)$ is independent of x by hypothesis (normal distribution $N(0, \sigma/\sqrt{n})$ but $f(z|y)$, as is shown in (16), although still normal and having the same variance, has non-zero prevision:

$$\mathbf{P}(Z|Y = y_0) = m_f - y_0 = (m_0 - y_0)/[1 + n\sigma^{-2}/\sigma_0^{-2}],$$

where y_0 was denoted in (16) by $n\bar{x}$. The term $m_f - y_0$ only vanishes if we take $\sigma_0 = \infty$; i.e. the improper uniform distribution over $\pm\infty$.

Objectivists will probably argue that, as a rule, really large errors do not occur, and that if they do one notices the fact and rejects the observation. However, the rejection of a complete, coherent formulation cannot be justified under the pretext that if something does not seem to work one can always get out of trouble by resorting to expedients which themselves cannot be justified (neither in the new, patched-up formulation, nor in the coherent one).

12.6.4. There are critics who occasionally attempt to ridicule this argument by pretending to interpret it as meaning that the difference between $\hat{x}$ and x_0 can be small at the same time as that between x_0 and $\hat{x}$ is large.† As we have stated above, we are not drawing a distinction between these two cases (there is none) but between conditioning on the hypotheses 'whatever x_0 may be' and 'whatever $\hat{x}$ may be'. Tracing this back to Bayes's theorem, what goes wrong is that those who do not wish to use it in a legitimate way—on account of certain scruples—have no scruples at all about using it in a manifestly illegitimate way. That is to say, they ignore one of the factors (the prior probability) altogether, and treat the other (the likelihood) as though it in fact meant something other than it actually does. This is the same mistake as is made by someone who has scruples about measuring the arms of a balance (having only a tape-measure at his disposal, rather than a high precision instrument), but is willing to assert that the heavier load will always tilt the balance (thereby implicitly assuming, although without admitting it, that the arms are of equal length!).

These same comments apply, essentially unaltered, to the case of hypothesis testing and to other topics (and so we shall not bother to repeat them), because they relate to the essence of the whole 'objectivistic' approach to statistics. One important consequence is the realization that objectivistic forms of significance test do not obey the likelihood principle. These are tests in which, for example, one rejects the null hypothesis $\theta = \theta_0$ because some given function $t(x)$ of the observed data (a statistic) has 'too large' a value (lying outside some given confidence interval; i.e. in the 'tails' of the distribution of t). The point concerning the likelihood principle is clear, because, for the objectivist, the confidence interval is one in which, with $100\beta\%$ probability, $t(x)$ must lie given θ_0 (and not vice-versa!). An example of this is given in Lindley, Vol. II, pp. 68–69.

These strictures do not imply, however, that the conclusions cannot, in practice, be satisfactory for most applications. Referring to our example, it will, in fact, be very rare for x_0 not to be close to $\hat{x}$. However, why should we blind our eyes to the possibility of it being otherwise? Why should we stick to the standard conclusion even in cases where we are suspicious? Why

† This is a rather imprecise objection, open to several interpretations. Only laymen (so far as this topic is concerned, anyway) could take it literally as providing evidence of an oversight. That we are dealing with two different things (cf. the explanation in the text) is clear, not only to the Bayesian, but also to objectivists of the Neyman–Pearson school. The difference is that the latter deliberately choose to base themselves on considerations of the form 'whatever x_0 may be', in order to avoid the Bayesian formulation (assuming arguments based upon the former considerations to be valid, despite the fact that they do not have the same meaning as those in the Bayesian framework, and, indeed going so far as to claim the former as 'modern', and the latter as 'old fashioned'). R. A. Fisher, on the other hand, attempted to create a fusion of the two. It seems to me that he felt the need for the Bayesian form of conclusion (although he expressed it in an illusory manner by means of an undefinable 'fiducial probability'), but wanted to approach the problem from the opposite direction (an approach rather like that of Neyman).

should we be forced to ignore facts which, if we do not wish to shut our eyes to them, should lead us to be suspicious?

In any case, a Bayesian analysis will indicate within what limits, and under what conditions, any particular method is approximately valid, and what needs to be done (following Lindley's example, perhaps) in order to turn it into an exact and acceptable procedure.

12.6.5. The method of *maximum likelihood* was developed in particular by R. A. Fisher, and, although it was known previously, it was through his work that it came to prominence.

In its crudest form, as a method of point estimation, it consists in taking the estimate of a parameter θ as the value (or vector, etc.) $\hat{\theta}$ which gives the (absolute) maximum of the likelihood for θ given by the observations x. One can give this a Bayesian interpretation as the estimate of θ given by the *mode* of the posterior distribution, assuming the prior to be uniform (since the posterior *coincides*† with the likelihood in this case, the point maximizing the former also maximizes the latter).

The most useful application of the concept is in providing a *normal* approximation to the posterior distribution (which can then, if one wishes, be used to give an interval estimate).

If we consider the standard case of exchangeability, i.e. repeated observations with the same density $f(x|\theta)$, the likelihood is the product, as we have seen many times before, and its logarithm is given by

$$L_n(x|\theta) = L(x|\theta) = \log p(x|\theta) = \sum_{h=1}^{n} \log f(x_h|\theta). \tag{31}$$

The logarithm is used simply for convenience, and the function L, to use the standard notation, is called the log-likelihood (the subscript n usually being omitted).

As n increases, the influence of the prior distribution π_0 on the posterior π_n becomes smaller and smaller, the fixed factor being overwhelmed by the n factors of the form $f(x_h|\theta)$. This was clear even in Section 12.2.5 (cf. equation (3′), and especially so in the examples we studied (cf. equation (15) of Section 12.3.2, etc.). This means that the more observations that are available, the more their influence is predominant in determining our posterior opinions, and the less significant the prior opinion becomes. This is what we would expect.

Because of this (a fact which, incidentally, has been appreciated for quite a while, and was well illustrated by Poincaré), the difficulties we mentioned

---

† Recall that it is not really correct to say 'coincides', because the likelihood is a point function, whereas the density depends on the point *and* on the measure (cf. the final remark of Section 12.3.4).

relating to the evaluation of the prior probabilities turn out to be less serious from a practical point of view. We are not saying that the problem disappears, but that it becomes possible to deal with it in a satisfactory manner by making precise the conditions and the limits within which it is possible to replace a given prior distribution by the uniform, for example, without causing any serious distortion.

In general, and under fairly weak conditions, the likelihood, for large n, is sharply peaked around its maximum, so that the maximum likelihood estimate $\hat{\theta}$ is, as it stands, quite informative (and this is true, in particular, in the case of the normal distribution). A point estimate on its own, however, is never very satisfactory, and it is fortunate that the maximum likelihood approach enables us to improve on this by also providing the variance, not of $\pi_n(\theta)$ itself, but of the normal approximation to it in a neighbourhood of $\hat{\theta}$. In fact, we have

$$(32) \qquad \sigma_n^{-2} = -\frac{\partial^2}{\partial\theta^2} L_n(x|\hat{\theta}).\dagger$$

A rough argument will suffice to show why this is so:‡ expanding $L_n(\theta) = L(x|\theta)$ around $\theta \simeq \hat{\theta}$, we obtain

$$L_n(\hat{\theta}) + (\theta - \hat{\theta})L_n'(\hat{\theta}) + \tfrac{1}{2}(\theta - \hat{\theta})^2 L_n''(\hat{\theta}) + \ldots,$$

and the density (again, in a neighbourhood of $\hat{\theta}$) is given by

$$(33) \qquad \pi_n(\theta) = K\pi_0(\theta)\,\mathrm{e}^{L_n(x|\theta)} \simeq K\,.\,\mathrm{e}^{+\frac{1}{2}(\theta-\hat{\theta})^2 L_n''(\hat{\theta})},$$

because: (a) $\pi_0(\theta)$ (in the small neighbourhood of $\hat{\theta}$ in which $L_n(\theta)$ is large) is practically constant (and equal to $\pi_0(\hat{\theta})$); (b) $L_n(\hat{\theta})$, which is constant, can be subsumed in K; (c) $L_n'(\hat{\theta}) = 0$, because L_n is maximized at $\hat{\theta}$; (d) we can neglect terms beyond those of second order.

The term $L''(\hat{\theta})$ is called the 'information' (but must not be confused with the concept as used in information theory; cf. Chapter 3, 3.8.5). The result can be extended to the case where θ is a vector, $\theta = (\theta_1, \theta_2, \ldots, \theta_s)$. The distribution is then multivariate normal, and a natural generalization of (32) defines the *information matrix* as the inverse of the variance–covariance matrix:

$$(34) \qquad I_{ij} = -\frac{\partial^2}{\partial\theta_i\partial\theta_j} L(x|\hat{\theta}_1, \hat{\theta}_2, \ldots, \hat{\theta}_s).$$

In other words, the I_{ij} give the coefficients of the $(\theta_i - \hat{\theta}_i)(\theta_j - \hat{\theta}_j)$ terms in the quadratic form $-Q$ appearing in the density $K\,.\,\mathrm{e}^{-\frac{1}{2}Q}$.

† Here, and in (34), the derivative of $L_n(x|\theta)$ is evaluated at $\theta = \hat{\theta}$.

‡ This is basically the same argument as that given in Chapter 11, 11.4.4.

12.6.6. Although our account has been very brief, it has dealt with several of the most important topics of mathematical statistics, both from the Bayesian and the objectivistic standpoints. Moreover, we have indicated the main points of departure of the two approaches.

In particular, we note that, in practical terms, the situation is altered by the fact of whether n is large (when we enter the realm of so-called *large-sample* theory), or small (*small-sample* theory). In the first case, practically any method works; whereas, in the second, different methods lead, in general, to very different conclusions. The Bayesian approach is part of a coherent, formal theory, which rules out any conceptual obscurity. On the other hand, in the case of small samples the conclusions are strongly dependent on prior opinions, and these may vary greatly from one individual to another. This is a genuine unavoidable fact, but it is not a drawback of the Bayesian approach. It would be a drawback if it were an unnecessary complication, but the fact is that if complications do actually exist the drawbacks and errors stem from ignoring them and providing pie-in-the-sky solutions which do not take them into account (as in the objectivistic approach).

Anyway, in concluding this summary I should like to quote the following words of Lindley (Vol. II, Preface, p. xii):†

> 'Most of modern statistics (*i.e. that of the objectivistic school*) is perfectly sound in practice; it is done for the wrong reason. Intuition has saved the statistician from error. My contention is that the Bayesian method justifies what he has been doing (*by reinterpreting and correcting it*) and develops new methods that the "orthodox" approach lacks.'

12.7 THE CONNECTIONS WITH DECISION THEORY

12.7.1. It is not our intention to discuss this topic at all thoroughly, nor would it be possible to do so within the limits of the present outline treatment. Had we wished to do so, however, we could have set the whole of this chapter within a decision–theoretic framework. What we shall do is to clarify some of the areas in which decision theory offers additional insights into certain of the problems of mathematical statistics, and into the comparison between the Bayesian and the more fashionable objectivistic approaches.

We mentioned this topic briefly at the end of Chapter 3 (and here and there in the sequel), where we observed that coherence required us to adopt the criterion of maximizing (expected) utility as the basis of decision making.

Basically, it tells us that we should arrive at a decision by first considering the individual increments of utility attached to the consequences of the various possible decisions, and then weighting these by the respective

† The explanations in parentheses, together with the quotation marks for 'orthodox', are not part of the original.

probabilities. A decision must therefore be based on probabilities; i.e. the posterior probabilities as evaluated on the basis of all information so far available. This is the main point to note. In order to make decisions, we first require a statistical theory which provides conclusions in the form of posterior probabilities. The Bayesian approach does this; other approaches explicitly refuse to do this.

Indeed, objectivistic approaches to statistics bend over backwards to give non-probabilistic answers to probabilistic questions, expressing them in YES–NO terms, as in the logic of certainty. More specifically, they talk in terms of 'accepting' or 'rejecting' a given hypothesis on the basis of some given test, and, although some hesitate to go this far, occasionally one hears that 'to accept an hypothesis' means 'to agree to behave as if it were certainly true'. This is nonsense. One should not behave 'as if an hypothesis were certain' unless it actually is regarded as certain. If it is not, then we cannot decide how to behave until we have attributed to it some probability p. The appropriate behaviour is then that which, on the basis of p, is calculated to maximize expected utility.

12.7.2. Some authors (notably R. A. Fisher) criticize the application of these ideas to problems of scientific inference, regarding them as essentially economic in nature and incompatible with pure research. We could object that even in the scientific field one cannot escape having to weigh up favourable and unfavourable consequences, but a more decisive reply stems from the fact that these 'economic' arguments reveal the necessity of making sure that opinions cohere. In particular, they show that one must pass from prior to posterior opinions in conformity with Bayes's theorem, and that this is the case no matter whether we are contemplating a bet, a business decision, or simply recording our conclusion for use in a scientific context.

There do not exist two entirely different forms of valid reasoning, one suitable in a commercial context, the other for pure research. No one working in the scientific field considers it beneath him to use the same arithmetic operations, or calculating machines, as are needed for commercial purposes. There is only one theory, and it does not matter whether it is used for utilitarian purposes, or for pure research, or simply studied for its own sake.

12.7.3. It is interesting to note that the movement towards a decision–theoretic point of view began within the framework of objectivistic theory. Above all, this was the result of Abraham Wald's introduction of the idea of associating a loss with an incorrect decision, taking, as an example of this, the acceptance of an hypothesis i given that hypothesis j is true (loss = L_{ij}, zero if $i = j$). However, this does not entirely remove the unsatisfactory identification of the decision as the 'acceptance of an hypothesis'. The necessary step involves singling out the individual possible 'actions'—choice among which corresponds to a probabilistic assessment—rather than acceptance of the various hypotheses. Some criteria of decision-making are taken over from other contexts, without examining closely their suitability for the problem under consideration (for example, the minimax criterion is considered acceptable, even though it corresponds to a different situation, that of competitive uncertainty—i.e. games theory). However, Wald's formulation did result in the explicit reintroduction of prior probabilities and hence Bayesian theory (albeit in a formal sense, without involving the subjectivistic interpretation).

Other movements in this direction have sprung from criticisms of various paradoxes and defects within the objectivistic framework itself. In order to remove these, it became clear that a Bayesian formulation was required. In this context, the contributions of I. J. Good, D. V. Lindley and L. J. Savage deserve explicit mention.

In addition there has been a great deal of research into the economics of uncertainty. Through the work of von Neumann and Morgenstern this gave new life and impetus to the study of utility theory, which had long been neglected (although various scholars—Daniel Bernoulli in the past, and F. P. Ramsey more recently—had shown an interest in it). These various strands of research found their culmination in the work of L. J. Savage, *The Foundations of Statistics* (1954) (so far, that is, as the theme of this book is concerned; the revision of statistical methodology from a Bayesian point of view is more recent, and is still continuing).

12.7.4. Finally, we should note that decision theory has very important things to say about questions relating to the planning of experiments for statistical purposes. In other words, planning experiments in order to improve the information on the basis of which decisions are to be taken.

One aspect of this involves the techniques of such experiments, these being studied in order to optimize the outcome; i.e. to obtain the most valid and useful information at the least possible cost. It would take us too long even to just mention the most important problems and methods, which have been extensively studied in the literature. It will suffice to simply point out that a vast amount of research has been done, and that its enormous contribution to technological progress cannot be properly appreciated unless one examines a number of examples.

So far as we are concerned, it is the more basic aspect of all this which interests us. We are referring to the fact that the reasons which make clear to us the correct form of argument for reaching useful, practical decisions, and that required for reaching conceptually valid conclusions, are the *same* in both cases.

In fact, the seemingly 'new' problem, '*what information is it most useful to obtain before making the decision?*', can be considered as it stands within our previous formulation, underlining yet again its general and comprehensive nature. It suffices to include as possible 'actions' not only those of the original formulation—i.e. those relating to the 'final decision'—but also the various possible choices of experimental procedure and model-building which lead up to it. The value of any piece of information (in the context of a particular decision problem) can be measured as the increment of expected utility deriving from it (or, in the simplest case, the increment of expected gain). This value is always positive (although it could be zero; if the worst comes to worst, we can always take the decision without taking into account the additional information), but there is usually some cost incurred (for labour, time, etc.). The net gain from the information (or, more precisely, from the decision to obtain it) is the difference between its value and its cost. The optimal decision (regarding what information one should seek to obtain) is given by that for which the difference between the value and the cost is maximized. In general, the process of collecting information may be quite complicated (performed sequentially, in a number of stages, with a built-in arrangement for subsequent choices to depend on the information obtained initially, and so on). From a conceptual point of view, our general approach can cope with all this without requiring any modification.

In this context, it is clear from a practical point of view that there is a need for coherence not only for each individual decision, but also at an overall level, linking the individual steps together. Such a requirement is perfectly obvious if problems are set out in a detailed fashion within their natural probabilistic setting, but it tends to be overlooked if one gets used to dealing with problems on a fragmentary basis.

Typical of the confusion that can arise is the statement that the 'minimax' procedure (in decision theory) is coherent. In actual fact, it is coherent for each individual application, because it turns out to be *Bayesian* under the choice of a particular prior distribution (and any one is free to choose it if they wish). It corresponds to the choice of the *least favourable* distribution; one that would be used by an opponent who wished to make things as difficult as possible for us. This analogy with games theory—more precisely, with two-person zero-sum games, i.e. those in which one person's loss is the other's gain—is often emphasized by referring to a statistical decision as a 'game against Nature'. The analogy only goes through, however, if one assumes 'Nature to be malevolent'.

Apart from any reservations one might have about this latter hypothesis,† we see at once that it cannot be applied in every case. In fact, if we simultaneously consider a number of decisions all depending on the same event,

† Some people attempt to justify it as a 'conservative policy' for anyone wishing 'to guard themselves against the risk of the worst happening to them'. The solution to this, if there is one, lies in choosing a very convex utility function, not in deliberately distorting one's opinions; this can only result in a worse decision, and is therefore unacceptable.

this approach will certainly lead to contradictions, because the least favourable distribution for one decision will not, in general, be the least favourable for the others. Nature (nor any other opponent for that matter) cannot be so evil-minded as to simultaneously adopt distributions—or '*strategies*' as they are called in games theory—which necessarily put us in a least favourable position for *any individual* decision problem that we might wish to consider.† As an obvious analogy, anyone being pursued by a number of hunters coming at him from different directions cannot escape in the opposite direction to all of them.

12.7.5. It might appear that these, our final considerations, have only been made possible by the long and wearisome journey that has gone before. In fact, this is not so. If one sticks to the approach that we have advocated throughout, all this—and let us repeat it once more, so that there is no doubt—is obvious. The time and energy was required for the long excursion which we made into objectivistic territory—a necessary journey, undertaken not as an end in itself, but in order to dispel the notion that an objectivistic formulation could constitute an acceptable, alternative approach. That journey is now over, and our work is done. Free at last from paradoxes and contradictions, we emerge from our sea of troubles.

† Anyone wishing to take seriously the hypothesis that Nature is ill-disposed towards him, should adopt a prior distribution which is least favourable *over the whole range of decisions* confronting him, and involving the circumstances under consideration. This would involve applying the minimax criterion to the single, compound problem (taking the entire complex of possible decisions as a single decision), or, alternatively (if the cases are independent), solving each individual decision one at a time, but in terms of what 'Nature's evil-minded strategy' would be over the whole complex of decisions (a very different situation from individual applications of minimax). This is the same as the distinction between minimizing a sum of functions $f(x) = \sum_n f_n(x)$—i.e. $f(\xi)$ at the value ξ where the sum obtains its minimum—and evaluating $\sum_n f_n(\xi_n)$, the sum at the individual minima (as if x could assume simultaneously—perhaps being evil-minded—the different values $x = \xi_n$).

Appendix

1 CONCERNING VARIOUS ASPECTS OF THE DIFFERENT APPROACHES

In every field, and in particular in the calculus of probability, there is scope, both hypothetically and in fact, for a number of axiomatic approaches, each of which, to a greater or lesser degree, differs from the others in various respects. It does not suit our purpose to choose just one of these, merely illustrating—even if exhaustively—that particular one; nor are we interested in presenting a somewhat wide and diverse collection from which each person makes his choice with the aid of a pin. The way that seems more appropriate, and that in any case we shall try to follow, consists in sticking to one preferred approach as a reference point, but at the same time illustrating both the variants within it that seem admissible, or necessary, and the approaches inspired by divergent views. This provides the framework for the necessary conceptual and formal comparisons.

From a conceptual standpoint our choice has already been made, and explained at some length, in Chapters 3, 4 and subsequently. At that time, we gave what might be called an axiomatic approach, but between then and now there is a difference in attitude which can be expressed (in the summary form of a single sentence) by saying that we must pass from an axiomatic approach to the *theory of probability*, to an axiomatic approach to the *calculus of probability*. This transition must not be taken as implying the existence of any distinction or separation between the two terms, or the desirability of creating such a distinction; we simply wish to draw attention to the different perspective that is obtained by emphasizing on the one hand the essential meaning, on the other the formal aspect.

The difference in perspective is the same as that which occurs when a given theory is viewed by a physicist and a mathematician. One concentrates his attention on the passage from the 'facts' to their mathematical translation; the other on the work involved in the latter. This then resolves itself into the difference between the axiomatization of a theory considered *from the point of view of its meaning*, and the axiomatization of a theory reduced *to its formal and abstract aspect*. In the first case, with reference to the example of a physical theory, the *axioms* encapsulate all those properties of an experimental nature that have been ascertained (or are assumed, perhaps hypothetically, to have been ascertained), and which suffice to give meaningful (i.e. operational) definitions of concepts and quantities, and to establish a mathematical theory to which they are subjected. In the second case, however, we omit the details and merely assume the result as our starting point: the *axioms*, independently of the meaning and validity of the physical interpretation from which one starts, are now nothing more than an expression of the mathematical nature of certain entities, and of the form of the relations among them. In this way, the mathematician can work with the axioms without worrying about those features which do not concern him *qua* mathematician. As always, the division of labour carries with it both advantages and disadvantages. A blind man with very acute hearing and a deaf man with very sharp eyesight will be able, in conjunction, to see and hear better than a normal individual, but they might 'understand' less owing to their inability to communicate. We will return to this point later.

The distinction we have just considered applies equally to the case of probability. As an axiomatic approach to the *theory of probability*, we understand the axiomatization made from the point of view of *meaning*. The latter consisted, for us, in the analysis of the conditions of coherence for bets (or something similar) on things we called 'events'; for others, it may consist of assertions either about symmetries, or frequencies, or things also called 'events', but which, perhaps, might be thought of as 'sequences of events', or whatever. In this way, one comes to impart meaning to certain words (quantities, etc.) and to establish relationships that must hold among them. As an axiomatic approach to the *calculus of probability*, we mean the axiomatization made from the *formal and abstract* point of view: we have rules with which to operate on symbols without the necessity of knowing which, if any, interpretation these rules and symbols have in the actual context.

Of course, such a contraposition is too crude to serve as anything other than a starting point; on no account must we gloss over the finer points (perhaps hidden to a superficial view, but nonetheless essential). In the choice of the mathematical axiomatization there is plenty of scope for choosing among formulations which are formally equivalent (but whose particular axioms, to those who recall the original meaning, might differ in their intuitive appeal); on the other hand, choices which are made concerning the more 'peripheral' aspects can appear rather arbitrary, and made simply for mathematical convenience.

The path we shall follow is motivated by our steadfast refusal to adopt this bad habit. In precise terms: *the axioms of the calculus of probability will be nothing more, and nothing less, than the translation into an abstract form of the conclusions which follow strictly from the practical exigencies brought to light during the preliminary discussions concerning the theory of probability.* It is useful, at this point, to clarify, in a summary and preliminary fashion, why this statement, so obvious in itself, is, instead, at odds with all those formulations, which, by following the same criterion *a little less strictly*, end up, in my opinion, by not following it at all. These clarifications will certainly not be enough to give a sufficient picture of the many factors to be taken into consideration and of their compass. However, they will enable those who bear them in mind to get to grips with the many considerations that will have to be worked out in detail, but without repeating too often, and tediously, these general motives.

We know that what we have to deal with in any case will be the characterization of certain functions **P** defined over the field of entities E, called 'events' (and then **P** is called 'probability'), or over the wider field of entities X called 'random quantities' (and then **P** is called 'prevision'). In order to carry out our task we will try to pose the formal questions concerning events in such a way as to reproduce, as faithfully as possible, the circumstances that can practically arise for events (together with variants—some important, some less so—to meet particular exigencies): similarly for random quantities. In order to define the functions acceptable as **P**, we will utilize only the conditions of coherence expressed in an abstract form.

In what way does this differ from the formulations more usually adopted at the present time? In the first place, the structure which is generally preferred is a closed, monolithic one. Rather than defining events in a general way, and then the functions **P** as extendible (in principle) to all events (either already conceived of, or conceivable in the future), one constructs on each occasion a definite, well-delimited (although possibly enormous) field of events with a particular function **P** attached to it once and for all. In terms of the standard image (in which events are thought of as sets in an abstract space), this means that the complete set-up (or 'probability space') is a *measure space* (i.e. a space with *one* particular, fixed measure). In contrast to this, the separate consideration of first the *space* (*without the measure*, or any other kind of structure), and then all the possible *measures*, not only, and most importantly, meets the needs of the subjective conception by providing $\mathbf{P}_i$ which are possibly different for each individual i ('tot capita, tot sententiae'), but also satisfies other more 'neutral' requirements (probabilities conditional on different hypotheses, or different states of information, or 'mixtures', and so on).

Moreover (independently of the previous objection, concerning **P**), this space–measure coupling gives rise to an unnatural, forced relationship between the two notions of event and probability, because it does not take account of the problems raised by the fixing of a particular function **P**. The current practice of reducing the calculus of probability to modern measure theory (countably or σ-additive, as in the Lebesgue theory)—apart from changes in terminology (set–event; measure–probability; function–random quantity; integral–expectation)—has resulted in the following:

probability is obliged to be not merely additive (as is necessary) but, in fact, σ-additive (without any good reason);

events are restricted to be merely a subclass (technically, a σ-ring with some further conditions) of the class of all subsets of the space (in order to make σ-additivity possible, but without any real reason that could justify saying to one set 'you are an event', and to another 'you are not');

people are led to extend the set of events in a fictitious manner (i.e. not corresponding to any meaningful interpretation) in order to preserve the appearance of σ-additivity even when it does not hold (in the meaningful field), rather than abandoning it.

Among other things, in the case of limiting processes and definitions of stochastic limits, this leads to the adoption of formulations that are unacceptable as they stand unless σ-additivity is imposed (at the cost of a great deal of inconvenience) as a necessary assumption at all times.

We should, of course, discuss these objections and reservations rather more fully, and go on to justify them; all the more so in that they will seem strange to those who are accustomed to the standard formulation. In fact, in the standard approach the points which do not seem to stand up to a critical examination are introduced either with the tacit suggestion that they are obvious, or they are couched in suitably seductive terms to overcome any initial reluctance to accept them.

There are other negative features of the space–measure coupling which are not related to the assumption of σ-additivity. An example is provided by the fact that *zero probability* is regarded as a property of the event in question (among other things, this sometimes leads to two events, $A \neq B$, being defined as 'equivalent' if their symmetric difference, $A\tilde{B} + B\tilde{A}$, or $A + B - 2AB$, has zero probability). Even more dangerous is the fact that *stochastic independence*—$\mathbf{P}(AB) = \mathbf{P}(A)\mathbf{P}(B)$—is considered as being a property of the events; and so on. One should beware of laying insufficient stress on the fact that it is a property of the function $\mathbf{P}$ (in relation to the events A and B), and not of the events as such (but this important distinction ceases to have any meaning if $\mathbf{P}$ is considered as given!).

In Chapter 2, we gave an account of what can be said about events from a logical standpoint (within the logic of certainty); in other words, concerning the events in themselves. We postponed until Chapter 3 anything which depended on the introduction of the function $\mathbf{P}$ defined on the events (without adding or altering anything concerning the notion of event, or the meanings of individual events). This separation was made in order to avoid any confusion early on; confusion which could have led to misunderstandings later.

Here we shall adopt the same policy, although, of course, in a deeper, more systematic and precise way. Certain distinctions which appear meaningful in other formulations no longer appear so in ours. Consider, for example, the distinctions between whether or not events are *atomic* (i.e. contain no events other than themselves and the empty one; in terms of sets of points, this reduces to those sets formed from the singletons), or between those events *belonging to either finite or infinite sets* of events, and so on. In the case of a random quantity X, having as possible values, for example, all real numbers between 0 and 1 (like X = 'percentage of time during which a given telephone link will be busy tomorrow between 9 a.m. and 5 p.m.'), let us consider the event $E = (X = 0{\cdot}4166666\ldots)$. It consists in obtaining exactly a given preassigned value, and could be regarded both as 'belonging to an infinite set' (i.e. of events $E_x = (X = x)$, $0 \leqslant x \leqslant 1$), and as an 'atomic event' (because a precise value, like $x_0 = 41\frac{2}{3}\%$, does not admit further refinement). It also belongs, however, to the field consisting of just the two events $E = (X = x_0)$ and $\tilde{E} = (X \neq x_0)$ (together with the events 0 and 1), and can be decomposed into $E = EA + E\tilde{A}$, by means of any event A not involving X (for example: A = 'it will rain tomorrow'; A = 'the party at present in government will not remain in power after the next election'; A = 'the azaleas in the window of the florist across the street will be sold today'). This can be extended to infinite subcases by considering other random quantities $Y, Z, \ldots$ (and, therefore, by considering as 'provisional atoms' the points (x, y), or (x, y, z), or $(x, y, z, \ldots)$ of $S_2, S_3, \ldots, S_n, \ldots$). It follows that any considerations put forward on the basis of these non-existent distinctions must be without foundation (an example of this is the assertion that an event E which is not impossible can only have zero probability, $\mathbf{P}(E) = 0$, if it 'belongs to an infinite set of events').

On the other hand, there exist real problems which arise in various connections with the notion of the 'verifiability' of an event; a notion which is often vague and elusive. Strictly speaking, the phrase itself is an unfortunate one because verifiability is the essential characteristic of the definition of an event (to speak of an 'unverifiable event' is like saying 'bald with long hair'). It is necessary, however, to recognize that there are various degrees and shades of meaning attached to the notion of verifiability. Some are more or less flexible: verifiable with a greater or lesser degree of *precision*; or within a shorter or longer period of *time*; or with a higher or lower level of *expenditure*; or with a greater or lesser *number* of partial verifications; and so on. Others are more precise: for example, we could consider 'absolute' degrees of precision, or 'infinite' time periods, and so on. The most precise and important, however, is that which arises in theoretical physics in connection with *observability* and *complementarity*. It seems strange that a question of such overwhelming interest, both conceptually and practically (and concerning the most unexpected and deep forms of application of probability theory to the natural sciences), should be considered, by and large, only by physicists and philosophers, whereas it is virtually ignored in treatments of the calculus of probability. We agree that it is a new element, whose introduction upsets the existing framework, making it something of a hybrid. We see no reason, however, to prefer tinkering about with bogus innovations rather than enriching the existing structure by incorporating stimulating refinements (disruptive though they may be).

It is our intention, therefore, to attempt to provide in this appendix an integrated view of questions of this kind which arise in connection with events. We should perhaps make it clear that our 'attempt' will be mainly concerned with the case of theoretical physics, and will consist of little more than a comparison of the positions adopted by various other authors, plus an indication of which position seems to us to be less open to criticism (as well as being better suited to deal with further problems concerning the verifiability of events).

There are other questions (already mentioned many times in passing) which concern the notion of 'possibility', and further aspects are revealed in cases where, either through haste or oversight (or because of one's own limitations, or because of the impossible nature of the task, or whatever), one has not drawn out all the logical implications contained† in the information in one's possession. The result of this is that the set of events considered 'certain' is not *closed* with respect to the logic of certainty.

Finally, we shall turn from the preliminary questions concerning events (and hence the logic of certainty) to the introduction of probability. It is the latter which is for us the real subject matter, the principle character as it were, and the rest is simply the setting of the scene.

We must now pass from the considerations which led us (in Chapter 3) to our basic formulation, to consider the *axioms* which constitute their translation into abstract form. The surest way of avoiding any kind of modification taking place during this translation is to directly express things in abstract form without any alterations. It suffices to preserve additivity and non-negativity. So far as the essential considerations are concerned, this rules out the attributing of a positive price (positive prevision) to a transaction (or bet) which will certainly lead to a negative outcome. From the abstract point of view, this obliges **P** to be such that we can never have

$$c_1\mathbf{P}(X_1) + c_2\mathbf{P}(X_2) + \ldots + c_n\mathbf{P}(X_n) > 0$$

† As an example of this, consider the matching problem (with n objects). It could happen that someone does not realize that the case of $n-1$ matchings is impossible (cf. Chapter 3, 3.8.4), either because he is not capable of arriving at this conclusion on the basis of the information available to him, or because it never occurred to him to doubt that all the values 0 to n were possible. It could also be that he had once known the result, but had subsequently forgotten it; or that he had not really forgotten it, but simply overlooked it at the time in question.

if

$$X = c_1X_1 + c_2X_2 + \ldots + c_nX_n \text{ is certainly} < 0.$$

These inequalities (imposed for every finite, linear combination) define (as the intersection of half-spaces) the convex set $\mathscr{P}$ of admissible functions **P** (and all that remains to be sorted out are a few details, like the possibility of substituting $\geqslant$ for $>$ in the inequalities, and so on).

On the other hand, we should point out that in expressing these conditions we have made use of, or at least made reference to, random quantities rather than events. In actual fact, writing $E_1 \ldots E_n$ in place of $X_1 \ldots X_n$ would have given practically the same condition† by introducing the X_h (which form a linear space) in an indirect fashion as linear combinations of the events (which do not form a linear space). To start directly with the linear space of the X_h (without giving any particular emphasis to events, which are, in any case, part of that space) not only enables one to deal with the whole set-up in one go, but also permits one to emphasize the adherence to the essential meaning.

Proceeding in this way, the axioms directly characterize **P** over its entire field of application: i.e. both over the field of events—where it can be given the name of *probability*—and over the field of random quantities—where it is called, more generally, *prevision* (or price, if we are dealing with practical situations).

This is a great advantage, not only from a formal point of view, but also because of the elegant simplification it provides. One avoids not merely the tiresome complication of having to consider two separate cases, but also a whole series of difficulties which stem from the fact that such complications are misleading as well as annoying. In the first place, one encounters a tiresome complication if one wishes to formulate the axioms in such a way as to deal only with events, excluding random quantities. A further complication then arises when one attempts to put right this exclusion and define prevision, taking into account that it has already been defined in the particular case of events, where it is called probability.

The obvious way, and the only possible way, of dealing with the exclusion would simply be to remove it—even though not straightforwardly—by means of some device that puts us back on the straight and narrow. It seems, however, that the first, unhappy step obliges us to continue with it in making the second step. In wishing to consider as a *definition* of prevision some relation connecting it with probability, one is led into an extremely unnatural position. In other words, one makes it appear as though the elementary notion of prevision presupposes a knowledge of something much more

† If one limits oneself to X_h having a finite number of possible values; from here one could proceed to the general case (with bounded X_h) by means of approximations from above and below.

complicated and delicate; that is, the probability *distribution* itself. Because it is unnatural, the situation is also dangerous, in the sense that it leads one to think that the definition to be made in this way, *ex novo*, allows a certain element of arbitrariness. In other words, that it requires, or permits, a choice of conventions, which are inspired by considerations of convenience.

In mathematical terms, expressed abstractly, all that we have said reduces to expressing a preference for, and then adopting, the first of the two paths open to us (which we indicate here by quoting the opening sentences of a more detailed description that can be found in Bodiou,† p. 5):

(1) Emphasis on linear functionals (Riesz, Bourbaki, L. Schwartz);

(2) Emphasis on measure (Borel, Lebesgue, Carathéodory, Fréchet, Kolmogorov).

The main thing, however, is not the conclusion we reach—i.e. the choice itself—but rather the *reasons* lying behind this choice. It is not a question of saying which mathematical formulation has the greatest merit from a mathematical point of view, but rather of saying which provides a means of interpreting most directly those things which are most directly significant, most directly important, and, above all, most directly *observable* (in a conceptual sense).

Our attitude towards the difference between the two approaches to the definition of $\mathbf{P}(X)$ (the prevision of X, usually denoted by $\mathbf{E}(X)$ = the mathematical expectation of X) can be clarified by means of an analogy (which is, in fact, exact, apart from the change in terminology). Given a solid body C, one can define its 'barycentre', $B(C)$, say, and also give an operational method of determining it, without formulae; but it will not normally be possible (nor will it be important) to discover the mass distribution of C. In particular, the notion of 'density' at a point is simply a convention, defined by a limit process which, given the structure of matter (molecules, atoms, particles), cannot, strictly speaking, make any sense. However, it can be said that if we assume the density ρ to be known, as a function of the point P, we are then able to say that the mass of the body, $m(C)$, and its barycentre, $B(C)$, must be given by:

$$m(C) = \int_C \rho(P)\,\mathrm{d}S, \qquad B(C) = \frac{1}{m(C)}\int_C P\,.\,\rho(P)\,\mathrm{d}S.$$

To summarize: the difference we referred to consists in choosing between those definitions which are direct and intuitive, and those expressed in formulae as in the example above. (Note that in the latter case we require a passage to the limit in order to define density, and then, to go back to the

† Georges Bodiou, *Théorie dialectique des probabilitiés englobant leur calcul classique et quantique*, Gauthiers-Villars, Paris (1964).

body itself, we have to do away with the density by integrating it. If there is any arbitrariness in the definition of the integral to be used, there is always the risk that some error is introduced.)

I find this undesirable habit of making simple things complicated to be very widespread at the present time (it is as if people go looking for trouble—and often they find it). I mention this not because I see it as my business to concern myself with it outside the confines of my own subject, but merely to point out that my noticing it and attempting to remedy it in the field of probability theory does not mean that I only see it as having taken root there. It happens more or less everywhere.

2 EVENTS (TRUE, FALSE, AND...)

By definition, an event must either be *true* or *false* (cf. Chapter 2, 2.3.4). It can be *uncertain* (for us, for the time being) only if, and insofar as, we do not possess the information required for establishing its truth or falsity. The same holds for any random entity; in particular, for random quantities. A 'random' quantity X is a quantity which has a well-determined value x; it could be, however, that we are not aware of what this value is (and it is because of this absence of information that it is, for us, for the time being, uncertain, and, hence, random). We can, in fact, limit our discussion to the case of events, because any information concerning X is simply information concerning some event of the form $X \in I$ (where I is any set).

But what does it mean to say that an event is either true or false? Two extreme interpretations would consist in making reference to an 'objective truth' or to 'immediate verifiability'. The latter is unobjectionable, but is extremely restrictive: it only holds in situations like that of a quiz where the answers can be found by turning to the next page. Even in this case, however, there are a number of implicit assumptions! We have to exclude the possibility of confusion or bewilderment such as would arise, for example, if every time one turned to the answers one found them different from when one last looked; or found them to be different according as one read them with the left eye or the right eye; and so on. Everyone will no doubt agree that these kinds of assumptions are ridiculous, but it should be noted that there is no logical reason for regarding them as such. One does so because they conflict with certain 'regularities' that 'objective reality' has accustomed us to. (Dually, from a solipsistic point of view, they conflict with certain 'regularities' that have guided us in our construction of our idea of 'objective reality'—in the image of what appears to us in our maybe-real-world-maybe-dream-world.)

Should we let ourselves be guided by the objective interpretation, the first of the two extremes we mentioned above? Up to a certain point this is inevitable (otherwise we would be forever in the grip of a 'ridiculous' scepticism, as in the examples above). It is necessary, therefore, to be constantly on the alert, with a critical attitude, remembering that many statements which appeared to a 'naïve' objectivism to be undoubtedly meaningful had subsequently to be modified and revised in terms of 'operational' definitions in order for it to be possible to give them a meaning.†

But when is 'objectivism' not 'naïve'? Unfortunately, the answer is far from reassuring: 'it is so up until the point when the unexpected occurrence of the contradictions or drawbacks to which it gives rise actually take place'.‡ When this happens, one has to seek a remedy, and this consists in moving as far as we can in the opposite direction. In other words, we cease to think of the 'objective' fact of something being either true or false, but rather of the fact of whether or not we can obtain the information that for us determines whether it is true or not (or, at least, whether there is a possibility'—in some sense or other—of obtaining this information).

This would lead us to regard some events as worthy of the name (since it actually makes sense to ask whether they are true or false), and others as requiring elimination (in that they are bogus—events in appearance only, non-events). If the possibility of a clear-cut separation of the two kinds of events existed (or, at any rate, were assumed to exist—possibly with some appropriate, simplifying hypothesis), there would be no problem. Everything could then remain as before (including the definition of event), with an additional warning that one should make sure that one really is dealing with events (i.e. with events that make sense—those that are verifiable).

Instead, it seems to be necessary to retain more flexibility. More specifically (although this suggestion might appear to be an unhappy expedient), it will be convenient to use the term 'event' quite freely, without any *a priori* selection and exclusion. In this way, the selectivity can be brought in later, case by case, taking into account the different requirements (sometimes clear-cut, more often vague) that arise in connection with 'verifiability', and interpreting them in the light of appropriate (albeit to some extent arbitrary) schematizations.

---

† At this point, in order to avoid confusion and misunderstandings, we should clarify the relationship between subjectivism in the field of probability, and subjectivism in relation to knowledge in general.

It is sometimes said that 'yes, of course probability is subjective . . . because *everything* is subjective'. Put this way, however, the statement is not in accordance with the subjectivistic conception of probability, and is, in fact, at odds with it. The fundamental point of the subjectivistic conception is that the notion of probability does not refer to something which is a property of the 'outside world' (and it does not matter whether the latter is regarded as an 'objective reality' or as a 'mental construct'). A solipsist, who considered all of so-called 'reality' to be 'subjective', in order to be self-consistent, and to correctly interpret the subjectivistic concept of probability, would perhaps be right in saying, instead, that probability is *objective*. It is objective in the sense that it expresses an autonomous judgment, and not something which is bound by 'external' circumstances to be interpreted in the sense of 'as if' (Veihinger's 'als ob').

‡ It is almost the same as saying that every individual must be regarded as immortal until he eventually happens to die.

It might be claimed that by adopting this approach we are begging the question, since the exclusion of that which *must* be excluded (because it is meaningless) becomes mixed up with the exclusion of that which *can* be excluded should we happen not to be interested in it. It is a fact, however, that our analysis (whether completely satisfying or not) does reveal the case of absolute unverifiability to be a limit-case of something more gradual (and, in a certain sense, 'economic'), involving different degrees of difficulty (of various kinds) in verifying whether an event is true or false. Nothing precludes one from evaluating this degree of difficulty in the light of the meaning and importance that such a verification would produce in practical terms.

One great advantage of proceeding in this way (and one which seems to me indispensable) is that our initial scheme remains the same: it includes all those things which we would have called events prior to embarking on these critical considerations, and permits us to carry out all the usual operations on them. When it comes to introducing a restriction (in a way which corresponds—within the given framework—to certain well-defined reasons), it will be sufficient to specify the subclass of events that one wishes to take into consideration (or to regard as making sense, or as verifiable, or whatever), and which other events one wishes to discard (regarding them as bogus, non-events, or as events whose meaning is unclear, or of little interest, or whatever). It may happen that in some circumstances certain given operations applied to verifiable events lead to verifiable events, but that in other circumstances they do not. There are various possibilities of this kind, and it is simply a question of noting what actually happens, rather than a theoretical question to be posed in abstract form as being an inherent feature of the concept of verifiability.

We have spoken thus far as if it were merely a question of distinguishing between genuine (i.e. verifiable) events, for which there are just two values, True and False, and bogus events, which are either not events at all, or are 'meaningless', or are 'intrinsically indeterminate'. There are cases, however, in which one discusses events for which there are three possibilities: True, False and Indeterminate (or Meaningless). This situation occurs above all in quantum mechanics in connection with the problem of complementarity, and hence of indeterminism. Having three possibilities could give rise to a three-valued logic (as, for example, in Reichenbach, 1942).

In considering (in Chapter 4) *conditional events* of the form $E|H$, we were, in fact, dealing with logical entities which could take on three values: True ($1 = 1|1$; i.e. both H and E true); False ($0 = 0|1$; i.e. H true and E false); and Void ($\varnothing = 1|0 = 0|0$; i.e. H false and hence the truth or falsity of E irrelevant). This is precisely the way in which the 'three truth values' of the above-mentioned three-valued logic are formed (with H = 'an *observation* made in order to verify whether E occurs or not'; and only after this does it make any sense to ask whether E is true or false). In actual fact, this reduction of the problem to the simple and familiar set-up of conditional events does seem to provide an adequate solution; moreover, it is especially satisfying in that it avoids any formulation which might appear to contain the germ of metaphysical infection.

There is no problem if only one considers the meaningful components making up the conditional event $E|H$ to be not H and E, but H and EH (cf. Chapter 4, 4.4.1). So far as the event E in $E|H$ is concerned, it is immaterial whether we take it to be E, or EH (the minimum possible), or $EH + \tilde{H}$ (the maximum possible), or any intermediate event $E = EH + A$, with $A \subset \tilde{H}$. To ask whether E is meaningful (and if so whether it is true or false), when H is assumed false does not make sense, when considered in relation to $E|H$. In this context, one would be considering the question of whether or not the possible residual part of the sentence made sense, or was true or false. If it were meaningful at all, this would represent the irrelevant A, which is outside the field of interest. One could, however, investigate whether, for *other reasons*, the E in the formulation adopted—i.e. its residual part A which is irrelevant for $E|H$—should be considered as meaningful and having interest outside of the hypothesis H. This is a separate problem, which concerns the event A as such, and does not have anything to do with the conditional event $E|H$, into which A enters only by the back door (like b in $5a - 0b + 2c$), or with its three logical values (for which, whatever A might be, there corresponds to $\tilde{H}$—and hence to A and $\tilde{H} - A$—always and only the same value; $\emptyset$ = Empty).

Let us now turn to a consideration of the mathematical representation of the field of events. We shall mention several variants, their appropriateness depending on the situation and on what is required. When it seems useful to do so, we shall also mention other possibilities which do not fit into our general framework of ideas. We have already provided a great deal of discussion in the text (Chapters 1–12) concerning the reasons for conflicting views on this subject; a few brief comments about certain peripheral topics should therefore suffice here. We shall try to give accurate accounts of those formulations which are not acceptable as such in terms of the approach adopted in the present work, and to bring out their worthwhile features, indicating how they might be applied in particular situations or as special cases.

3 EVENTS IN AN UNRESTRICTED FIELD

The basic set-up with which we shall begin is that already described in a summary fashion in Chapter 2, and which we have adhered to ever since, despite the occasional reservation. It serves our purpose in two ways: firstly, it is useful as it stands, in that it provides for a suitable representation and interpretation of the case which we shall regard as the most general, and, apparently, the simplest; secondly, with appropriate modifications, it provides a means of obtaining schemes for representing the other cases of interest (and some aspects of these might, in fact, appear simpler than the first case).

The simplicity of this first case lies precisely in the fact that no restrictions of any kind are imposed when it comes to forming the events: the latter could always be thought of as *arbitrary subsets* of a set of 'elementary possible cases' of a partition admitting indefinite refinements. In formal terms, we could express this more precisely as follows: 'at any given moment', the field $\mathscr{E}$ of events under consideration corresponds to the entire collection of subsets (or subdivisions) of the set (or *partition*) $\mathscr{Q}$ of the 'elementary possible cases', Q, which, 'at that moment', one wishes to single out. The partition $\mathscr{Q}$ must be considered as having no structure whatsoever, and, moreover, it must be considered as 'provisional', 'not once-and-for-all' (and this is what the references to a 'given moment', etc., are intended to convey). This implies that we can only consider as meaningful those notions and properties which are, in a certain sense, invariant with respect to 'refinements' of $\mathscr{Q}$. In more precise forms the latter are represented by subsets $\mathscr{Q}'$, for which each set of $\mathscr{Q}$ consisting of a single point Q is replaced by a set containing many 'points' Q' (in general, there could be an infinite number). To summarize: we can provisionally identify the events E of $\mathscr{E}$ with the subsets $\mathfrak{P}(\mathscr{Q})$ of $\mathscr{Q}$, and also with the corresponding subsets $\mathfrak{P}(\mathscr{Q}')$ of $\mathscr{Q}'$;— i.e. without taking too seriously the temporary interpretation of the $Q \in \mathscr{Q}$ as 'points'.

We shall give an example straightaway, in order to clarify this. Let E be the event $X^2 + Y^2 \leqslant a^2$ (where X and Y are random quantities). If $\mathscr{Q}$ is the (x, y)-plane, the event E corresponds to the disc of 'points' (x, y) with $x^2 + y^2 \leqslant a^2$; whereas if $\mathscr{Q}'$ is the three-dimensional space of points (x, y, z) (or four-dimensional, (x, y, z, t), etc.) the event corresponds to the cylinder of points (x, y, z) (or x, y, z, t), etc.) such that $x^2 + y^2 \leqslant a^2$. By considering not only the random quantities X and Y, but also others like Z and T, etc., we change the field $\mathscr{Q}$ and the notion of point (to each point (x_0, y_0) there corresponds the infinity of points on the line (x_0, y_0, z), or on the plane (x_0, y_0, z, t), etc.). The set to which E corresponds—or, conventionally, with which it is considered identified—changes, but this is an irrelevant contingency, arising from the form of representation; what does not change is the meaning of the proposition itself, which is completely contained in the inequality $X^2 + Y^2 \leqslant a^2$.

All this could have been expressed in a better way had we eliminated completely the notion of point, but it appears to be more instructive to put it forward and then to present the arguments against it. In this way, we underline the contrast between the more usual formulations on the one hand, and the refusal to accept 'closure'—as is otherwise inevitable—on the other.† On the other hand, it turns out to be useful to accept the 'points' as indicating the limit of subdivision beyond which it is not necessary to proceed (at a given 'moment'; i.e. with respect to the problems under consideration). There is just one condition: that we always bear in mind that this is only useful insofar as it helps to 'fix ideas' at the time in question. If one were to attribute to it some absolute meaning, it would lead to a confusion of the ideas, and to a tangle of misunderstandings.‡

The field $\mathscr{Q}$ (or, more generally, a field $\mathscr{S}$ of which $\mathscr{Q}$ is just a part) is often obtained by starting from some given set—which we shall call a basis $\mathscr{B}$—of events E_h, or, more generally, of random quantities X_h (this was the case in the previous example, where we started with X and Y, and then considered adding in Z and T). To think of the field $\mathscr{Q}$ (and therefore of the field $\mathscr{E} = \mathfrak{P}(\mathscr{Q})$) as having been generated from a basis $\mathscr{B}$ is completely irrelevant; it is convenient, however, to refer to this case in order to take up the theoretical discussion again, and to develop it in a more expressive manner.

In Chapter 2, and also in the example above, we have already seen how, given n random quantities X_h, the whole picture could be summed up by

---

† To approach the formulation of a theory by starting off with a preassigned, rigid and 'closed' scheme seems to me a tiresome and cumbersome procedure, wherever it is followed. (It is true that it serves to guarantee one against antinomies and suchlike, but this is not a good reason for always having recourse to it; in the same way as it is not necessary to shut oneself inside a tank in order to journey through a peaceful and friendly country.)

In connection with the use of 'points', and their abandonment in geometrical representations, we refer the reader back to our remarks in Chapter 2, 2.4.3 (especially to the quotations of von Neumann and Ulam).

Further discussion, closely relevant to this point, and (insofar as the present topic is concerned) upholding precisely the same position, can be found in Bodiou (1964, p. 3). Abstracting from the space of points, and describing the events directly as the elements of a Boolean algebra, or a Boolean lattice, he observes that 'apart from the formal simplification thus obtained, the new axiomatization is more directly interpretable in terms of the logic of the attributes.... If one assumes that each element of the Boolean algebra is a union of "atoms", one proves equivalence to the Kolmogorov axioms; *the emphasis, however, is more directly on the essential feature—the global lattice, and not the set of its atoms*'. (We should add—although it is not necessary at this point—that Bodiou does, however, retain the usual conditions, which admit σ-additivity.)

‡ The most serious such misunderstanding likely to arise is the idea that a conditional event $E|H$ has some special significance when H is 'atomic': in other words, when H corresponds to a 'point' in some given representation (although this would obviously be incomplete, since both EH and $\tilde{E}H$ must make sense, and $H = EH + \tilde{E}H$). In this way, one would be led to think that $\mathbf{P}(E|H)$ has an absolute meaning, unchanged even if some further information can be added to that expressed by H (here, $\mathbf{P}(E|H)$ is the probability of E, 'knowing all the circumstances that can influence E—and, one might add, determined up to the present moment—as expressed by H').

Considering the early stage we are at in our present attempt at a systemization, the above brief comments may seem premature. However, it is perhaps useful to have some idea of the arguments we shall have to consider, even though we shall only come across them later, while developing this treatment.

considering a single 'random point' Q in the cartesian space $\mathscr{S} = S_n$ (with coordinate system x_h), where Q is the point defined by $x_h = X_h(h = 1, 2, \ldots, n)$. Not all the points of $\mathscr{S}$ are, in general, *possible*, but only those of a subset $\mathscr{Q}$ (obtained by eliminating the cases $\mathscr{S} - \mathscr{Q}$ which, on the basis of the data of the problem, turn out to be impossible). In particular, if the X_h are events, $X_h = E_h$, $\mathscr{S}$ reduces to the set of the 2^n vertices of the hypercube (with coordinates 0 or 1), since we can only have $x_h = 0$ or $x_h = 1$. In this case, $\mathscr{Q}$ is the subset of possible vertices; in other words, the constituents (Chapter 2, 2.7.1). In the general case, nothing really changes, except that the E_h, or the X_h, may be infinite in number; the indices h will then run through some infinite set H (not necessarily countable), and even if we write the more familiar $h = 1, 2, 3, \ldots$, or simply say 'all the E_h (or X_h)', we shall mean $h \in H$.

In this way, the preceding (cartesian) representation will hold without any alteration, except that the number of dimensions (of axes, of coordinates) is infinite,† and $\mathscr{S}$ will be S_H (the cartesian space with an infinite number of coordinates, x_h, $h \in H$). In the case of events (i.e. if all the X_h reduce to events E_h) the vertices of the hypercube (in infinite dimensions) are characterized by indicating for which h we have $E_h = x_h = 1$ (for the others, $E_h = x_h = 0$). In other words, they correspond to subsets of $\mathscr{B}(\mathscr{S} = \mathfrak{P}(\mathscr{B}))$ or, equivalently, to the functions $f(\cdot)$, elements of $\mathscr{S} = 2^{\mathscr{B}}$, which to some subsets of $\mathscr{B}$ assign the value 1, and to the others 0. One easily recognizes the identical form of procedure to that which led to constituents in the case of finite n; it is a question of stating that out of the events of the basis $\mathscr{B}$, a certain subset are true, and the others false. Of course, some of the products will, in general, be impossible; i.e. demonstrably false on the basis of the data. We shall need to remove these from $\mathscr{S}$ in order to obtain $\mathscr{Q}$. If we wish, we can always reduce the case of random quantities to that of events: it suffices to substitute for each X_h the events $E_{h,x} = (X_h = x)$, for all the values x possible for X_h. Calling $\mathscr{B}'$ the modified basis which arises from this substitution for all the X_h, we can always write $\mathscr{S} = \mathfrak{P}(\mathscr{B}')$, or $\mathscr{S} = 2^{\mathscr{B}'}$.

If, having constructed $\mathscr{S}$ in this way (using one variant or another), we preserve, even if implicitly, through the x_h, the record of how $\mathscr{S}$ was generated from the basis $\mathscr{B}$, a linear space structure (or that of a subspace) remains as a trace of this in $\mathscr{S}$ (and hence in $\mathscr{Q}$). On the other hand, we might actually be dealing with a problem of geometrical probability (even of geometry, in the sense of ordinary, physical space), and hence we inevitably have the geometric structure (one could think, for example of $\mathscr{S} = \mathscr{Q}$ = surface of the earth, Q = point at which a lost—or stolen—object is located). It does not matter.

† Note (although this is not really important here) that the concept of the number of dimensions—when this is infinite—could be understood in a different way as the number of non-zero linearly independent elements (and this is actually intrinsically more meaningful): this notion no longer coincides—as in the case of finite n—with the 'number of coordinates'.

physical phenomena. More specifically, in the study of phenomena where for certain aspects the particle interpretation is appropriate, and for others the wave interpretation holds. Neither interpretation can lay claim to being universally acceptable, nor can the two be considered simultaneously without leading to 'contradictions'.

The question has been, and continues to be, a live topic of discussion; many-sided, and requiring special competence in several fields. The arguments put forward have offshoots in many directions, making it exremely difficult both to encompass them all (even if one restricts oneself to the essential points), and to single out with sufficient clarity either one topic, or a small group of them, on which one would like to concentrate attention.

The aspect which concerns us here is the logical–probabilistic one (and, in fact, for the time being, just the logical aspect, although with a view to the probabilistic side of things, for which it will serve as support). The study of this aspect could not be carried out, however, without touching upon points relating to other aspects, and without indicating the position taken up with respect to them; a position which appears to correspond to that underlying the proposed choice of approach in the logical field.

Let us, without further ado, indicate which works we shall be referring to most frequently in what follows: on the one hand, that of von Neumann, and, in particular, the exposition and development given by Bodiou, whose formulation is in the field of direct interest to us; on the other hand, that of Reichenbach, who seems to me to present the questions most lucidly from the logical and philosophical point of view.† The solution that I will put forward is a different one, but it could, in a certain sense, be seen as a simplified version of those set out by these authors.‡

† John von Neumann, *Mathematical Foundations of Quantum Mechanics*, Princeton University Press (1955) (a translation, by Robert T. Beyer, of *Mathematische Grundlagen der Quantenmechanik*, Springer, Berlin (1932)); J. von Neumann and G. Birkhoff, 'The logic of quantum mechanics' *Annals of Math.* (1937); G. Bodiou, *Théorie dialectique des probabilités* (etc.), Gauthiers-Villars, Paris (1964); Hans Reichenbach, *Philosophical Foundations of Quantum Mechanics*, University of California Press (1944).

References to these works in the present section will be indicated by means of initial and page number; R, p. 238, for example, for Reichenbach.

‡ My attitude had previously consisted in rejecting *ad hoc* interpretations in relation to quantum physics in order to reduce everything, essentially, to familiar situations (to facts which were 'complementary' in the sense that they were conditional on mutually exclusive experiments; like the behaviour of an object in two different destructive testing situations; or the victory of a tennis player in two different tournaments taking place at the same time in two different countries). A mention of this can perhaps only be found in the CIME (Centro Italiano Matematico Estivo) course given in Varenna, 1959. This solution seems to coincide with that of B. O. Koopman, *Quantum Theory and the Foundations of Probability* (1957).

Subsequent reflection (after a good deal of reading—the most relevant being that mentioned above), has not changed my original view, but rather made it more precise. In any case, it is, of course, simply an attempt at explanation (as we remarked at the beginning of the chapter) given the many issues involved, some of which may have escaped my notice.

The different solutions, or interpretations, relating to the points we have to consider here, are concerned, explicitly, with the systematization in the logical domain of those kinds of statements which, because of their association with 'anomalies' like those mentioned above, lead to confusion. In order to incorporate them, it is suggested that, in general, we must have recourse to new logical structures, different from the usual structures, such as many-valued logics, or logics with modified operations and rules (in particular, 'non-modular' logics).

The very starting points on which the analysis of these problems is based differ, however, one from the other. This difference is mainly between those who consider the problems as strictly peculiar to quantum physics, and who therefore pose the problems directly in terms of its technicalities, and those who see the problems as problems of thought in general. In the latter case, these problems could still appear more or less bound up with quantum physics, but only for contingent reasons; i.e. because they satisfy needs which actually arise in that theory (some would say 'exclusively' so, some 'mainly').

The formulation of von Neumann (vN, pp. 247–254) is strictly in terms of quantum theory, and takes as its starting point a Hilbert space (of functions ψ) in which the (linear Hermitian) operators correspond to quantities. An *event* is a quantity capable of assuming only the two values 0 and 1,† and therefore represented by a *projection-operator* E (which is idempotent, $E^2 = E$; i.e. having possible values—and eigenvalues—either 0 or 1); i.e. by a closed linear manifold $\mathcal{M}$ (that onto which E projects orthogonally). The event E is certain or impossible according as ψ belongs to $\mathcal{M}$ or is orthogonal to it; in all other cases, E has probability equal to the square of the projection of ψ onto $\mathcal{M}$. Two events are incompatible if they are orthogonal; they are simultaneously verifiable (not 'complementary') if they are commutative (in which case the logical product and logical sum are meaningful); and so on. To quote von Neumann (p. 253),

'As can be seen, the relation between the properties of a physical system on the one hand, and the projections on the other, makes possible a sort of logical calculus with these'.

The study of this kind of logical calculus (in terms of projections) has led (vN and B) to the identification of *non-modularity*‡ as the characteristic

† Let me just mention, as an interesting curiosity, that this is the same convention as I had adopted (in a paper of 1964, and now here) after much hesitation, considering it novel and perhaps unacceptable. I subsequently realized that, far from being new, it had been in use since 1932 (together with all its developments). I wonder if the fact of its not being taken up confirms my doubts about its unacceptability?

‡ See, for example, L. Lombardo-Radice, *Istituzioni di algebra astratta*, Feltrinelli, Milan (1965), pp. 332 ff.

property which distinguishes the lattice of this logic from that of standard (Boolean) logic.

A development which is inspired by the trend towards studying, in a more autonomous manner, or even completely separately, logic (and probability) on such a lattice—or on similar structures, also referred to many-valued logics—can be found in the work of Bodiou. His intentions are clearly summarized in the following passage (B. p.7): 'The primary motivation for our work, quantum theory, might appear contingent and particular, and capable of disappearing by the wayside if, by chance, quantum theory should come to be incorporated within a classical theory, which eliminates its "anomalies". This is what contemporary probabilists seem to believe and to expect. We shall attempt to show that they are wrong, and that the quantum calculus is simply a special case, imposed by necessity, of a general calculus of probability, which we call *dialectic*. This latter, far from being an unnatural growth on the body of the classical calculus, in fact subsumes it.'

Discussions of which statements and interpretations 'are or are not meaningful' are more directly considered, and more rigorously set out in Reichenbach, in a form which makes specific reference to quantum mechanics (and compares, in this context, the work of various authors), but which, from a conceptual point of view, can be adapted to any context whatsoever. For this reason, we shall develop our own analysis by using his (Reichenbach's) remarks as a guideline, putting forward our remarks as comments on his. In any case, the object of the analysis is that of finding the logical constructions which will prove suitable for resolving the difficulties in which we find ourselves; a topic which has attracted many currents of ideas from many different sources. This goal does not appear to have been achieved, nor does it appear that the efforts to reach it have opened up any promising avenues. I have the feeling that (as I said in my preliminary remarks in Section 6) the correct path is straightforward and simple, but it is my belief that it is obscured precisely by preconceived ideas about what it is that constitutes a necessary prerequisite for any logic.†

This would also appear to be a move in the direction of a natural continuation of a natural process; that of eliminating the drama from the initial state of confusion brought about by the appearance of something new, in contrast to one's accustomed way of seeing things. This has already happened for the Copernican system and for non-Euclidean geometry, for logical

† A comment seems called for at this point. My agreement or disagreement with the opinions of various authors concerns the individual points which necessarily arise in the course of an argument, and does not indicate any general position for or against. In every work there are inevitably a number of points with which the reader agrees or disagrees, either strongly or to some extent, or is in doubt about, or indifferent to, or simply does not understand. This also holds true for those works which I value to the extent of making them the basis for a discussion, an indication in itself of the stimulation I derived from them.

paradoxes and for relativity theory, and was bound to happen for the 'anomalies' of quantum physics. It seemed as if either *logic* itself was on trial, or had fallen apart completely. Reichenbach makes it clear, however, that logic, including probabilistic logic, is not to blame; on p. 102 he says:

'The rules of logic cannot be affected by physical experiences. If we express this idea in a less pretentious form, it means: If a contradiction arises in physical relations, we shall never consider it as due to formal logic, but as originating from wrong physical interpretations.'

The attributing of 'anomalies' not to the *structure of the physical world*, but rather to the 'structure of the languages in which this world can be discussed' is even more decisive. 'Such analysis expresses the structure of the world indirectly, but in a more precise way' (R, p. 177). These are the languages which, by means of *definitions*, introduce into the world of observable phenomena something which we might call 'interphenomena' (non-observables). As examples of such 'definitions', consider those which attach to the 'observed value' the meaning of being the value of the quantity *before and after*, or only *after*, and so on. As examples of such languages, consider the particle language, the wave language, and a neutral language: The first two '... show a deficiency so far as they include statements of causal anomalies, which ... can be transformed away, for every physical problem, by choosing the suitable one of the two languages. The neutral language is neither a corpuscle language nor a wave language, and thus does not include statements expressing causal anomalies. The deficiency reappears here, however, through the fact that the neutral language is three-valued; statements about interphenomena obtain the truth-value *indeterminate*' (R, p. 177). The same situation is described by Bodiou as the existence of several 'coherent formal systems', like the mechanics of points, and the wave theory of light. Two 'attributes' pertaining to different systems, like a statement in particle form and one in wave form, 'might be *incoherent* without being *contradictory*' (B, p. 11).

The appearance of the word 'indeterminate', or of the distinction between 'incoherent' and 'contradictory', indicates that in order to find something suitable for our purpose we must bring into the world a new logic. There is no difference, in principle, between the approaches of the two authors cited. The one introduces straightaway a third 'logical value', and then goes on to define the logical operations by means of 'truth tables'; the other defines the operations axiomatically, and could (it seems to me—in fact, he does not, although I think he should do so) define the 'truth values' on the basis of them.

Reichenbach distinguishes two variants according as a statement which is neither *true* nor *false* is called *indeterminate* or *meaningless*. The different names do not correspond to different meanings of the partitions into the three cases; the change in name corresponds to the case in which 'it is necessary to make an observation H in order to know whether E is true or false'. One agrees then to say that

E is *true* if observation H has given the result E,

E is *false* if observation H has given the result not-E

$$E \text{ is } \left\{ \begin{array}{c} \textit{meaningless} \\ \text{or } \textit{indeterminate} \end{array} \right\} \text{ if the observation } H \text{ has not been made.}$$

Using the notation introduced for conditional events, this turns out to be exactly what we agreed to say by writing $E|H$ instead of E, and putting $E|H = 1$ (true), or $= 0$ (false), or $= \varnothing$ (void) (and, if we wish, we could call it 'meaningless' or 'indeterminate' instead of 'void').

The meaning of the trichotomy does not depend at all on which words we use; the way in which it is defined is the only thing that matters. It might be conceded, however, that it does make some difference whether we use 'meaningless' instead of 'indeterminate'. It is a difference of philosophical attitude—an acceptance of the Bohr or Heisenberg interpretation. And there are formal consequences if one believes that a meaningless statement cannot even be mentioned, whereas by calling it *indeterminate*, and considering this, as '*an intermediate truth-value*' (R, p. 145) lying between true and false, it becomes permissible to speak of it and *above all to work with it*.

This is, in fact, the requirement that must be satisfied if something is to be called a mathematical structure or, in particular, a logical structure. It is very easy to construct such a structure. Considering the tableau on the left, there are 3^9 ($= 19{,}683$) different ways of substituting the letters T, F, I (True, False, Indeterminate) in place of the asterisks, and one can choose a subset of these to which to assign the title of operation and a symbol (e.g. $+$) to replace the symbol $\circ$ between A and B at the corner.

$A \circ B$	B: T	F	I
A: T	*	*	*
F	*	*	*
I	*	*	*

$A + B$	B: T	F	I
A: T	T	T	T
F	T	F	I
I	T	I	I

The table headed $A + B$ is thus filled with entries which are either T or F or I, placed in the 1st, 2nd, or 3rd row according as A is true, false, or indeterminate, and in the 1st, 2nd or 3rd column depending on the value of B (an example is given in the tableau on the right). Reichenbach (p. 151) introduces seven such binary operations (some of which are taken from the work of Post): disjunction and conjunction (extensions of logical sums and products), three forms of implication (the standard one, an alternative and a form of quasi-implication), two forms of equivalence (the standard one and an alternative), and three unary operations of negation (cyclical, diametrical and complete). Four of these operations are due to Reichenbach himself, but he leaves out 'some further implications' defined by Post. Variants due to other authors are also mentioned.

Without going into a more detailed discussion (which would lead on to more substantial objections), we note that all this could be expressed in terms of two-valued logic by thinking of a 'three-valued event' (in our terminology a 'conditional event', but nothing is altered if one refers to it—or thinks of it?—in a different way), E, say, expressed in the form $E'|E''$, or in that of the partition into three cases in which it is either true, false or void:

$$E^T = (E = 1) = E'E'', \qquad E^F = (E = 0) = \tilde{E}'E'', \qquad E^I = (E = \varnothing) = \tilde{E}''.$$

An E whose logical value depends on the logical values of two other 'three-valued' events, A and B, is obtained by defining as logical functions of $A^T, A^F, A^I, B^T, B^F, B^I$, both E' and E'', and E^T, E^F, E^I (should they turn out to be exhaustive and exclusive), and putting $E = E'|E''$ or $E = E^T|(E^T + E^F)$.† In this way, one avoids creating a number of symbols and names of operations and consequent rules (which are difficult to remember and sort out, and difficult to use without confusion arising). Above all, one avoids creating the tiresome and misleading impression that one is dealing with mysterious concepts which transcend ordinary logic.

Bodiou (like, of course, many previous authors whose approach and notation he follows) does not base his work on three truth-values (although, in his notation, the three possibilities for a proposition a seem to me to be expressible in the form a, Ca, $C(a \vee Ca)$, where C denotes negation). He does not even have a symbol for the 'third value' (whereas he uses u, True, and $\varnothing$, False, corresponding to our use of 1 and 0). This takes one even further from an immediate understanding of the meaning. There is also a section (B, pp. 30 ff) devoted to many-valued logics in which there are M 'truth-values', which can be denoted by $k/(M-1)$, $k = 0, 1, 2, \ldots, M-1$, but not even here is there a value equivalent to 'Indeterminate'. The work, in fact, proceeds in an entirely different direction, in which the value $W(a)$ of a proposition should have a meaning something 'similar' to 'probability' (but with its own

† Note that E'' is the same thing as $E^T + E^F$, whereas for E' it does not matter whether we take E^T or $E^T + E^I$, or any event in between ($E' = E^T + D$, with D contained in E^I).

'rules': $W(a \vee b) = W(a) \vee W(b)$, with the same for $\wedge$; fortunately, we have $W(Ca) = 1 - W(a)$).†

With this, the time has now come (for two reasons which are formally identical, but as far as interpretation goes are totally unrelated) to examine the real meaning of these questions; no longer just formally, but in depth. And it is at this point that, in our examination, we must take into consideration, together with the notion of indeterminacy, the notion of *complementarity*.

10 VERIFIABILITY AND 'COMPLEMENTARITY'

The essential problem, the basic doubt which came to the surface in the previous analysis, can be expressed formally in the following way. Suppose we have two or more 'three-valued events' (we shall consider them as conditional events, but it does not make any difference), and let us denote them by $E_1 = E'_1|E''_1, E_2 = E'_2|E''_2, \ldots, E_s = E'_s|E''_s$. Can it be meaningful and interesting to define another such, an $E = E'|E''$, whose meaning is related to the meaning of the others? And, in this case, will its 'truth-value' be a function of those of the others?

If we think of the general case (for example, of s conditional bets) it is likely that a few events (simple, two-valued ones) will be of interest; like $E_1^T E_2^T \ldots E_s^T, E_1^T + E_2^T \ldots + E_s^T$, and the similar forms with E_h^F or E_h^I (expressed by means of simple events), which express the fact that either all the events, or at least one, are won, or are lost, or are called-off (because the conditioning event did not occur).

If we think of cases of actual interest, however, it appears that we have to reconsider our whole approach to the problem, since some Es may arise which are connected with the E_h in a meaningful way, although not necessarily with their logical values. As a trivial example—but, for this very reason, instructive—let us begin by considering the (two-valued) events E'_h, and for each of them construct the (two-valued) event E''_h = 'I know (at this moment) whether E'_h is true or false'. By this means, we have transformed our field of events into a field of three-valued events, in which the third value stands for 'unknown', whereas True and False stand for 'known as true' and 'known as false'. One often emphasizes—and for good reason—that 'Indeterminate' is not to be confused with 'unknown' (see, e.g., R, p. 142). The same thing can be underlined in a rather better way by saying that the two notions coincide only in this particular example. One might say that this example corresponds to thinking of what 'I know at this moment' as frozen (I will no

† Such rules, proposed by other authors, are changed by Bodiou in a way which brings them closer to probability theory (but then, of course, *one no longer has operations on logical values*).

longer be able to learn anything about that which is now unknown to me, and I have no further interest in it; for me it will be for ever indeterminate, or even 'meaningless').

In this example it is natural to call the logical sum of two conditional events E_1 and E_2, $E = E_1 \vee E_2$, the conditional event corresponding to the logical sum $E'_1 \vee E'_2$ of the corresponding events E'_1 and E'_2. We have, therefore, $E = E'|E'' = (E'_1 \vee E'_2)|E''$, where E'' = 'I know (at this moment) whether $E'_1 \vee E'_2$ is true or false' (i.e. if at least one of E'_1 and E'_2 is true), and this does not coincide with $E''_1 \vee E''_2$ (although it necessarily contains it) since one might well know, for instance, that someone 'arrived yesterday or today', but not know which. It follows, therefore, that the event $E = E_1 \vee E_2$ thus defined is *not* a logical function of E_1 and E_2 in the sense we have seen so far (function of their logical values). E is certainly true if at least one of the E_h is (certainly) true, certainly false if they are both known to be false; but if they are both indeterminate (unknown) it could be either indeterminate or (known to be) true. (And if there are more than two of the E_h, one has the latter case if at least two are indeterminate and the others false.)

This example, as we have said, is trivial; but the cases in which considerations of this kind find an actual important application are precisely (I would even say exclusively) those modelled on the same scheme (except that 'I know ... ' is replaced by 'it has been verified that ... ', or 'it will be verified that ... '—within a certain time period, for example—and so on).

Quantum theory provides an obvious example of a case in which everything is more clear-cut. If E_1 and E_2 denote two events (or, equivalently, the respective projection-operators), an observation by means of the operator $E_1 \vee E_2 = E_1 + E_2 - E_1E_2$ is an observation for the event-sum, but not for either of them individually. Similarly, one can make an observation of $X + Y$ or XY, etc., without making observations of the two separate random quantities X and Y. The concept is an analogous one, but, if we wish to confine ourselves to projection-operators representing events, we must restrict ourselves to considering the events consisting of whether or not X (or Y, or $X + Y$, or Y, etc.) belongs to a given interval, $E = (a \leqslant X \leqslant b)$; i.e. $E = 1$ if X lies between a and b, and $E = 0$ otherwise.

When we turn to verifiability (in the various senses considered in previous sections) the situation is similar. Leaving aside the details and the finer points (we do not have to repeat them here, having dwelt upon them—perhaps at too great a length—already), it will suffice to refer to the 'trivial example' considered above, taking as 'indeterminate' that which will be 'known (not now, but) after a certain time, or after certain checks have been made, or after having obtained some given information, etc.' Here, too, 'E is indeterminate' is the statement of an objective fact; the lack of the information required in order to decide the truth or falsity of E within the specified time period, and according to the rules laid down for doing so. The difference is that the assignment of the value 'indeterminate' (and the same for 'true' and 'false') is, in this sense, not immediate (as it was in the trivial example): it is not excluded, however, as in two-valued logic (where 'unknown' is always considered as a temporary state of affairs, pending knowledge of the truth—even if the period of waiting should turn out to be in vain, or to last for ever).

Within this context, it appears to be possible to pose the problem of complementarity and to discuss its real meaning. Let us first of all observe that, in the sense we have just used it, the qualification 'indeterminate' might be attributible to an event *as of now*: i.e. when, on the basis of what we already know about the events, and about the possibilities and known means of obtaining information, we are in a position to exclude the possibility of getting to know whether E is true or false within the given time period, and according to the specified mode of doing so. Clearly, only a few of the 'unknown' events (or *possible* events, as we used to call them) will be 'as-of-now indeterminate' (in exceptional cases all of them might be, and this would reduce to the trivial example considered above). We could say, in order to be more precise, that the division (at a given moment) of the events into certain, impossible and possible (i.e. with values known, as of now, to be true, false or unknown) could be pursued further, subdividing the possible (unknown) events into five subcases depending on the situation considered as 'final'. Specifically, we can distinguish the events which will eventually be certainly indeterminate (the case I already mentioned), or those for which there is doubt between the outcomes T–I (true–indeterminate), I–F (indeterminate–false), T–F (true–false), T–I–F (true–indeterminate–false).

From these conclusions concerning single events, we can pass to the properties of two or more possible events. When we were restricting ourselves to True and False, for example, we were able to say whether two or more events were incompatible or exhaustive. This meant that although it was possible for any of the events to occur, not more than one of them actually could. In the same way, it is possible that in the case of indeterminacy similar exclusions can be made. It may be certain, as of now, that, from among two or more events, each of which might or might not in the end turn out to be indeterminate (i.e. they are all either T–I, I–F or T–I–F; none of them are I or T–F), at least one remains indeterminate, or at least one does not, and so on, and so forth.

The interesting case in practice is that of two events, one at least of which remains certainly indeterminate (but it is not known which; otherwise it is easy to couple it with another one). Two such events are called *complementary* (and a similar definition holds for quantities, as we shall shortly see). The purpose of the more general discussion given above was merely to show how the notion corresponds to a natural examination of the possibilities which present themselves when we extend the classification by introducing 'indeterminate' as a third logical value.

Complementary events arise, for example, when establishing: whether or not a tennis player wins if he takes part in one or other of two tournaments taking place at the same time in two different countries;

whether a coin will show Heads or Tails the next time it is tossed, assuming that the next toss is performed by either Peter or John;

what are, fixing them in one's mind, the registration number and the features of the driver of a suspect car that flashes past (assuming that it is at best possible to observe one or other of the two items);

what is the behaviour of one and the same object when it is subjected to one or other of two destructive tests;

whether a given building (e.g. the Tower of Pisa) will remain standing until some specified date under the assumption that some kind of repair work is carried out (or assuming some other project); and so on.

As well as events, one could equally well speak of *complementary quantities*. Referring to the above examples, we could consider the remaining life of the Tower of Pisa conditional on one or other of the hypotheses considered (of which only one will be observable—that conditional on the course of action actually chosen). Two random quantities X and Y are, by definition, non-complementary (i.e. simultaneously measurable) in the strict sense, if this condition holds for all the events $X \leqslant x$, $Y \leqslant y$, for arbitrary x, y (and, in quantum theory, this reduces to the spectral decompositions of the corresponding operators; see vN, Chapter II, and, for non-complementarity, p. 254). An amusing example, but one which well conveys the idea (in a nutshell), is the complementarity of the two measurements that a tailor would have to make simultaneously when one of them requires the client to hold his arm straight downwards, and the other requires that he hold it parallel to the floor and with the elbow bent to give a right angle.

The most celebrated example is undoubtedly that of complementarity in quantum mechanics, and there is no doubt that this is the most important case, because of the profound nature of the implications regarding our conception of the nature of phenomena and the knowledge of them that we can attain. There is also a more 'technical' and precise way of expressing the condition of complementarity for events in this case. As we have already mentioned, E_1 and E_2 are non-complementary events if, as projection-operators, they commute. Does this (together with related factors) provide sufficient justification for the idea that one has to make a *distinction of a logical nature* between complementarity in the realm of everyday affairs and in that of quantum physics? The answer would seem to be no. Otherwise, why should we not say that incompatibility—corresponding as it does in the quantum theory formulation to orthogonality—should be considered in that context as something completely different, even though it is exactly the same thing? This comment is certainly not sufficient to settle the argument; nor is the much more basic fact that we have up to now presented the notion of complementarity without having encountered any need to introduce such distinctions. We shall have to be more specific, albeit in a summary fashion, about the physical meaning of the problems under consideration, and, on the other hand, we must examine, from a critical standpoint, the arguments put forward on the basis of these physical considerations to support the opposite point of view.

As a first step, we shall simply develop the *description* of what we shall need as background for our purposes. This can then be used as a basis for the understanding of the logical situation, and also by those who are not familiar with the physical and mathematical interpretations of the schemes on which we shall base our considerations. We shall take as our starting point the remarks of Section 9, concerning the interpretation of events as projection-operators, and, in fact, we shall restate this, for the convenience of the reader, and provide an integrated and extended version by including the case of quantities. Notwithstanding the inevitable fact that many things will be glossed over, this should be sufficient, and the resulting picture should turn out to be clear and precise enough for the purposes for which it is intended.

11 SOME NOTIONS REQUIRED FOR A STUDY OF THE QUANTUM THEORY CASE

As the fundamental notion, we take a space $\mathscr{H}$ which, by means of its points, or rather vectors, provides a suitable representation of the 'states' in which a given physical system S can find itself. The space is like ordinary three-dimensional space, with all the affine and metric properties (those of analytic geometry), but is infinite-dimensional (*Hilbert* space). Its points, or vectors, represent functions (functions defined on the space of possible configurations of the system; for example, of 3 coordinates x, y, z in the case of a single free particle, and of $3N$ in the case of N distinct particles). The 'state' of the system, at a given instant, is characterized by one of these functions, its ψ-function (or ψ-point, or ψ-vector, as we call the point, or vector, which represents it in the space $\mathscr{H}$); ψ is such that the vector has modulus $\|\psi\| = 1$.

The way the system evolves in time is described by the variation of ψ as a function of time (deterministically set out by equations similar to those of classical mechanics). The difference is that these equations no longer tell one how the configuration of S varies, as it is and as it is observed, but only how the probability of finding it in this or that configuration varies if one submits it to an 'observation' at some future time t.†

We cannot 'see' the vector ψ; it is a mathematical abstraction in a space that is a mental fiction. But, by starting with the results of the last observations made (and assuming that the system S has not, in the meantime, been subjected to any external disturbance), and knowing the laws governing its evolution, we can, in principle, determine it. In any case, assuming that we knew the vector ψ, little or nothing would be known about the actual configuration of the physical system S in which we are interested, and about its evolution.

More precisely—in order to make clear what ψ is or is not sufficient to explain—let us consider an arbitrary event E, i.e. any statement whatsoever concerning the space S at a given instant (this must always be understood, even if not mentioned explicitly; two events or two quantities of the same kind, but relative to different times, are two distinct events or quantities). What we can say about E is that it is certainly *true* if the vector ψ belongs to some given linear manifold $\mathscr{M}$ (associated with E), and certainly *false* if it belongs to the linear space of all vectors orthogonal to $\mathscr{M}$. In these two cases it is superfluous to make an observation, because the answer would certainly be the one we have just given (but it would also be innocuous because it would not disturb the system at all). If, on the other hand, the vector ψ is neither contained in, nor orthogonal to, the space $\mathscr{M}$, an observation concerning E gives an unforseeable result. Knowledge of the state, which is all contained‡ in the vector ψ, does not determine this result in

† Let us just mention some of the omitted details. The functions ψ (and, in general, all the functions considered) are complex (roughly speaking, for the same reason as it is convenient to express oscillations in terms of $e^{i\omega t}$ rather than $\cos \omega t$), and, as such, considered as vectors, they have bounded moduli ($\|f\|^2 = \int |f|^2 \, dS < \infty$). The space of these functions is the *Hilbert space* with the *Hermite* inner-product ($f \times g = \int f . g^* \, dS$, where the asterisk denotes the complex conjugate; we always have $(f \times g) = (g \times f)^*$ and $|f \times g| \leqslant \|f\| \, \|g\|$). One can directly characterize an infinite-dimensional linear metric space as a *Hilbert* space by adding the properties of completeness and separability. With a system of (orthogonal, etc.) cartesian coordinates, it is the space of points defined by sequences of coordinates $x_h (h = 1, 2, \ldots, n, \ldots)$ such that $\sum_h |x_h|^2 < \infty$ (and this expression gives the modulus of the vector with coordinates x_h; the Hermitian inner-product of two vectors is given by $\sum_h x_h y_h^*$). The linear operators that we shall come across are also Hermitian (or self-adjoint: $A^* = A$, where A^*, the adjoint of A, is defined by $A^* f \times g = f \times Ag$); the operators can be represented by matrices (with reference to an orthogonal cartesian system) with entries A_{rs} (and then $(A^*)_{rs} = (A_{sr})^*$; A is Hermitian if $A_{sr} = (A_{rs})^*$). See vN, especially pp. 34–46, and for a more direct exposition and interpretation, E. Persico, *I fondamentali della meccanica atomica*, Zanichelli, Bologna (1936).

Let the above serve to give an idea of the various detailed specifications which, like the present one, would be out of place in the main text if they were to give the reader the impression that he has to acquire a knowledge of these notions, or to refresh his rusty memory, or to worry about the details, in order to understand those few points on which his attention would be better focused. And let it serve also as a warning for those who were tempted to accept the present formulations literally, or to be put off at finding them incomplete.

‡ Cf. the comments which are made later, following the discussion of the possibility of explanations introducing 'hidden parameters'.

advance, but it is not without value, because it gives all that can be given: that is, the *probability*. The details are as follows: we decompose the vector ψ into two components, one parallel to $\mathscr{M}$, the other orthogonal; i.e. into $\psi = E\psi + \tilde{E}\psi$ (in this way indicating the projection E onto $\mathscr{M}$, and the orthogonal projection, $\tilde{E} = 1 - E$, onto $\mathscr{H} - \mathscr{M}$). The probabilities of the E and $\tilde{E}$ which result are given by the squares of the respective projections:

$$\mathbf{P}(E) = \|E\psi\|^2, \qquad \mathbf{P}(\tilde{E}) = \|\tilde{E}\psi\|^2.$$

Note that, instead of $\|E\psi\|^2 = E\psi \times E\psi$ (1st form) we can also write $E\psi \times \psi$ (2nd form; which equals $E\psi \times (E\psi + \tilde{E}\psi)$, whereas $E\psi \times \tilde{E}\psi = 0$), which is also valid even if E is not a projection operator, and has a similar meaning even in this case (as we shall see).† In the case where E is a projection-operator, one verifies immediately that $\mathbf{P}(E) + \mathbf{P}(\tilde{E}) = 1$, as was necessary. In fact (because of the orthogonality of the two components, and hence by Pythagoras), we have

$$\mathbf{P}(E) + \mathbf{P}(\tilde{E}) = \|E\psi\|^2 + \|\tilde{E}\psi\|^2 = \|\psi\|^2 = 1.$$

The interpretation of E as a projection-operator turns out to be even more expressive, however, in the light of the following. The most important fact is that the vector ψ is not restricted to a passive rôle of indicating the probability of the required result being observed. The observation itself is forced to choose whether to fall into the space $\mathscr{M}$ corresponding to E, or into the orthogonal space $\mathscr{H} - \mathscr{M}$ corresponding to $\tilde{E}$ (and it does so with the probabilities indicated). The outcome simply constitutes the information about the choice made. From the position taken up after the jump we have obliged it to take, the system returns to an evolution according to the previous laws until there is a new disturbance.

This picture (anthropomorphic, but this perhaps helps one grasp the ideas in the absence of a more detailed technical exposition) contains 'in a nutshell' all that is required for a complete treatment. It will suffice, essentially, to consider simultaneous questions about several events, rather than a single one (this will also hold for measurements of quantities), and to distinguish the cases which give rise to the circumstances to be discussed.

Instead of a partition into two opposite events (E and $\tilde{E}$) we can think of a 'finer' partition, still 'complete', into some number n of incompatible events, $E_1, E_2, \ldots, E_n$, or even into an infinite number. Each E_h will be defined by the corresponding (closed) linear space $\mathscr{M}_h$, and all these spaces must be taken to be orthogonal to each other, and such that taken altogether they

† The equivalence between the two forms does not hold in the general case. There, in fact (cf. the last footnote but one), we have $A\psi \times A\psi = A^*A\psi \times \psi$, and, in the case which interests us (the Hermitian form), this equals $A^2\psi \times \psi$. In order that it equals $A\psi \times \psi$, we must have A idempotent; i.e. A = projection-operator (with all the eigenvalues idempotent, $\lambda^2 = \lambda$; that is, $\lambda = 1$ or $\lambda = 0$).

form the entire space $\mathscr{H}$ (i.e. there must not exist a vector in $\mathscr{H}$ orthogonal to all the $\mathscr{M}_h$, and then there will not exist vectors which are not linearly dependent on the vectors of the $\mathscr{M}_h$). The total dimension is countable, being coincident with that of $\mathscr{H}$. It follows that, in the case of a finite partition, at least one of the $\mathscr{M}_h$ must be infinite-dimensional (in particular, note that for E and $\tilde{E}$ either one is infinite-dimensional or both are). In the case of a countable partition, this is not necessarily the case, and we can even consider the extreme case where all the spaces $\mathscr{M}_h$ are one-dimensional.

This is the fundamental case in terms of which the discussion of all the others is framed. In other words, it is the case in which we have a system of orthogonal cartesian axes corresponding to an infinite set of events $E_h(h = 1, 2, \ldots, n, \ldots)$, interpretable, for example, as the (distinct) values, λ_h, which a quantity Z can assume: $E_h = (Z = \lambda_h)$. On the other hand, we clearly have $Z = \sum_h \lambda_h E_h$ (as a random quantity), since one and only one of the E_h will occur (and will take the value 1; all the others will be 0), and the sum will reduce to the corresponding value λ_h. Defining Z as an operator (associated with the quantity of the same name) in the same way, it seems clear from the identity of the written forms that one is dealing with the operator formed by multiplying the axis vectors (functions) E_h (the eigenvectors or eigenfunctions) by the λ_h (the eigenvalues). This gives the prevision (or mathematical expectation) of the quantity Z by means of the same formula used for E (2nd form):

$$Z\psi \times \psi = (\textstyle\sum_h \lambda_h E_h)\psi \times \psi = \sum_h \lambda_h (E_h \psi \times \psi) = \sum_h \lambda_h \mathbf{P}(E_h) = \mathbf{P}(Z).$$

In a similar fashion, one obtains, immediately, the distribution function of Z: putting $E_z(\lambda) = \sum E_h(\lambda_h \leqslant \lambda)$, one has the event $E_z(\lambda) = (Z \leqslant \lambda)$, or the related projection-operator; it follows that

$$E_z(\lambda)\psi \times \psi = \mathbf{P}(E_z(\lambda)) = \mathbf{P}(Z \leqslant \lambda) = F_z(\lambda),$$

where we denote by F_z the distribution function of Z.

The collection of projection-operators $E_z(\lambda)$ (or the set of their linear spaces, each of which, if we proceed in the direction of increasing λ, contains all the preceding ones) defines the spectrum of Z; in this case, a discrete spectrum (there are a countable number of values of λ_h).

Going back to the physical problem, we can repeat what was said in the case of an event. If ψ belongs to one of the axes, and only then, the corresponding event E_h will certainly be true; i.e., thinking of the quantity Z, its value will be λ_h with certainty. It is unnecessary (but harmless) to make an observation. In all other cases, the vector ψ will be forced (if we make an observation on all the E_h together, i.e. on Z) to choose along which axis it is to lie: the result (E_h, or λ_h) indicates which choice was made, and, after the jump, we go back to the normal evolution.

There is one difference, and a very important one, with respect to the previous case. Now we know exactly,† after the observation, the position chosen by ψ (whereas, beforehand, we knew only that it belonged to $\mathcal{M}$, or alternatively to $\mathcal{H} - \mathcal{M}$). This happens only in the case now under consideration; that of a partition which gives rise to spaces which are all one-dimensional. When one defines on them a quantity Z, it is necessary that the λ_h values (the eigenvalues) are all distinct (simple); otherwise, the refinement of the subdivision we have reached will in part be destroyed (one has the case of 'degeneracy'). For the case we are dealing with (the non-degenerate case) one says—for obvious reasons—that a 'maximal observation' has been made.

We now turn to the problem of complementarity of observations.

Can we ask for the simultaneous verification—i.e. with one and the same observation—of two or more events? Or for the measurements of two or more quantities? (And we note that 'simultaneously' can only mean 'with one and the same observation'; another observation made immediately afterwards would already find ψ changed by the effect of the first one.)

The answer is obvious if one thinks of two 'maximal observations', like the measurements of two quantities Z' and Z'', to be performed simultaneously. In doing so, we force ψ to lie on one of the axes of the first system and also on one of the second system. Now ψ obeys any order whatsoever, but cannot accept contradictory orders—and this would be the case if the two systems of axes do not coincide. In such a case, Z' and Z'' cannot be measured simultaneously; i.e. they are *complementary*.‡ If the axes do coincide, the result is trivial, because Z' and Z'' are functions one of the other (if Z' assumes the value λ'_h, it means that the hth axis has been chosen, and hence that Z'' takes on the value λ''_h). The coincidence of the system of axes implies commutativity (in terms of operators, the condition is $Z'Z'' = Z''Z'$), and the same also holds in the case of events, $E'E'' = E''E'$, or of non-maximal quantities, $XY = YZ$. Non-maximal quantities X, Y relating to one and the same system of axes (suppose it to be that of Z) are obtained by taking eigenvalues μ_h and ν_h, which are not all distinct (so that X and Y as functions of Z, $X = f(Z)$, $Y = g(Z)$, are not invertible). Conversely, if X and Y are not complementary, and have as possible values the μ_i and ν_i, respectively, one can construct a Z (corresponding to a maximal observation) of which X and Y are functions, and having distinct values λ_h corresponding to all the compatible pairs (μ_i, ν_j). In particular, in the case of events, non-complementarity, $E'E'' = E''E'$, means that the corresponding spaces $\mathcal{M}'$ and $\mathcal{M}''$ are mutually orthogonal

† The ψ is, in fact, uniquely determined, because a multiplicative constant (real or complex) is irrelevant.

‡ Think of the example of the tailor: complementary measurements are those which to be made simultaneously would require the client to simultaneously assume several different, incompatible positions!

(i.e. if we call $\mathcal{M}$ the intersection, $\mathcal{M} = \mathcal{M}'\mathcal{M}''$, then two vectors, one from $\mathcal{M}' - \mathcal{M}$, and one from $\mathcal{M}'' - \mathcal{M}$—i.e. from $\mathcal{M}'$ and $\mathcal{M}''$, respectively—and orthogonal to $\mathcal{M}$, are always orthogonal to each other; then, in fact, and only then, is the product of the two projections the projection onto the intersection $\mathcal{M}$, and does not depend on the order). As special cases, one has the case of inclusion (if $\mathcal{M}' - \mathcal{M} = \{0\}$, we have $\mathcal{M}' \subset \mathcal{M}''$, $E' \subset E''$), and that of incompatibility ($\mathcal{M} = \{0\}$, $E'E'' = 0 =$ impossible).

Non-complementarity between X and Y can be interpreted in the same way because it is equivalent to the non-complementarity of each of the events (projection-operators) $E_x(\mu)$ and $E_y(\nu)$; in other words,

$$E_x(\mu)E_y(\nu) - E_y(\nu)E_x(\mu) = 0$$

for any μ and ν whatsoever. Geometrically, this is equivalent to the orthogonality of the spaces $\mathcal{M}'_\mu$ and $\mathcal{M}''_\nu$ (i.e. orthogonality between the vectors of $\mathcal{M}'_\mu - \mathcal{M}'_\mu\mathcal{M}''_\nu$ and of $\mathcal{M}''_\nu - \mathcal{M}'_\mu\mathcal{M}''_\nu$). One could consider the same condition in a weaker version (limiting ourselves to checking the validity for certain values of μ and ν instead of for all of them), but we shall consider this in the context of 'continuous spectra', where it is more interesting.

The case of the 'continuous spectrum' arises with a quantity which can assume any value (between $-\infty$ and $+\infty$, or in an interval, etc.), rather than just a finite or countable set of values as considered so far. In quantum physics one deals with non-quantized quantities (like the coordinates) in addition to the quantized ones (like energy).

In this case, too, in considering quantities $X, Y, \ldots$, everything can be expressed by $E_x(\mu)$, $E_y(\nu), \ldots$, except that an $E_x(\mu)$ will actually vary for all increments in μ (and not just in going through certain values of μ; the eigenvalues $\mu = \mu_i$), and the distribution function

$$F_x(\mu) = \mathbf{P}(X \leqslant \mu) = E_x(\mu)\psi \times \psi$$

will, in general, turn out to be continuous.† A decomposition into a finite or infinite number of incompatible events could be obtained by dividing the axes of the μ in some fashion into intervals $\mu_i < \mu < \mu_{i+1}$ ($i = 0, \pm 1, \pm 2, \ldots$, in order to denote them in increasing order, and letting them be, in general, unbounded in both directions). Only in this way can we have a partition

$$E_i = (\mu_i < X \leqslant \mu_{i+1})$$

that gives us, in the way we indicated previously, an 'approximate measurement' $\hat{X}$ of X, defined by choosing a subdivision μ_i, and in each interval a value $\hat{x}_i$, and then setting $\hat{X} = \sum_i \hat{x}_i E_i$. In other words; $\hat{X}$ is the function of

† We ignore the mixed case of probability in part concentrated, in part continuous, etc. (cf. Chapter 6, 6.2.2–6.2.3).

X defined by the step-function

$$f(X) = \sum_i \hat{x}_i(\mu_i < X \leqslant \mu_{i+1}).$$

From this point of view, X is not a measurement with absolute precision, but with arbitrarily high precision if we substitute for it an $\hat{X}$ defined by a function with arbitrarily small steps. An observation on $\hat{X}$ forces ψ to lie in one of the spaces $\mathscr{M}_i$ (corresponding to $E_i = E_x(\mu_{i+1}) - E_x(\mu_i)$), and it is never maximal because the subdivision could always be made finer.

If we now consider X and Y (both with continuous spectra), the condition for non-complementarity is still the commutativity of X and Y as operators, $XY - YX = 0$; in other words, commutativity between the $E_x(\mu)$ and the $E_y(\nu)$. If the condition holds, then, apart from obvious complications, what was said in the case of the discrete spectrum also applies here: for example, it is also true in this case that one can construct a Z of which X and Y are functions, but that one can only obtain it (clearly) by means of procedures based on the Peano curve or something similar (vN, p. 178).

If, however, we content ourselves with approximate measurements, $\hat{X}$ and $\hat{Y}$, then the orthogonality of $E_x(\mu_i)$ and $E_y(\nu_j)$, relative to the points of subdivision chosen for the μ and ν, is sufficient for non-complementarity, and this weaker condition may also hold if X and Y are complementary. In other words, complementarity does not necessarily exclude the possibility of simultaneous approximate measurements; i.e. of two suitably chosen $\hat{X}$ and $\hat{Y}$.

And at this point we arrive at the special case of quantum mechanics, where complementarity often arises in the particular guise of non-commutativity, expressed by

$$XY - YX = h/2\pi i \quad (h = \text{Planck's constant}).$$

This holds where X and Y are coordinates, and for a conjugate impulse, or, more generally, in the terminology of classical mechanics, for 'canonically conjugate' quantities.

From this relation of non-commutativity (and hence complementarity between X and Y) we can derive a justification of Heisenberg's Uncertainty Principle, which indicates the way in which the precision of the measurements of X and Y—which can be made arbitrarily high if performed separately—turn out to have a reciprocal relationship under a simultaneous observation.†

The following is just a brief development of this crucial point. From $XY - YX = a$, it follows that

$$(XY - YX)\psi \times \psi = a\psi \times \psi = a\|\psi\|^2 = a \quad (\text{if } \|\psi\| = 1).$$

† The procedure which is briefly outlined here is taken from vN (p. 230, ff.), and is there (note 131, p. 233) attributed to ideas of Bohr, and work of Kennard and Robertson.

We note also that

$$XY\psi \times \psi = Y\psi \times X\psi \leqslant \|Y\psi\| \,.\, \|X\psi\|,$$

with a similar result for YX. It follows (by the triangle inequality!) that the bound for the difference is given by

$$(XY - YX)\psi \times \psi \leqslant 2 \,.\, \|Y\psi\| \,.\, \|X\psi\|;$$

i.e.

$$\|Y\psi\| \,.\, \|X\psi\| \geqslant \tfrac{1}{2}|a| = \frac{h}{4\pi}.$$

The inequality holds no matter where we take the origin for X and Y, and, in particular, if we take it at the mean value. The two moduli then have an interpretation as standard deviations, and we have the uncertainty principle in its usual form: $\sigma_x\sigma_y \geqslant h/4\pi$.

This means that it is impossible to approximate X and Y by means of some $\hat{X}$ and $\hat{Y}$ by choosing arbitrarily small 'steps' for both of them: their order of magnitude must be such that the product (as an order of magnitude) is not less than h. Geometrically, the subdivision into rectangles in the (X, Y)-plane, a consequence of the subdivision by means of the μ_i and ν_j, respectively, on the x-, y-axes, cannot be so fine as to give rectangles whose areas have orders of magnitude less than h (the choice of the ratio of height to width remaining arbitrary). These rectangles are the regions for which it can be 'verified' whether the pair of measurements fall inside or not: $(\hat{X} = \hat{x}_i)(\hat{Y} = \hat{y}_j)$ is, in fact, equivalent to $(\mu_i < X \leqslant \mu_{i+1})(\nu_j < Y \leqslant \nu_{j+1})$. Observe that this is precisely one of the conditions of 'bounded precision' considered in Section 8 (cf. Figure A.1).

On the basis of the digression, which we are now about to end, we might, at this point, take up again the discussion of the logical aspects of indeterminacy. However, let us first take advantage of the opportunity which has grown out of the discussion concerning the precision of a measurement in the quantum theoretic field, in order to examine the question in relation to the considerations we put forward about the subject in general (in Sections 7 and 8). We have just pointed out the similarity between relations of indeterminacy and bounded precision in terms of area; more fundamental, however, is the analogy between the case of 'unbounded precision' (considered in Section 7) and the situation presented (following von Neumann) for the measurements of non-quantized quantities (with continuous spectra). These quantities (vN, p. 222) '... could be observed only with arbitrarily good (but never absolute) precision', in contrast to what happens with the '... introduction of an eigenfunction which is "improper", i.e., which does not belong to Hilbert space', a procedure which '... gives a less good approach to reality than our treatment here. For such a method pretends the existence of such states in which quantities with continuous spectra take on certain values exactly, although this never occurs.' These critical comments are directed towards procedures which make use of the Dirac function (and they are repeated very frequently). In this connection, I think it appropriate to indicate its relation to the point of view that we are following, both to avoid misunderstanding and to make things clearer. I sympathize with von Neumann's attitude, not when he seems to be inspired by scruples of mathematical rigour, and attacking imprecise definitions (because, generally speaking, formal imperfections can always be removed), but when he shows care in not attributing absolute certainty and precision to a quantity without really good reason. I approve even more strongly of the observation (as he adds in note 126, p. 222) that not even attributing X to one of the intervals of the subdivision can be considered as certain, except as an idealization: 'Nevertheless,' he concludes, in an admirably undoctrinaire manner, 'our method of description appears to be the most convenient one mathematically at least for the present.' 'I sympathize', also, in the sense that I would like quantum physics to make room for this elegant example of contraposition; of quantities which are or are not quantized, which are, respectively, precisely measurable or not. In any case, it cannot, of course, be a matter of taste, whether mathematical or philosophical, and if the opposite formulation should, on a closer examination, turn out to correspond more closely to a meaningful physical interpretation then it must be welcomed with open arms.

12 THE RELATIONSHIP WITH 'THREE-VALUED LOGIC'

Let us now go back to three-valued logic and to the related conceptual questions that are raised by quantum mechanics.

So far as the nature of 'three-valued logic' is concerned, we came to the conclusion that the 'three values' correspond well to the requirements of applications—quantum-theoretic or otherwise—but that they do not give rise to a 'logical calculus', because the most meaningful considerations are not connected with operations which could be performed on such 'values'. If one examines the actual situations directly, the preconceived choice of a machinery consisting of formal operations similar to those of ordinary logic does not appear to be appropriate as the unique way of constructing a formulation which is to replace it.

The best proof is provided by the many-valued logic—'similar to the calculus of probability'—which Bodiou mentions and develops. In order to have the satisfaction of finding a true relationship (in the calculus of probability), he has to put together two things which are falsely defined (in the calculus of probability), and which, on the other hand, cannot be modified if one wants them to be expressed as 'logical functions' (and the final consolation lies in the observation that they only work if one has either probability

0 or 1, in which case formal, two-valued, logic, without probability, is essentially sufficient).†

The most important point to be examined is the reason behind the different attitudes (already mentioned in Section 10) of those who regard 'Indeterminism' as a concept specifically and exclusively belonging to quantum physics, and those who see no distinction *of a logical nature* between this case and the world of everyday affairs (although nobody denies the very important and significant differences which derive from the physical and mathematical structures peculiar to the quantum-theoretic set-up).

Von Neumann, in speaking about the 'sort of logical calculus' to which the projection-operators gives rise, says of this calculus that '... in contrast to the concepts of ordinary logic, this system is extended by the concept of "*simultaneous decidability*" *which is characteristic for quantum mechanics*' (vN, p. 253).

It seems that such sentences have no practical implication and, therefore, no actual content. So far as von Neumann is concerned, it may be that he never examined the possibility of examples of a different kind. This is not the case, however, with Reichenbach (as we shall see); indeed, one might think that he was constantly preoccupied with such dilemmas (like waves versus particles, values before and after, etc.). These dilemmas were, instead, clearly resolved by von Neumann, as is shown, for example, by the following remark (note 148, p. 282);

'In contrast with this, however, it is to be noted that quantum mechanics derives both "natures" from a single unified theory of the elementary phenomena. The paradox of the earlier quantum theory lay in the circumstance that one had to draw alternately on two contradictory theories (electromagnetic theory of radiation of Maxwell–Hertz, light quantum theory of Einstein) for the explanation of the experience.'

The attitude of Bodiou—see the previous quotation (B, p. 7)—also seems to stem from having overcome these distinctions. On the other hand, it seems

† Perhaps this 'many-valued logic' might be useful in other cases, and in other senses, without reference to probability. For example, by giving a proposition a a certain value $V(a) = k$, if, in a system with a given set of ordered axioms $A1, A2, \ldots, An$, the proposition is decidable (true or false) on the basis of $A1, A2, \ldots, Ak$, but not using only $A1, A2, \ldots, A(k-1)$. The scheme as it stands does not even work in this case, but, in a certain sense, we get closer.

So far as non-modularity is concerned, one can observe that modularity no longer holds in our scheme, $(E|H)$ (or in similar ones), when the 'truth-value' (Void or Indeterminate) is considered greater than 0 (False) and less than 1 (True), and a scale of intermediate values (in various possible senses, e.g. probability), which are considered not comparable with $\varnothing$, are inserted between these two values. The most natural convention would be that of taking $\mathbf{P}(E|H)$ as the value, putting $\mathbf{P}(E|H) = \varnothing$ if $H = \varnothing$; possibly using 0^* and 1^* to distinguish the certainly False and certainly True cases (obtaining the partially ordered set of values $0^* < 0 \leqslant p \leqslant 1 < 1^*$, $0^* < \varnothing < 1^*$, $\varnothing$ not comparable with values $0 \leqslant p \leqslant 1$). It is not clear to me whether this has any connection with the appearance of non-modularity—in many different ways, some not immediate—in the treatments given by von Neumann and Birkhoff, and Bodiou.

an inevitable progression to attain more and more comprehensive views, which remove from their isolation those things which, when they first appeared, seemed abnormal.

What is the difference then, from a logical point of view, between the complementarity or non-complementarity of measurements in the case of a physicist and in that of the tailor (whom we met above in our trivial example)? Or among the examples given previously—like that of the coin whose next toss could be made by either Peter or John—and an example of a quantum-theoretic nature? It is precisely in the context of such an example that Reichenbach has developed his arguments (R, pp. 145–146, and p. 168), basing himself upon an absolutely rigid division between the indeterminacy of the quantum world and the determinacy of the macroscopic world. This division is so complete that Reichenbach says the following concerning the outcome of *that toss which John might have made* (but which instead was made by Peter). Since it is a question of 'a macroscopic affair, we have in principle other means of testing', by making precise measurements of the state of John's muscles before or after the toss made by Peter, and in many other ways:

'... or let us better say, since we cannot do it, Laplace's superman could. For us the truth value of John's statement will always remain unknown; but it is not *indeterminate*, since it is possible in principle to determine it, and only lack of technical abilities prevents us from so doing.'

In discussing the merits of the question, one might object that the 'determinism' of the macrocosm—to which Reichenbach makes explicit reference—has a merely static character, and this renders completely unpredictable those facts for which numerous microscopic circumstances might prove decisive (not to mention the fact that even the result of a single collision between particles, as recorded on a photographic plate, can cause macroscopic phenomena like the publishing of papers, the holding of lectures and conferences, and endless indirect consequences and repercussions). Moreover, not even Laplace (so far as I know) ever suggested that his 'superman' was capable of predicting not only everything that is going to happen, but also what would happen if... something that is not going to happen were to happen. How could it come about that the state of the muscles, and so on, could inform us about the result of the toss that has not been performed (and why not the text of a conversation that has not taken place; the adventures of a journey not undertaken; etc.), rather than informing us directly that it is predetermined that the toss, or the conversation, or the journey, will not take place (or did not take place)? To my knowledge, no-one, even in theological discussions, has ever claimed to have decided whether divine omniscience includes the knowledge of what exactly would have happened to the world conditional on every imaginable hypothesis about the form of Cleopatra's nose (or any other fact, either substantial or irrelevant, concerning the world's history).

In my opinion, however, there is no point in entering into the merits of such questions, physical or metaphysical as the case may be, because logic can only be *neutral* and *anterior* with respect to any contingent circumstance of scientific knowledge, or hypothesis, concerning the world of phenomena.† Logic has to be applied to the wider field of everything that is imaginable, and the inevitable circumstance that fantasy is of so little use in extending the field beyond what has already been observed or realized is already too restrictive. Science fiction itself has rarely anticipated reality by more than a few decades. To make use of new ideas or discoveries can be legitimate for the purpose of bringing up to date points of view in logic by including in its domain new areas of what is conceivable, areas which had previously been ignored (and this is what we are attempting to do). The approach which consisted, instead, in making every logical theory restrictive and ephemeral, by reducing it, moment by moment, to a reflection of current scientific views, would have got things upsidedown.

Before turning to another topic, it would perhaps be appropriate to clarify certain views on the theme of determinism, given the connection with discussions pertaining to the present theory, and given that we have commented upon it (even though in order to decide that it was not relevant). In my opinion, the attachment to determinism as an *exigency of thought* is now incomprehensible. Both classical statistical mechanics (or Mendelian hereditary) and quantum physics provide explanations—in the form of coherent theories, accepted by many people—of apparently deterministic phenomena. The mere existence of such explanations should be sufficient to give the lie for evermore to the dogmatism of this point of view. What I mean is that the fact that such theories exist, or are conceivable, should be sufficient (no matter if they are wrong, or even if they are merely successful mental constructs, clearly science-fictional in character).

It is a rather different matter to pose oneself similar questions from what one might call a psychological–aesthetic angle, rather than a dogmatic one. As a result of our own individual tastes and habits, each one of us will have a propensity to find one or other of a deterministic or indeterministic formulation of a law or theory more or less simple and convincing. In particular, we evaluate with greater or lesser sympathy (*a priori*—i.e. before some possibly deeper knowledge or examination of the detailed reasons for and against) the ideas which tend to characterize probabilistic-type quantum theory as merely a partial explanation, unsatisfactory and provisional, and requiring replacing sooner or later by something deterministic.

† On the other hand, this has been perfectly expressed elsewhere by Reichenbach himself.

Personally, I am of the opinion that nothing should ever be excluded *a priori*: tomorrow's notions will almost certainly be as inconceivable for us today as today's notions would have been for a man of the nineteenth century, or for Neanderthal man. This is, however, a distant prospect; the foundations of physics are those we have today (perhaps for many decades, perhaps centuries), and I think it unlikely that they can be interpreted (or adapted) in deterministic versions, like those that are apparently yearned for by people who invoke the possible existence of 'hidden parameters', or similar devices. I hold this view not only because von Neumann's arguments against such an idea seem to me convincing (vN, pp. 313–328), but also because I can see no reason to yearn for such a thing, or to value it—apart from an anachronistic and nostalgic prejudice in favour of the scientific fashion of the nineteenth century. If anything, I find it, on the contrary, distasteful; it leaves me somewhat bewildered to have to admit that the evolution of the system (i.e. of its functions ψ) is deterministic in character (instead of, for example, being a random process) so that indeterminism merely creeps in because of the observation, rather than completely dominating the scene. This can actually lead one to search for some meaning which makes the function ψ objective, although this notion is the very least suited to appear to be capable of such a transformation.

In any case, for what concerns us as human beings, interested in foreseeing the future with some degree of confidence on the basis of our scanty, imprecise and uncertain knowledge of the present and the past, all arguments about determinism are purely academic, and have no more meaning than would a discussion about the number of angels that can dance on the head of a pin. No matter how the world's history develops, nobody could disprove either the assertion that everything is determined by the past through iron laws (but we can foresee either nothing or very little because we are too ignorant both of the past and of the laws), or the assertion that everything occurs 'by chance' (and this does not exclude the possibility that 'by chance' things might develop according to some 'law'). In the final analysis, it seems to be of very little consequence or assistance to us whether we take up a position for or against the plausibility of the hypothesis that Laplace's superman could work out the entire future *if only he knew the entire present in every detail*. Such a statement, in the sense we have just examined, must, in fact, be said to be neither true nor false, but instead indeterminate, the hypothesis being, without any doubt, illusory, and therefore false.

13 VERIFIABILITY AND DISTORTING FACTORS

As the final part of our survey of the various factors which are important when we attempt to determine and verify the outcome of an event, it remains to consider the most troublesome of them. These are the factors which, for reasons relating to the individuals involved, or to their self-interest, are capable of modifying the outcome, or of influencing its verifiability, or of simply raising doubts about the possibility of such distortions.

Many examples of this are well-known, and we shall just quote them without having anything useful to say about overcoming the difficulties. The deepest discussion (which may well be new) will centre, however, on the events of the three-valued logic illustrated in Sections 10–12 above, where it seems impossible to give a complete definition without encountering similar difficulties regarding possible distortions. Let us begin, however, with the most well-known and obvious cases.

These are events for which the will of the individual concerned enters in directly (and, in this respect, there is nothing to distinguish this case from that of events which depend on animals, or on other natural factors). There is a difference—a distorting factor—when such a will can be influenced by facts which are objects of our study, and which therefore alter this very object of study. This happens in evaluating a probability if the circumstances upon which the evaluation itself is based are modified by the evaluation, or the knowledge of this evaluation, or a contract drawn up on the basis of this evaluation, and so on.

This much is true: although we are again speaking of probability before the appropriate time, we have to do so in order to present the examples; the object under consideration is, however, the difficulty of avoiding the problems by means of detailed specifications in the description of an event.

The evaluation of the probability of an event can influence its occurrence. If someone, at some given moment, perhaps because of a vague feeling, or even for no reason at all, considers the danger of a traffic accident to be higher than usual, he will try to be more careful, and the risk will diminish. If, on the other hand, we are dealing with an event whose positive outcome is desired—like succeeding in a business deal, an examination, or a race—it can happen that a greater feeling of confidence leads to one being in a better position to succeed.

The knowledge of someone else's evaluation of probability can have a marked snowball effect, as a result of the confidence that tends to be placed in the opinions of experts. If in circles which are considered well-informed the expectations are pessimistic (or optimistic), and an increasing number of people, when informed of these opinions, behave as if they correspond to reality, the expectations will end up by being borne out by reality—even if they were initially without foundation.

Nevertheless, the most direct example is provided by the influence of a person's self-interest on the outcome of events. In the case of insurance, it can lead to faked or fraudulent accidents; but we are still dealing here with the kind of case which is, to some extent, identifiable. Much worse (from a logical point of view, since it constantly eludes one's grasp) is the effect of the insufficient precautions which an individual might take, knowing that he is insured. Similar influences are at work if a prize is attached to the occurrence of an event (e.g. an additional bonus for each goal, either for the player who scores, or for the whole team to share out), or even if it arouses admiration, or merits reproach.

It is even easier for a person to have an influence on the process of verification rather than on the outcome itself. An individual who is interested in proving that an event has occurred will devote a lot of energy to obtaining information to this end, and will take care to collect the necesary documentation and to send it to the appropriate authority. On the other hand, someone interested in concealing such news will be more or less negligent, or may even attempt to suppress it, or to destroy the evidence.

In order to avoid all this, one should provide a description of the event which is sufficiently detailed to preclude the possibility of distortion. In fact, the clauses of an insurance policy abound in detailed specifications of the obligations of the person concerned, the risks excluded, and so on (although it is clearly not possible to extend the specification beyond those cases which are easiest to define and to pick out†).

An even more entangled situation is to be found in the theory of games. In the simplest case, one has two players, each of whom must make a decision (without knowing the decision of his opponent), and the result (one player's gain, the other's loss) depends on the two decisions made. Each would like to know the decision of the other in order to adjust his own decision accordingly; not knowing it, he could evaluate the probabilities of the various decisions the other might make, and in order to do this he would need to go through a similar reasoning process by putting himself in the other's shoes.

This, and other more complicated situations, are objects of study in games theory. But all the aspects of distorting factors which we have mentioned so far are only intended as remarks in passing, merely to put the reader on guard against the difficulties one encounters when dealing with cases where they arise (the difficulties may or may not be serious, but they are virtually impossible to eliminate).

Above all, these examples serve as an introduction, in order that it should not appear (misleadingly) to be a rather special feature which arises when one delves more deeply into the study of 'three-valued' events. There is, in fact, a particular, novel feature in this case, but it arises later, and is not related to the distorting factor (which derives from the choices that can affect verifiability). We shall examine this latter aspect first.

A conditional event $E|H$ presents no problem of this kind if E and H turn out to be known with certainty—as true or false—within the time and

† A discussion, together with useful examples, can be found in H. M. Sarason, 'Come impostare e applicare le statistiche assicurative', *Giorn. Ist. Ital. Attuari*, **I** (1965), pp. 1–25.

manner specified. In fact, if we think in terms of having made a bet—our guideline—we will then know, without any room for doubt, that it is called off if H turns out to be false, won if both H and E turn out to be true, lost if H but not E turns out to be true. But what if H or E, or both, turn out to be non-verifiable (in some preassigned manner; for example, within a given time period during which the bet has to be decided)? As a first step, we must decide what is to happen to the hypothetical bet in such circumstances. It seems natural—and, in any case, this is what we shall do—to make the following convention: it is either won or lost, respectively, only in the cases of H and E true, H but not E true, respectively; it is called off both if H is false, and if H is indeterminate, and also if H is true but E is indeterminate. In formal terms, considering E and H as three-valued events, $E = E'|E''$, $H = H'|H''$, the conditional event $E|H = (E'|E'')|(H'|H'')$ would correspond (in Reichenbach's terminology) to *quasi-implication* (as introduced by him), with the following truth table (in Reichenbach's notation, $E|H$ corresponds

$E\|H$		H: T	I	F
E	T	T	I	I
	I	I	I	I
	F	F	I	I

to $H \ni\!\!- E$). In terms of the four simple (two-valued) events E', E'', H', H'', this becomes

$$E|H = (E'|E'')|(H'|H'') = \begin{cases} 1 & (T) \text{ if } E'E''H'H'' \\ \varnothing & (I) \text{ if } \sim(E''H'H'') \\ 0 & (F) \text{ if } \tilde{E}'E''H'H''; \end{cases}$$

in other words,

$$E|H = (E'|E'')|(H'|H'') = E'|(E''H'H'').$$

Distorting factors enter in here, too, as soon as one allows the possibility that somebody might influence the outcome, or the knowledge of the outcome, of $E|H$. The particular case of greatest specific interest is that in which H represents the performing of the experiment—or, more usually, experiments—from which information about the outcome of E is drawn (either in fact, or potentially). This covers the cases of all measurements and experiments in both classical and quantum physics, and of all the investigations which are appropriate for the ascertaining of the truth of any assertion concerning practical matters. This situation arises most clearly when E consists just of the result of an experiment which must be expressly performed (e.g. H = the toss of a coin, or the launch of a satellite, and E = Heads, or entering into orbit, respectively). In such a case, it would make no sense at all to enquire whether E was true or false without assuming that H were true; but the case in which E is thought of as being true or false independently of an experiment H for ascertaining it no longer appears different when one is concerned with the actual ascertainment of E. We can very well imagine that E = A. N. Other has gone down with a certain illness, or E = the residue of a substance contains poison, are statements which are true or false in themselves, independently of the fact of our knowing whether they are true or false. If we, or anyone, wish to know whether E *is* true or *is* false (and not merely to say that it is one or the other) then E should be replaced by $E|H$, where H denotes the performing of an act leading to its ascertainment. We could say, for example, that H = A. N. Other undergoes tests to establish whether or not he is infected with the given disease, or H = the residue of the substance is analysed in order to ascertain whether or not it contains poison, and then E = the outcome is positive. But what tests and analyses should be performed?

Let us exclude the possibility that an experiment (e.g. the tests or analyses mentioned in the above examples) could give a wrong answer: this would by no means be absurd, because experiments concerning facts related to the one we wish to ascertain can only lead us to increase or decrease the probability we attribute to it. Everything stems from our convention of considering that a question has been answered only if it is certain, and we call E indeterminate if the ascertainments which have been made have not proved sufficient to resolve the doubt. (Just as, in the case of 'insufficient evidence', it would be inadmissible to claim that a suspect is both guilty and not guilty.)

However, it is only rarely that by performing an experiment H one obtains an answer with certainty. What usually happens (at least in cases which are sufficiently complicated for it to be worthwhile to apply considerations of this kind) is that H may give an answer (in which case it is an exact answer), but, on the other hand, may not (and then E remains indeterminate). To be precise, H should not simply denote the performance of some given experiment, but rather the successful performance of it (in the sense that, with respect to E, the answer is either YES or NO, and not MAYBE). If we wished to split hairs, we could put $H = K'|K$, where K denotes the experiment in the sense of its performance, K' the fact that K was a success, and thus H is the successful performance of K (i.e. the hypothesis that ensures the ascertainment of the truth or falsity of E).†

† Only thus can one avoid having to consider E itself (as well as $E|H$, which becomes, in this notation, $E|K$) as a three-valued event (indeterminate, notwithstanding the—unsuccessful—performance of the experiment). We have our doubts about the actual utility of such notation, other than for a once and for all explanation, and we avoid insisting on, or taking up a position, regarding the desirability of more or less logically perfect forms of notation.

In general, however, there will be no one unique experiment K which we can (or cannot) perform in order to ascertain E. There will exist various possibilities $K_1, K_2, K_3, \ldots$ (and even if there were only one 'type' of experiment available, we could always vary the time, the apparatus, or the experimenter, etc.), and they might or might not be compatible for various reasons (and possibly not repeatable in the case of failure), ranging from physical incompatibility to contingent limitations (e.g. lack of time, apparatus or available personnel, funds, raw materials, etc.). In order not to further complicate the notation, we can suppose that the list given by the K_i includes not only the individual experiments (e.g. let $K_1, K_2, \ldots, K_{38}$ denote the performance of just one of 38 possible different experiments), but also all possible combinations or strategies (e.g. the one consisting in first performing K_5, K_{19} and K_{22}, and then, if none of these succeeds, K_7, and, if this is still not sufficient, K_9 and K_{31} together, and then stopping whatever happens, is a strategy which will be denoted by a number greater than 38—e.g. by K_{728}). For each K_i, K'_i will mean that K_i was successful; i.e. that K_i succeeded in establishing whether E was true or false. In the case of an individual experiment, K_2, say, K'_2 will mean that this experiment was successful; in the general case, for example for K_{728}, K'_{728} will mean that at least one of the component experiments of the strategy was successful (and—in the case of a sequence of experiments, as in the example of K_{728}—the experiments following will then not even be performed). By going into more and more detail (like the time and manner of performing the experiments, possible repetitions, etc.), the number of distinct strategies could be increased without limit. So far as our notation is concerned, however, it is simply a question of extending the list of the K_i.

In this way, the problem of ascertaining E is translated, in practice, into the problem of ascertaining one of the conditional events $E|K_i$, depending on the choice of K_i, which is arbitrary within the limitations imposed (of time, money, etc.). And it is thus that the arbitrariness has its effect on the verification of E. In extreme cases, there might be one-way experiments; i.e. experiments which either prove that E is true, or prove nothing (or vice-versa). Suppose that one experiment shows whether or not a given liquid is pure water, and another experiment shows whether or not it contains strychnine. If the question is whether or not the liquid is poisonous, the first experiment can only return a negative answer, and the second only a positive one (because to know that it is not pure water, or that it does not contain strychnine, neither proves nor disproves the presence of poison). Even without taking these extreme cases into consideration, any method might present, by its very nature (and taking previous experiences into account, according to the evaluation of each individual), different characteristics in its functioning, and different probabilities of breaking down, depending on whether E is true or false.

Up to this point, we have merely been dealing with cases involving distorting factors, just like the others considered previously (even if these cases deserve special attention because they are less obvious than most other examples). The most important specific factor in the present case is, however, quite a different one, which—as we mentioned at the beginning—only 'arises later', after the analysis given above of the experiments K_i, their successes K_i', and the consequent realization of the corresponding hypothesis H_i.

The new factor is the following: to be realistic, one should also substitute E_i for E. Let us explain right away what we mean by this. If we perform the experiment K_i, its success K_i' does not give us directly the answer 'E is true', or 'E is false'; it does not make directly visible to us the fact that we wish to affirm or deny by these phrases. Neither, if we are dealing with the more general question of measuring a quantity, does it enable us to realize what the value is by making it visible or tangible. The answer reduces to a signal (a movement, a light, a noise, a colour, etc.; in the case of quantities, the position of a pointer on a dial, the reading of a counter, the height of a column of mercury, etc.). For an event, we shall have one of two signals, E_i or $\tilde{E}_i$, as possible outcomes of the experiment K_i (as well as the absence of any answer at all—or, if one prefers, $\varnothing$, or $\tilde{K}_i'$), and these may differ from experiment to experiment. *But*, it could be argued, *this is irrelevant*, *because we know that they correspond to E being true*, *or E being false*.†

Agreed . . . , but what does this mean exactly? The last sentence, so simple, clear and straightforward, is admirably suited to an hypothesis which is equally simple and clear; the hypothesis which assumes that one of the many experiments has been taken as the *definition* of E (i.e., if the one chosen is K_{13}, then E means E_{13}, or, better, $E_{13}|K_{13}$), and that this experiment is always possible and always successful. It follows that the statement that E is true because a different experiment K_4 has given the signal E_4, can be strengthened by the remark 'since it is certain that the answer E_{13}—i.e. E—will be obtained if we perform experiment K_{13}, it is unnecessary to do so, because of the trouble, expense, etc.; but, if you do not believe it, try it, and you will see!'. In this way, for example, if I derive the height of a distant tower by trigonometric methods, or by observing how long it takes for stones dropped from the top to reach the ground, I can say to someone who does not believe it 'go up and measure it'. And one might allow that the argument is considered in general to be valid, even when the invitation becomes rather less realistic (distance from the centre of the earth, distance between two stars, or two galaxies). But what if someone does not believe it?

In the previous hypothetical case, there was a criterion which could appropriately be assumed as a definition, both because of its meaning,

† Cf. the discussion of H. Jeffreys that we have already quoted (Chapter 11, the end of 11.1.1).

and because it could always be applied (at least in principle) with a guarantee of success. In particular, it could be applied to statements about things which are not directly observable, and possess disconcerting properties (as in the case of 'waves' and 'particles'). What do we do in the absence of such a criterion? We could define all the experiments K_i (and the simple ones will suffice, it is not necessary to consider strategies) by means of the respective answering signals, and observe that, in any case, no matter which K_i are applied, and no matter how many of them, in all successful cases they give a concordant answer; either always E_i, or always $\tilde{E}_i$. This, practically speaking, assures the meaning and uniqueness of the notion related to E (or, more generally, to a quantity), provided that the coincidence of answers for any two methods, K_i and K_j, could be verified experimentally by applying both of them in precisely that same situation (or at least indirectly, by means of a chain of equivalences, each link of which could be verified experimentally).

But what if we come across cases where it is not possible to perform more than one experiment in a given situation? Any statement of the form 'having observed the outcome E_i of the experiment K_i, we know that *had we performed* the experiment K_j we *would have* obtained E_j', is entirely without content, since the assumption is false. It is the same situation as the one we already considered, albeit light-heartedly, at the beginning of Chapter 4 (Section 4.1), when we asked 'whether or not it is true that had I lived in the Napoleonic era and had participated in the Battle of Austerlitz I would have been wounded in the arm'.

It might help to place one's confidence in a more speculative kind of generalization, such as one makes in passing from the direct verification of indirect measurements of length on the 'human' scale to admitting the same thing for inaccessible distances. The generalization required would have to admit the coherence and validity (justified by numerous indirect proofs) of the entire set of concepts, arguments and calculations which constitute the scientific view of the world.

It is a fact, however, that, so far as the 'ordinary man in the street' is concerned, the only reason he believes in these things is a lack of appreciation of the fact that they are more abstruse and delicate than he imagines. The logical situation for him is, under the worst assumptions, the following: he is given the explanation that the fact of E being true (to fix ideas, think of E as an event on which he would like to bet) can be verified in one and only one of the many possible ways provided by performing an experiment K_i and receiving a corresponding answering signal E_i (the choice being made by the experimenter; in any case, *this tells him about the same fact*). But this leading statement conveys nothing to the man in the street, who has no idea of what 'fact' it is that one is dealing with. (For the scientist, too, it is very much an intellectual conviction; but we are not interested in him.) The man in the street only knows, naturally enough, that he might bet on the outcome of an experiment whose choice is in the hands of his opponent. He may therefore think (in theory, of course—not in practice, because it is not nice to be suspicious) that the choice will be made to his disadvantage: e.g., he may end up trying to draw a white ball from an urn containing only black balls, this being one of the possible choices open to his opponent.

Leaving these more or less picturesque illustrations aside, it would seem that the conclusion—a negative and disturbing one—cannot be other than the following: *one does not succeed in giving an operational meaning to a statement E (or to a quantity X) by means of a collection of statements $E_i|H_i$, which do have operational meaning, without introducing the statement that all the obtainable results E_i are necessarily conformable* (*and this does not have an operational meaning if the H_i are incompatible*).

14 FROM 'POSSIBILITY' TO 'PROBABILITY'

The logic of certainty only distinguishes events which are either true or false, and which can only be *possible* (uncertain) rather than certain or impossible† (for us, in our more or less temporary state of ignorance). We have discussed questions of a critical nature by remaining within this ambit as a preliminary exploration of the field into which we have to introduce and apply the logic of the probable.

The time has now come to deal with the main critical questions which specifically concern the subject of direct interest to us: the theory of probability.

We do not want to repeat ourselves by going back to the beginning and starting from scratch: we have already dealt with many questions in the text, and many comments were necessarily made as we developed our approach. It will be more appropriate to refer back to these, to draw the threads together, and to go more deeply into them, in order to provide a synthesis, and, finally, what will, hopefully, turn out to be a sufficiently integrated view of the entire subject.

Let us begin by sketching a broad outline, including both those topics for which we shall limit ourselves to recalling our previous remarks—or just adding the odd word here and there—and those which we shall take up again later because they require further analysis, or more thorough discussion.

Without further comment, we shall take as axioms those already established as the basis of our subjectivistic formulation. This will in no way prejudice the (technically neutral) possibility of comparison; our starting point, in fact, makes comparison easier, because it represents the minimal set of

† We do not have to worry about 'indeterminacy', considering it (see Sections 9–12) reducible to the case of two-valued logic by means of conditional events.

conditions common to all formulations. The subjectivistic formulation, as we have said repeatedly, is, in fact (and deliberately so), the *weakest* one; its only requirement is coherence, and in no way does it seek to interfere with an individual's freedom to make an evaluation by entering into the merits of it on some other grounds.

In discussing these concepts, we shall provide a comparison with other points of view, which differ in various respects (in the interpretation of the notion of probability, the mathematical details, and the qualitative formulation).

Those interpretations of the notion of probability in a (would-be) objective sense which are based on symmetry (the classical conception; equally likely cases), or on frequency (the statistical conception; repeated trials of a phenomenon), provide criteria which are also accepted and applied by subjectivists (as, to a considerable extent, in this book). It is not a question of rejecting them, or of doing without them; the difference lies in showing explicitly how they always need to be integrated into a subjective judgment, and how they turn out to be (more or less directly) applicable in particular situations. If one, instead, attempts to force this one or that one into the definitions, or into the axioms, one obtains a distorted, one-sided, hybrid structure.

The mathematical details remain those which derive from the positions we adopted concerning zero probability, countable additivity, and the interpretation of asymptotic laws (points which we have already encountered, and commented on, many times). In this regard, we shall have to consider many further points, which we glossed over in Chapters 3, 4 and 6, in order not to overcomplicate the exposition (prematurely), and to add some details concerning a number of new features. These considerations, together with some others, will enable us to sort out, and comment upon, the differences between the axiom system we have adopted here, and that given by Kolmogorov (1933), the formulation which, broadly speaking, has been adopted by most treatments of the last few decades.

Finally, under the heading of 'qualitative formulations', we will have to mention two separate topics. The first concerns the possibility of starting from purely qualitative axioms—i.e. in terms of comparisons between probabilities of events (this one is more probable than that one, etc.)—without introducing numerical probabilities, but eventually arriving at them by means of comparisons of this kind. The second deals with the thesis which several authors have recently put forward, namely that probabilities are intrinsically indeterminate. The idea is that instead of a uniquely determined value p one should give bounds (upper and lower values, p' and p''). That an evaluation of probability often appears to us more or less vague cannot be denied; it seems even more imprecise, however (as well as being devoid of any real meaning), to specify the limits of this uncertainty.

15 THE FIRST AND SECOND AXIOMS

The entire treatment that we have given was based on a small number of properties, which were justified in the appropriate place in the text as conditions of coherence. In order to develop the theory in an abstract manner, it will now suffice to assume these same properties as axioms.

There will be two axioms (the first and the second) dealing with previsions, and a third dealing with conditional previsions. The third one—which is needed in order to extend the validity of the first two to a special case—will be dealt with later (Section 16): we concentrate for the time being on the first two.

Axiom 1: *Non-negativity*: if we *certainly* have $X \geqslant 0$, we must have $\mathbf{P}(X) \geqslant 0$;

Axiom 2: *Additivity* (finite):

$$\mathbf{P}(X + Y) = \mathbf{P}(X) + \mathbf{P}(Y).$$

From these it also follows that

$$\mathbf{P}(aX) = a\mathbf{P}(X), \qquad \inf X \leqslant \mathbf{P}(X) \leqslant \sup X,$$

as well as the (Convexity) condition, which includes Axioms 1 and 2:

(C) any linear equation (or inequality) between random quantities X_i must be satisfied by the respective previsions $\mathbf{P}(X_i)$; in other words,

$$\text{if we } \textit{certainly} \text{ have } c_1X_1 + c_2X_2 + \ldots + c_nX_n = c \ (\text{or} \geqslant c)$$

$$\text{then } \textit{necessarily } c_1\mathbf{P}(X_1) + c_2\mathbf{P}(X_2) + \ldots + c_n\mathbf{P}(X_n) = c \ (\text{or} \geqslant c).$$

By taking differences, (C) can be written in an alternative form:

(C′) No linear combination of (fair!) random quantities can be uniformly positive; in other words, the $\mathbf{P}(X_h)$ must be chosen in such a way that whatever be the given $c_1, c_2, \ldots, c_n$, *there does not exist a* $c > 0$ *such that*

$$c_1(X_1 - \mathbf{P}(X_1)) + c_2(X_2 - \mathbf{P}(X_2)) + \ldots + c_n(X_n - \mathbf{P}(X_n)) \geqslant c$$

certainly holds.

We could put forward as a further (possible) axiom one which consists in excluding the addition of other axioms; i.e. one which considers *admissible*, as prevision-functions **P**, all those satisfying Axioms 1 and 2, or equivalently, condition (C).† On the other hand, this is implicit, since nothing is said to the contrary. In any case, we shall say that every function **P** satisfying Axioms 1 and 2 is *coherent*.

As we have already mentioned (Chapter 3, 3.10.7), a coherent function **P**, defined on some given set of random quantities $\mathscr{X}$ (an arbitrary set, in general infinite), can always be extended, preserving coherence, to any other random quantity, X_0, say. From any inequality of the form (C'), one can obtain, by solving it with respect to one of the summands (let us assume $c_0 = \pm 1$, and take it to be the one corresponding to X_0; were this not the case, it suffices to divide through by $|c_0|$), an inequality for $\mathbf{P}(X_0)$ of the form

$$\mathbf{P}(X_0) \leqslant \inf\{X_0 + \sum_{h=1}^{n} c_h(X_h - \mathbf{P}(X_h)) - c\} \quad (\text{or} \geqslant \sup\{\ldots\}).$$

As a result, we obtain $x' \leqslant \mathbf{P}(X_0) \leqslant x''$, where x' denotes the greatest lower bound, and x'' the least upper bound. If $x' = x''$, the extension will turn out to be uniquely defined;

$$\mathbf{P}(X_0) = x' = x'',$$

i.e. $\mathbf{P}(X_0)$ will be determined by the values given over $\mathscr{X}$. If $x' < x''$, the admissible values for $\mathbf{P}(X_0)$ will consist of all those in a closed interval (as is obvious by convexity). The extension would be impossible if $x' > x''$, but this is ruled out by the observation that there would then exist a linear combination $X_0 + \sum_i c_i(X_i - \mathbf{P}(X_i))$ always $> x'$, and another one

$$X_0 + \sum_j c_j(X_j - \mathbf{P}(X_j))$$

always $< x''$; their difference ($\sum_i - \sum_j$; X_0 cancels out) would then turn out to be $> x' - x'' > 0$. But this would mean that there was a contradiction of (C') already contained in $\mathscr{X}$, contrary to the hypothesis.

It follows immediately from this that one can always define a $\mathbf{P}(X)$ for all the X belonging to an arbitrary set of random quantities (in particular, one can always define a $\mathbf{P}(E)$ for every event in an arbitrary collection of events—e.g. those corresponding to all subsets of a given space), even assuming $\mathbf{P}(X)$ as already assigned in some given field, and extending it. It is sufficient,

† Note that we are not dealing here with the basic issue of whether under given circumstances all the coherent evaluations **P** are admissible (subjectivistic conception), or whether only one of them corresponds to reality (objectivistic conceptions). For the objectivist also, it is a question of knowing which **P** are formally admissible (e.g. the **P** which he can adopt when he has the information he is now lacking—about composition of urns, frequency of statistical phenomena, etc.), or even that he judges to be possible with respect to the abstract scheme without knowing which concrete events are represented by the symbols E_1, E_2, etc. On the other hand, this is the attitude adopted by the supporters of all points of view when they are faced with the notion of '(abstract) probability space'.

as we have done here, to carry out the extension for new X one at a time, by means of *transfinite induction* (assuming, of course, the Zermelo Postulate, in order to well-order the X_h; the indices, h etc., will be transfinite ordinals). One has to be a little careful that nothing goes wrong for the X_k which have no antecedent (such as X_ω, where ω denotes, as usual, the first ordinal which comes after the natural numbers).† In our case, however, the contradiction would derive from the comparison between two *finite* linear combinations, and should have occurred at the last of the steps corresponding to the X_h which appear (and the fact that there are an infinite number of steps between this X_h and our X_k does not enter into the argument).

Let us return now to the problem of the extension, in order to consider when it turns out to be uniquely defined. One obvious case is that of a random quantity X_0 linearly dependent on those of the original field $\mathscr{X}$; i.e. belonging to the linear space $\mathscr{L}$ generated by the X belonging to $\mathscr{X}$. In this case, the uniqueness of the extension holds for any **P**.

Condition (C), however, reveals what the situation is in terms of a particular **P**. Instead of linear relations, we have, in general, linear inequalities, $\sum_i c_i X_i \geqslant c$, which, solved in terms of X_0 (as above for $X_0 - \mathbf{P}(X_0)$), give random quantities X' and X'', linear combinations of random quantities belonging to the field $\mathscr{X}$ (and hence belonging to $\mathscr{L}$), which bound X from below and above: $X' \leqslant X$ and $X \leqslant X''$, respectively. We observe that the problem is the same one that we already encountered in a special case (Chapter 3, 3.12.4), and by passing, as here, to the general and abstract case, we also reached essentially the same conclusions. As we vary **P** (defined over $\mathscr{X}$, and hence on $\mathscr{L}$, and, in particular, for X' and X''), the X' for which $\mathbf{P}(X')$ is a maximum, $\mathbf{P}(X') = x'$, will also vary (or, if x' is an upper bound rather than a maximum, the X' to be chosen in order to obtain $\mathbf{P}(X')$ arbitrarily close to x' will vary): similarly for X''. Having chosen X and X'' in this way, we have $X' \leqslant X \leqslant X''$, with $\mathbf{P}(X'' - X') = x'' - x'$ (or $x'' - x' + \varepsilon$, with $\varepsilon > 0$ arbitrary, in the case when they are not the maximum and minimum). In general, therefore, one has a uniquely defined extension if upper and lower bounds of X exist for which the difference $\Delta = X'' - X' \geqslant 0$ has prevision $\mathbf{P}(\Delta) = 0$, or such that $\mathbf{P}(\Delta) < \varepsilon$ (for arbitrary, fixed $\varepsilon > 0$).

In order to denote what can be said about the probability (or prevision) outside some given linear space $\mathscr{L}$ in terms of the prevision function **P** defined over it by the evaluations of probability (or prevision), it is convenient to use the same notation (*mutatis mutandis*) as we used in Chapter 6, 6.4.4.

† Lebesgue measure, too, can be extended, preserving countable additivity, to an arbitrary non-measurable set, and hence to an arbitrary number of such sets, one at a time. In this case, however, an infinite number of steps can lead to a contradiction without any single step doing so (in the same way as a convergent series remains such if we replace the 1st, 2nd, 3rd, . . . , terms with 1, and so on for any finite number, but not if we replace an infinite number of terms).

We thus denote by

$$\mathbf{P}^-_{\mathscr{L}}(X) = x', \qquad \mathbf{P}^+_{\mathscr{L}}(X) = x'', \qquad \mathbf{P}^{\pm}_{\mathscr{L}}(X) = x,$$

the minimum and maximum (previously indicated in the text by x' and x'') of the values $\mathbf{P}(X)$ which are compatible with the knowledge of $\mathbf{P}$ over $\mathscr{L}$, and, respectively, their common value (if they are equal).

We shall say more about this—for other reasons—in Section 19.4.

16 THE THIRD AXIOM

Conditional probabilities $\mathbf{P}(E|H)$, or conditional previsions, $\mathbf{P}(X|H)$, are expressible, in cases where H has non-zero probability, in terms of the unconditional probabilities by means of a formula which, in an abstract, axiomatic treatment, can be taken as a *definition*:

$$\mathbf{P}(E|H) = \mathbf{P}(EH)/\mathbf{P}(H), \qquad \mathbf{P}(X|H) = \mathbf{P}(XH)/\mathbf{P}(H).$$

In this case, there is nothing much to add, apart from noting that here, too, an extension (in the sense of $\mathbf{P}_{\mathscr{L}}$) gives rise to an interval of indeterminacy:

$$\mathbf{P}^-_{\mathscr{L}}(X|H) \leqslant \mathbf{P}(X|H) \leqslant \mathbf{P}^+_{\mathscr{L}}(X|H).$$

To see this, suppose that $\mathbf{P}_1$ and $\mathbf{P}_2$ are two extensions of $\mathbf{P}$ as given over $\mathscr{L}$, and that these give to XH and H the values

$$\mathbf{P}_1(XH) = x_1, \qquad \mathbf{P}_2(XH) = x_2, \qquad \mathbf{P}_1(H) = h_1, \qquad \mathbf{P}_2(H) = h_2.$$

In addition to $\mathbf{P}_1$ and $\mathbf{P}_2$, their convex combinations,

$$\mathbf{P}_\lambda = \lambda\mathbf{P}_1 + (1-\lambda)\mathbf{P}_2 = \mathbf{P}_2 + \lambda(\mathbf{P}_1 - \mathbf{P}_2) \; (0 \leqslant \lambda \leqslant 1),$$

will also be extensions of $\mathbf{P}$, and will give

$$\mathbf{P}_\lambda(X|H) = \mathbf{P}_\lambda(XH)/\mathbf{P}_\lambda(H) = \frac{x_2 + \lambda(x_1 - x_2)}{h_2 + \lambda(h_1 - h_2)}.$$

Since the denominator does not vanish (for $0 \leqslant \lambda \leqslant 1$; or at most at one of the end-points if one of the h_i is zero, a case that we shall not consider now, however), the hyperbola increases or decreases monotonically between the extreme values

$$\mathbf{P}_1(X|H) = x_1/h_1 \quad \text{and} \quad \mathbf{P}_2(X|H) = x_2/h_2.$$

In the extension, the set of possible values for $\mathbf{P}(X|H)$ is thus an interval as asserted.

If $\mathbf{P}(H) = 0$, we have a new situation. Does it make sense to consider this case? And, if so, for what purpose? If one were to take the formula, with $\mathbf{P}(H)$ in the denominator, as the actual, unique definition of conditional probability and prevision, then the concept, in this case, would become meaningless. If the meaning were to be assigned in some other, direct, way—for example (as was done in Chapter 4, in line with the subjectivistic point of view), by means of conditional bets—then the meaning would be retained.

But the theorem which expresses coherence, connecting it to the non-conditional **P** (the theorem of compound probabilities), no longer holds (and neither does the criterion of coherence) if its formulation (Chapter 4, Section 4.2) has to be in terms of the existence of a 'certainly smaller' loss. In order to extend the notions and rules of the calculus of probability to this new case, it is necessary to strengthen the condition of coherence by saying that *the evaluations conditional on H must turn out to be coherent conditional on H* (i.e. under the hypothesis that H turns out to be true). This is automatic if one evaluates $\mathbf{P}(H) \neq 0$, in which case we reduce to the certainty of a loss in the case of incoherence. The loss for $\tilde{H}$ (Chapter 4, Section 4.3) is, in fact, the sum of the squares of $\mathbf{P}(H)$ and $\mathbf{P}(EH)$; but if $\mathbf{P}(H)$, and therefore $\mathbf{P}(EH)$, are zero, this loss is also zero in the case $\tilde{H}$ (which has probability $= 1$, and is, in any case, possible).

Although this strengthening of the condition of coherence might seem obvious, we had better be careful with it. There are several other forms of strengthening of conditions, often considered as 'obvious', which have consequences that lead us to regard them as inadmissible. In this case, however, there do not seem to be any drawbacks of this kind; moreover, the 'nature' of the strengthening of the condition seems more firmly based on fundamental arguments (rather than for conventional or formal reasons, or for 'mathematical convenience') than others we have come across, and to which we shall return later. In any case, we propose to accept the given extension of the notion of coherence, and to base upon it the theory of conditional probability, without excluding, or treating as special in any way, the case in which one makes the evaluation $\mathbf{P}(H) = 0$.

If we wish to base ourselves upon a new axiom, we could express it in the following way:

Axiom 3: *The conditions of coherence* (*Axioms* 1 *and* 2) *must be satisfied, also, by the* $\mathbf{P}_H$ *conditional on a possible* H, where

$$\mathbf{P}_H(E) = \mathbf{P}(E|H), \qquad \mathbf{P}_H(E|A) = \mathbf{P}(E|AH)$$

is to be understood.

This means that $\mathbf{P}_H$ is the prevision function that we may have ready for the case in which H turns out to be true, and the axioms oblige us to make this possible evaluation in such a way that if it is to have any effect it must be coherent. This is implicit in the previous definition if one makes the evaluation $\mathbf{P}(H) \neq 0$. Axiom 3 obliges us to behave in the same way, simply on the grounds that H is possible and we might find ourselves actually having to behave according to the choice of $\mathbf{P}_H$—even if, in the case in which we attribute probability 0 to the hypothesis H, the sanction provided by the losses does not apply outside of the case $\tilde{H}$.

Axiom 3 permits us to define the ratio of the probabilities of two arbitrary events—even if they have zero probabilities—in the manner already introduced in Chapter 4, 4.18.2. In Section 18.3, we shall expressly return to the topics concerning zero probabilities, topics previously dealt with in Chapter 4, 4.18.3–4.18.4.

17 CONNECTIONS WITH ASPECTS OF THE INTERPRETATIONS

The axioms of an abstract theory are, as such, arbitrary, and independent of this or that interpretation (at this level, interpretations do not, strictly speaking, exist; or, to put it a little less strongly, one might say that they are ignored).

It goes without saying, however, that the choice of axioms is influenced by the interpretation they will have when the theory is applied in the field for which it has, in fact, been constructed, and on which one would like it to turn out to be adequately modelled.†

In the case of the theory of probability, any judgement about the adequacy of the axioms depends on one's concept of probability, and, in addition to the subjectivistic concept, which we have adhered to throughout, we shall also have to consider the 'classical' and 'statistical' concepts.

From the subjectivistic point of view, the axioms are valid in that they are a translation of the necessary and sufficient conditions for coherence (our starting point in Chapters 3 and 4). It follows that no other axioms can be admitted (since these would introduce further restrictions).

Mention should be made of a formulation which is subjectivistic in a purely psychological sense, and in which no axioms would be acceptable. This is the approach in which one simply thinks of evaluations of probability—in general, incoherent—being made by some, arbitrary, individual. It is clear that without sufficient preparation and thought everyone would give incoherent answers in every field (e.g. by estimating distances, areas, speeds, etc., in an incoherent manner). This does not imply, however, that there exists—albeit only in the individual's own mind—a different theory (e.g. a non-Euclidean geometry) to be made an object of study. The object of study could only be the extent of his intuitive inability to understand the conditions of coherence, and to avoid breaking them. Otherwise, one would have to say that, in a system of bets, he deliberately chooses to behave in such a way as to lose.

From the classical point of view—probability 'defined' as the ratio of favourable cases to possible cases, all considered 'objectively' equally likely

† As someone rather neatly put it—Frechet attributes the remark to Destouches—a book which starts off with axioms should be preceded by another volume, explaining how and why these axioms have been chosen, and with what end in view.

for reasons of symmetry—the axioms are true by virtue of the laws of arithmetic (sums of fractions, together with certain other details which are required to achieve the necessary rigour). There is, for any given application, just one admissible **P**. It would appear to be valid to consider infinite partitions into equally probable cases (by virtue of symmetry).

As an extension of this point of view, one might consider the 'necessary' conception, which takes probabilities of a collection of events, possibly outside the range of cases considered in the 'classical' approach, to be uniquely defined for *logical* reasons. A typical example—one which accepts the possibility of 'an infinite number of equally likely cases'—is provided by Jeffreys' admission of improper initial distributions (e.g. uniform in X, or in $\log X$, etc.). It appears that Carnap's point of view is similar to this.†

From the statistical point of view—where probability is regarded as 'idealized frequency'‡—additivity always holds for arithmetic reasons (as in the classical case). According to this conception (again as in the classical case), there should be a unique admissible **P**. It is difficult to attempt to venture hypotheses about the interpretation of more delicate cases (e.g. zero probability).

An attempt to make the statistical conception more precise consists in defining probability not as an 'idealization', but rather as a *limit* of the frequency (as the number of trials, n, tends to ∞). In throwing a die, the limit-frequency of 'evens' is without doubt the sum of the limit-frequencies of '2', '4' and '6', if these limits exist (and this is assumed to be the case in the scheme we are considering). It seems equally clear that such additivity does not (necessarily) hold for infinite partitions: a 'die with an infinite (countable) number of faces' could well turn up each face with limit-frequency zero§

† It is always difficult to judge whether similarities are real or apparent (particularly between authors with different backgrounds, working in different fields).

‡ This phrase does not really convey anything, but it is the only way to refer to the many confused explanations given by the supporters of this conception, and it may be that, in fact, there is nothing of substance to 'understand' (alternatively, it may be me who lacks the resources necessary for success in this toilsome venture).

§ A 'reasonable enough' example might be obtained by saying that each face of the die has probability $p_h = e^{-1}/h!$ of occurring h times (in an infinite series of trials). Alternatively, if one prefers, one could say that out of 10,000 faces occurring 'at random' we will have 'in prevision' (or, less appropriately, 'on average')

$10{,}000\ p_h =$	3679	3679	1839	613	153	31	5	1
where $h =$	0	1	2	3	4	5	6	7

(i.e., in the infinite series of trials 0, 1, 2, ..., 7 repeats will occur, but 8 or more will not occur even once—on average about 0·10 times).

This is the Poisson distribution with $m = 1$, which holds asymptotically for the game of matching n objects ($n \to \infty$), or for that of n drawings with n balls (e.g. 90 drawings, with replacement, of the 90 numbers in Bingo), again as $n \to \infty$. We have remarked that this example is 'reasonable enough', but it is no more than this, because the choice of this scheme from among infinitely many others is arbitrary. One could, for example, vary the scheme for drawing the balls, by assuming that as n, the number of balls, increases, the number of drawings is not n, but $2n$, or n^2, or $\sqrt{n}$, etc.

(indeed, *it should* do so, if we continue to admit, in some shape or form, the assumption of equal probabilities for all the infinite faces). But this seems to get overlooked (cf. Chapter 3, 3.11.6, the case $C9 = N9$).

Although it is really going beyond our aim of giving a critical analysis of particular attempts at axiomatization, it seems necessary to spend some time on the following case in order to make some comments (we shall refer to the version given by von Mises, which is the most developed, and which was in favour for a time).

On the one hand, it appears that the hypothesized sequences with well-determined limit-frequencies should represent 'idealizations' of problems of 'repeated trials'. This would seem to be so in view of introductory remarks alluding to ideas like the 'empirical law of averages', and because of an additional restriction ('Regellosigkeits-axiom', or axiom of 'non-regularity'), which is intended as a summary, in objective and descriptive terms, of the apparent effects of the *independence* of successive trials. Should one exclude periodicities? The grouping of the results in blocks (e.g. each 'colour' at least 3 times in a row)? The sequences definable in terms of simple mathematical formulae, or sentences not exceeding 100 words? On each occasion one would probably answer yes; but in actual fact there is never any reason to call a halt before having excluded all the possibilities, nor, conversely, any justification for absolutely excluding any given case.

On the other hand, if one wishes to consider the actual case of a sequence of trials (in general unlimited, but to begin with let us assume it to be limited to a finite number, n), independent, and with equal probabilities—according to the concepts derived from such a formulation—one must not think in terms of the 'sequences' of the previous model. One must think of n parallel sequences; i.e. a sequence of n-tuples representing (let us say) a fictitious infinity of 'copies' of the actual sequence of n trials (for $n = \infty$, a fictitious infinity of copies of the whole actual sequence). Only in this absurd supermodel do the simple and obvious ideas of independence and equal probabilities make any sense, and is one able to (correctly) conclude that in the actual (Bernoulli) sequence one has stochastic convergence (weak, strong, in mean-square), but not definite convergence (as was postulated in the original scheme).

The original scheme is therefore a sham. For the purpose of winning over the unwary—who do not notice the sleight of hand—properties are attributed to it which look like the probable properties of actual sequences of 'repeated trials', but which are in fact misleading and incompatible with them. By mysterious manipulations of an infinite number of such shams, one finally succeeds in saying those things which could have been said directly anyway (i.e. that the trials are independent and equally probable). The fruits of these labours are that one now does not understand the (subjective) meaning of the words, and that the flurry of sophistical acrobatics has created the illusion that one has established or produced an 'objective' something or other.

In bringing to a close our summary of the various interpretations, we repeat that the subjectivistic conception is not in opposition to any of them, but rather that it utilizes all of them. It is simply a question of rejecting the claims of exclusiveness which lead to incomplete and one-sided theories, of correcting the distortions made in order to make them appear objectivistic, of considering them as methods whose appropriateness varies with the situation, and of seeing them as having one and the same function: that of aiding the individual in his task of evaluating the probabilities (always subjective) to be attributed to events of interest.

18 QUESTIONS CONCERNING THE MATHEMATICAL ASPECTS

18.1. We now turn to an examination of those aspects of a purely mathematical or formal nature. In a certain sense, we will be considering the properties of the function **P**, and the meaning of the implications of these properties. We refer here to meaning in a formal sense; without reference—except incidentally, and for purposes of clarification—to the different interpretations and assumptions which precede the choice of axioms (concerning which see Section 17).

In order to provide an overall perspective, it will be convenient to present the various questions—including those we have already dealt with—in the context of a comparison with the Kolmogorov axiom system,† a formulation which is well-known to everyone.

The basic differences are as follows:

(1) we REJECT the idea of 'atomic events', and hence the systematic interpretation of events as sets; we REJECT a ready-made field of events (a Procrustean bed!), which imposes constraints on us; we REJECT any kind of restriction (such as, for example, that the events one can consider at some given moment, or in some given problem, should form a field);

(2) we REJECT the idea of a unique **P**, attached once and for all to the field of events under consideration; instead, one should characterize *all* the admissible **P** (the set $\mathscr{P}$); $\mathscr{P}$ turns out to be closed (Chapter 3, Section 3.13), and thus any **P** adherent to $\mathscr{P}$ in fact belongs to it (a property which does not hold in the Kolmogorov system);

(3) our approach deals directly with *random quantities* and *linear* operations upon them (events being included as a special case); we thus avoid the complications which arise when one deals with the less convenient Boolean operations;

(4) we REJECT countable additivity (i.e. σ-additivity);‡

† A. Kolmogorov, *Grundbegriffe der Wahrscheinlichkeitsrechnung*, Springer, Berlin (1933). The first time I developed a systematic discussion in the context of a comparison with this theory was in 'Sull'impostazione assiomatica del calcolo delle probabilità', *Annali Triestini*, **XIX**, University of Trieste (1949).

‡ It is worth mentioning, incidentally, that if one decides to proceed in the direction of assuming things for 'mathematical convenience', then not even countable additivity appears to be sufficiently restrictive. Several authors, including Kolmogorov himself, have recently proposed axioms ('perfect' additivity, and such-like) which make the principles of probabilistic reasoning, essential to every human being, completely dependent on the abstruse subtleties of set theory at its most profound. See, for example, D. Blackwell, 'On a class of probability spaces', in *Proc. 3rd Berkeley Symp.*, **II**, pp. 1–6 (1956), and other works referred to therein.

(5) we REJECT the transformation of the theorem of compound probabilities into a definition of conditional probability, and we also REJECT the latter being made conditional on the assumption that $\mathbf{P}(H) \neq 0$; by virtue of the exclusions we have made in (4) and (5), the construction of a complete theory of zero probability becomes possible;

(6) Kolmogorov's proof of the compatibility of his axioms is open to criticism (cf. the paper of mine quoted in the footnote at the beginning of this section); this is, however, a problem that can be resolved, and it has no substantive implications.

To a greater or lesser extent, all these matters have been touched upon already, either in the text or in this Appendix. We shall only concern ourselves now with those aspects which require further analysis or more detailed discussion.

18.2. *Zero probabilities.* Let us first of all go back to our earlier discussion (at the end of Section 16), and let us repeat the proofs and definitions that we gave (in Chapter 4, 4.18.2), basing ourselves now on Axiom 3.

Axiom 3 permits us to define the ratio of the probabilities of two arbitrary events, A and B, by observing that for all H which contain A and B (i.e. $H \supset A \vee B$) the ratio $\mathbf{P}(A|H)/\mathbf{P}(B|H)$ does not change (except possibly to become indeterminate, 0/0). Suppose, in fact, that H' and H'' are events containing $A \vee B$, and that they do not give rise to the case of 0/0, and let $H = H'H''$ be their product, which also contains A and B (or one could take H to be $A \vee B$). Since $\mathbf{P}_{H'}$ and $\mathbf{P}_{H''}$ must be coherent, we can write

$$\mathbf{P}_{H'}(A) = \mathbf{P}_{H'}(AH) = \mathbf{P}_{H'}(H) \,.\, \mathbf{P}_{H'}(A|H).$$

But $\mathbf{P}_{H'}(A|H) = \mathbf{P}(A|HH') = \mathbf{P}(A|H)$, because $H \subset H'$, $HH' = H$. Finally, we have $\mathbf{P}_{H'}(A) = \mathbf{P}_{H'}(H) \,.\, \mathbf{P}(A|H)$, and $\mathbf{P}_{H'}(B) = \mathbf{P}_{H'}(H) \,.\, \mathbf{P}(B|H)$, and hence it follows that

$$\frac{\mathbf{P}_{H'}(A)}{\mathbf{P}_{H'}(B)} = \frac{\mathbf{P}_{H'}(H) \,.\, \mathbf{P}(A|H)}{\mathbf{P}_{H'}(H) \,.\, \mathbf{P}(B|H)} = \frac{\mathbf{P}(A|H)}{\mathbf{P}(B|H)}.$$

The same holds true for every H'', and, finally, in order to obtain the ratio, it suffices simply to take $H = A \vee B = A + B - AB$; in this case we certainly have $\mathbf{P}_H(A) + \mathbf{P}_H(B) \geqslant 1$, and 0/0 cannot occur.

In this way, the formula $\mathbf{P}(E|H) = \mathbf{P}(EH)/\mathbf{P}(H)$ *is always meaningful and valid*, and the same is true for every application of the theorem of compound probabilities, and, more generally, for any operation involving probability ratios, so long as they make sense (i.e. so long as one does not introduce the nonsensical, indeterminate expressions, $0/0$, $0 \,.\, \infty$, ∞/∞, which must be avoided by means of the procedure used for defining the ratio in each case).

There are, therefore, different orders, or layers, of zero probability (as we have already noted in Chapter 4, 4.18.3, where we also saw how very rich and complicated structures of such layers could be constructed). We shall see later, in 18.3, what the situation would necessarily be in this respect were we to assume the axiom of countable additivity (or, at least, if the condition were assumed to hold in some specific example or other). For the present, however, let us return to the general case.

The theorem of total probability will have to be interpreted in the following extended sense (which includes the case of zero probabilities): given n incompatible events, $E_1, E_2, \ldots, E_n$, the probability of the sum-event E is the sum of the non-zero probabilities, if there exist any, and, if not, it is *the sum of the zero probabilities of maximal order*. If, for example, E_3 is of maximal order (i.e. $\mathbf{P}(E_h)/\mathbf{P}(E_3) < \infty$, $h = 1, 2, \ldots, n$), the sum-event has probability $\mathbf{P}(E) = \mathbf{P}(E_3)$ if for all $h \neq 3$ the preceding ratio is not only $< \infty$, but in fact $= 0$. In general, it is given by

$$\mathbf{P}(E) = \mathbf{P}(E_3) \sum_{h=1}^{n} c_h,$$

where $c_h = \mathbf{P}(E_h)/\mathbf{P}(E_3)$ ($c_3 = 1$; the other c_h may be zero, in which case they do not count, or they may be greater than, or less than, 1).

The introduction of conditional probability freed from the restriction that the 'hypothesis' have non-zero probability, and the consequent possibility of comparing zero probabilities, is important both from a conceptual and from a practical point of view. This importance derives not so much from the fact that we can see the potential usefulness in interesting applications, but rather from the warning it provides against inaccurate ways of approaching—or, at least, of expressing—certain questions. We have in mind methods of approach which either lead to confusion, or to an over-hasty choice of the path and interpretation to be followed, because of the absence of a precise meaning to which one can refer.

No doubt some will regard discussion of this kind as rather artificial and academic; nothing more than hair-splitting *ad infinitum*. They may be right, and they will do well to pose their problems in such a way as to avoid the difficulties. But in order to do this, they must first be able to recognize the difficulties as such, so as to overcome them without lapsing into naïvity or contradiction. In any case, since there do exist differences of opinion in this respect, and since the one which I consider to be correct, and which I uphold, differs from that which forms an integral part of the theory currently most in favour, there is no alternative, in the present context, but to consider the matter more deeply.

But why worry about events with zero probability? Are they not, for this very reason, eventualities which can be ignored?

From time to time, someone imagines that he had discovered the way of eliminating the problem altogether, by *establishing* that the values 0 and 1 must be reserved for the probabilities of the impossible event and the certain event, respectively. Every possible event should have positive probability (strictly less than 1). It is easy to see—and we shall do so presently—that this leads in our case to the same kind of absurdities as one encounters when trying to invent a measure which only assigns zero to the empty set. It is only in the very simplest examples (i.e. those where we only meet finite or countable partitions) that it may happen, *by chance*, that there are no possible events with zero probability, or that, if there are an infinite number of them, their union still has zero probability (a case in which, to some extent, one might regard them as eventualities that can be ignored). If, on the other hand, we considered a non-denumerable partition, we would have to conclude that it was impossible to consider a non-denumerable partition into *possible* cases (because at most a countable number can have positive probability if their sum is not to become infinite $\sum p_h = \infty$). We need go no further than the logic of certainty to see the absurdity of this statement.

The major difference between events of zero probability and impossible events is the following: the union of an infinite number of the former can have a non-zero probability (and may even be the certain event), whereas any union of the latter can only be an impossible event.

It is in this setting that one comes across the most controversial question of them all; that of 'countable additivity'. If we limit ourselves to a discussion of it in the context of events having zero probability, countable additivity implies that taking a countable union will never yield an event with positive probability (and certainly not the certain event). One has to examine the specific question of whether it is possible and appropriate to assume this property as an axiom of probability (the property holds, as is well-known, for Lebesgue measure, where the nature of the definition excludes the cases for which it would not hold†). The majority opinion is that the answer is yes. In my opinion, this is a consequence of external factors, which, generally speaking, are not examined in order to check whether or not they correspond to the essential nature of the problem.

Certain aspects of conditional events also involve us in a consideration of the problems which derive from the presence of events with zero probability. If an event is possible, then—independently of the probability attributed to it, even if it is zero—events conditional on it, and bets related to these events, can always be considered. In this way (by means of the above-mentioned formulae, which we need not consider here), it becomes possible to compare all zero probabilities. They may be of the same order (i.e. having

† As in the example given by Vitali (quoted in Chapter 6, 6.5.9). See G. Vitali and G. Sansone, *Moderna teoria delle funzioni di variabile reale*, Zanichelli, Bologna (1935), part I, pp. 56 ff.

a finite ratio); or of a different order (ratio equal to zero, or, conversely, to infinity). For the purpose of providing an analogy, we have a situation similar to that which arises in comparing two geometrical objects of zero volume; they can be compared by considering the ratios of their areas, or of their lengths, depending on whether they are both two-dimensional, or one-dimensional. On the other hand, we would say that one was of smaller order if it were a line segment while the other were part of a plane.†

We are not concerned with pushing the analogy too far, because the geometrical case has certain special features of its own. What is common to both is the idea of measures of different orders (or, if one prefers, of non-Archimedean quantities). However, the example is to be understood in a purely illustrative sense, with a warning that one should not take into account notions like dimension, distance, volume, limit, cardinality, etc.

Let us just mention that the consideration of probability as a non-Archimedean quantity would permit us to say, if we wished, that 'zero probabilities' are in fact 'infinitely small' (actual infinitesimals), and only that of the impossible event is *zero.* Nothing is really altered by this change in terminology, but it might sometimes be useful as a way of overcoming preconceived ideas. It has been said that to assume that

$$0 + 0 + 0 + \ldots + 0 + \ldots = 1$$

is absurd, whereas, if at all, this would be true if 'actual infinitesimal' were substituted in place of 'zero'. There is nothing to prevent one from expressing things in this way, apart from the fact that it is a useless complication of language, and leads one to puzzle over 'les infiniment petits'.

Despite all that has been said, some readers may still be of the opinion that all these things are pointless hair-splittings anyway—and, in a certain sense, I would like to reply YES. The fact remains, however, that, paradoxically, the only way of dealing with these things is to think about them and analyse them in detail, carefully studying the most valid and appropriate way of setting them aside, case by case. Even in those cases where approximate answers are preferable to exact ones (because of the illusory nature of the exactness), it is especially important to be doubly precise in one's arguments, in order to know which things remain valid, and which require modification, when the reasons for, and the degree of, this illusory exactness are taken into account.

---

† A more systematic method of comparison—for simplicity, we shall always refer to ordinary, three-dimensional space—would be to consider, for each set I, the set of points I_ρ whose distance from I is less than ρ, and the function $V_I(\rho) = \mathrm{Vol}(I_\rho)$ (volume of I_ρ). One can now define the *ratio of the measure of two sets* I' and I'' to be the limit as $\rho \to 0$ of the ratio $V_I'(\rho)/V_I''(\rho)$ (if it exists). It does not always exist, but in the most 'regular' cases we have $V(\rho) = k\rho^{3-d}(1 + 0(\rho))$; in other words, $V(\rho)$ is comparable with a power ρ^α $(0 \leqslant \alpha \leqslant 3)$, and, in particular, volumes, areas, lengths, the number of isolated points are given by the coefficients k in the cases for which α turns out to be 0, 1, 2, 3 ($d = 3 - \alpha$ is the number of dimensions, $d = 3, 2, 1, 0$).

To this end, we shall make use, among other things, of the ideas that were developed concerning the 'precision' factor, and we shall arrive at conclusions which (hopefully) will appear reasonable, sensible and, perhaps, obvious. But this feeling will only be justified when we have arrived at the conclusion by means of an accurate evaluation of alternative suggestions, which clarifies just what is, and what is not, really significant and well-founded.

18.3. *Countable additivity*. An extensive treatment of this topic was given in Chapter 3, Section 3.11, and we have also referred to it on many subsequent occasions. Let us recall the main points as a prelude to making some further critical comments.

The property of additivity, which we have assumed as an axiom, says that in a *finite partition* the sum of the probabilities must equal 1. In other words, if $E_1, E_2, \ldots, E_n$ are exclusive and exhaustive, the probabilities $p_1, p_2, \ldots, p_n$ attributed to them must be non-negative with sum equal to 1. In fact, this is not merely a necessary condition for the evaluation to be coherent and admissible, but it is also sufficient.

In the case of an infinite partition into events E_h ($h \in H$, where H is arbitrary), we can only say, on the basis of our axiom, that the sum of every finite number of the p_h must be $\leqslant 1$: in other words, that at most a countable number of them can be positive ($\neq 0$), and that for such values† we must have $\sum p_h \leqslant 1$. If, in particular, the set of positive p_h has sum $= 1$, then the E_h with zero probability also have zero probability when taken together: i.e. it turns out that the union $E = \sum E_h$ $(p_h = 0)$ also has zero probability, $\mathbf{P}(E) = 0$. If, on the other hand, we obtain $\sum p_h = P < 1$, i.e. if a probability $1 - P$ is *missing* in the partition, then the possibilities are as follows: if there are a finite number of events with non-zero probability, then this missing probability is necessarily that of $E =$ the union of the events with zero probability; otherwise, it may be attributed arbitrarily to E and to $\tilde{E} =$ the union of the events with positive probability;

$$\mathbf{P}(E) = P', \qquad \mathbf{P}(\tilde{E}) = P + P'', \qquad P' + P'' = 1 - P.$$

† Even if there are an infinite number of them one can speak of a 'sum' in the sense of 'upper bound of the sum of a finite number of terms' (if one thinks of the 'sum of the series' no conclusion would be legitimate). If we denote by $\sum$ the upper bound (possibly $+\infty$) of the sums of a finite number of terms, we may denote in this way the sum of an arbitrary infinite number of non-negative numbers, and, in particular, of events. For example, $\sum E_h(h \in K)$ will denote the number of successes among those E_h for which $h \in K$, and we observe that the standard convention—cf. Chapter 1, Section 1.9—in which $(h \in K) = 1$ or $(h \in K) = 0$ according as h belongs, or does not belong, to K, allows one to write $(h \in K)$ as a factor, on the same line, instead of as an index written below the $\sum$ sign (cf. the application, which follows shortly, with $(p_h = 0)$ for $(h \in K)$ where $K =$ 'the set of the indices for which $p_h = 0$'). If the E_h are incompatible, the sum is necessarily either 0 or 1.

To summarize: given the probabilities p_h of the events E_h of a partition, if their sum is $=1$ then the probabilities of all the events depending on it—i.e. sums of a finite or infinite number of events in the partition—are uniquely determined. In this case, we shall say that the probability $\mathbf{P}(E)$ is countably additive on the partition $\{E_h\}$. Otherwise, this only holds for event-sums of a finite number of the E_h, or for their complements: in any other case, a margin of indeterminacy equal to $1 - P = 1 - \sum p_h$ remains;

$$p' = \sum p_h(E_h \subset E) \leqslant \mathbf{P}(E) \leqslant 1 - \sum p_h(E_h \subset \tilde{E}) = p'' = p' + (1 - P).$$

Let us be clear that 'indeterminacy' simply means that the extension is not, in general, uniquely defined; one only has bounds. There is no 'indeterminacy' in any specific sense; such as being 'barred' from attributing a well-defined value to $\mathbf{P}(E)$. It is simply that E is not one of the events whose probability has already implicitly been evaluated by virtue of our evaluations for the E_h; it is just one of the many for which our choice is more or less open. We are completely free in our choice (i.e. can give $\mathbf{P}(E)$ any value between 0 and 1) in the particular case in which all the events of the partition have been attributed zero probability (and E is not the sum of a finite number of the E_h, nor the complement of such a sum). This happens in the case of a continuous distribution (on the line, or in the plane, or in ordinary space, ...) for which we have established only that it is 'without concentrated masses', or for a countable number of exclusive (and exhaustive) events of zero probability.

Conclusions of this kind may be hard to accept, or perhaps may even appear paradoxical. At least, the way in which many authors bend over backwards to avoid them—by introducing some new axiom (or 'strengthening' the existing ones)—seems to suggest that this is the case. The following are some of the kinds of restrictions which could be imposed:

(Z) denying that it is legitimate to attribute zero probability to a possible event;

(Za) denying that a union of events with zero probability can have non-zero probability;

(Zb) as (Za), but only considering a countable number of events;

(Ka) assuming countable additivity for arbitrary partitions;

(Kb) as (Ka), but only for countable partitions.

We have introduced the letters Z, Za, Zb, Ka, Kb, in order to facilitate references to these 'axioms'; in what follows, we shall, of course, argue against them.

The first and the last have actually been proposed; Za and Zb are pro-progressively weaker versions of Z, and are also special cases of Ka, Kb, respectively; the inclusion of intermediate possibilities would only serve the purpose of pointing out these connections.

First of all, we should draw attention to the lack of any real arguments on the part of those who support such a restriction. It is usually presented as a 'natural' extension of the theorem of total probability as $n \to \infty$ (as, for example, in Cramèr); or as a 'natural' property by analogy with Lebesgue measure (and this is the most common idea); or by Baire extension by continuity (like, for example, in Feller). In other words, for 'mathematical' reasons, and not for reasons relating to probability theory.

One mathematical consequence of this is that it becomes impossible to think of a $\mathbf{P}(E)$ defined for all the events which could be formed on the basis of a non-denumerable partition. An example is provided by the power set of any set whose cardinality is that of the continuum (by virtue of the results of Vitali, Lebesgue, Banach, Kuratowski and Ulam, concerning the impossibility—except in trivial cases of 'concentrated mass' at a finite or countable number of points—of extending σ-additive measures to all the subsets of a non-denumerable set†). To admit σ-additivity is to contradict the basic idea that one can attribute to any uncertain event whatsoever a probability—without any, logically inexplicable, discrimination between one event and another. Of course, it could happen that this 'basic idea' itself gives rise to conflicts with other requirements: for example, were it true that there does not always exist an extension of a finitely additive $\mathbf{P}$, we should have had to re-examine the whole question of whether, and in what way, a mathematical theory of probability was possible (with goodness knows what weakening of the axioms‡). The fact that such a disaster does not occur for finite additivity, but does occur if one attempts to replace it with σ-additivity, clearly indicates that the substitution is entirely inappropriate.

If one accepts the subjective concept of probability, the conclusion becomes even more obvious.

In order to reach this conclusion, it was not even necessary, in fact, that the contradiction of not being able to find an arbitrary $\mathbf{P}$ came to light. It was sufficient that the choice was restricted in a way which appeared to

† From our point of view, it suffices that this has been established for some sets. It appeared preferable, therefore, not to weigh down the text with details of how the result has been proved (Ulam) provided that the cardinality of the set is not 'inaccessible'.

‡ A more 'minor' difficulty may serve as an example. A paradox, due to Haussdorff, says that a spherical surface can be divided into three sets, A, B, C such that each is superposable both on each of the others and on their union; it follows that any 'measure' which was finitely additive and invariant under rigid motions would assign to these sets both $\frac{1}{3}$ and $\frac{1}{2}$ (and $\frac{2}{3}$ and so on; as is well-known, one can logically deduce anything if one starts from something ridiculous). This contradicts geometric intuition, but not the idea of probability (nor the 'axiom of choice'). As Paul Lévy said, in order to refute this interpretation, the simple fact is that 'the continuous in higher dimensions is even more complicated than we thought'. (Cf. E. Borel, *Les paradoxes de l'infini*, Gallimard, Paris (1946); Paul Lévy, 'Les paradoxes de l'infini et le calcul des probabilitiés', and a note by Borel, in *Bull. Sci. Math* (1948), pp. 184–192.)

preclude each individual being permitted an unfettered evaluation. And this occurs even in the case of a countable partition. Let us suppose that there are a countable infinity of 'possible cases', and—in order to avoid thinking of points or sets on the real line which appear 'special' in some way—let us imagine that they are represented by points on the circumference of a circle, whose distance apart is a rational multiple of 2π (i.e. by taking the origin at an arbitrary one of these points, with $\theta = 2k\pi$, k rational, $0 \leqslant k \leqslant 1$). That an axiom should not permit me to attribute probabilities which are negative, or have sum greater than one, is something which can be clearly understood as a condition of coherence; it does not impose any restriction on my freedom of opinion. But suppose that an axiom (like *Zb* or *Kb*) prohibits me from attributing the same probability, $p_h = 0$, to all the events; or even (like *Kb*) forces me to choose some finite subset of them to which I attribute a total probability of at least 99 % (leaving 1 % for the remainder; and I could have said 99·999 % with 0·001 % remaining, or something even more extreme). If I do not happen to hold these opinions, and have no reasons for adopting them, then this is no longer a question of coherence; it is a direct interference with my judgement!†

Moreover, to permit the assignment of zero probabilities to all the events (of a countable partition) is a much less restrictive idea than, as in the finite case, considering them as 'equally probable' ($\mathbf{P}(E_h) = 1/n$, $h = 1, 2, \ldots, n$). The equivalent of this would be to consider the $\mathbf{P}(E_h)$ equal, not in the sense that $\mathbf{P}(E_h) = \mathbf{P}(E_k) = 0$ as real numbers, but in the sense that $\mathbf{P}(E_n)/\mathbf{P}(E_k) = 1$ (as a ratio of zero, or 'infinitely small', probabilities). For the first condition ($p_h = 0$) to hold, much less is required: in terms of ratios, it is sufficient that there do not exist probabilities of maximal order (for example, that give, for each h, $\mathbf{P}(E_{h+1})/\mathbf{P}(E_h) = \infty$), or that, if they do exist, their sum (taking one of them to be unity) is infinite. (It is also sufficient—but this cannot be derived from the ratios alone—that the probability of all the cases be infinite in the given scale; and it may happen that this occurs for the union of cases with probability of smaller order without occurring for those of maximal order.)

Assuming (in line with Axiom 3) that, if accepted, axioms *Zb* or *Kb* should also hold for probabilities conditional on an arbitrary possible event H, they would imply even more restrictive conditions for the probabilities of individual 'possible cases', in order to avoid—pulling out a countable number of them—the possibility of a case of probabilities all zero. There could be at most a countable number with the same order, so that, given a non-

† It is strange that the very same people who, in general, would encourage one in the finite case to accept a judgement of equal probabilities, on the grounds that a person 'knows nothing', seek to prohibit someone who, on the grounds that he 'knows nothing', would like to make the same judgement in the countably infinite case.

denumerable infinity of 'possible cases', we should have a non-denumerable infinity of different 'orders' of probability.†

We should also mention that, from time to time, problems which, explicitly or implicitly, run counter to the assumption of countable additivity are also considered by authors who insist on the latter as an axiom. Sometimes the case of 'an integer chosen at random' is regarded as 'meaningful' but 'breaking the rules' (the probability taken to be the limit density; for example, the probability that the integer is a multiple of $k = \lim$ [(number of multiples of k between 1 and $n)/n] = 1/k$). At other times (see, for example, Rényi, Chapter 3, 3.18.5), one considers conditional probabilities; for instance, a distribution inside a circle, which is then made larger and larger, so that the probability of each finite region tends to zero. Countable additivity is prescribed for the conditional probability (i.e. within any circle), but it is not made clear that this no longer holds for the limit distribution (which is not explicitly dealt with in its own right, the passage to the limit being merely a device).

This seems to provide further evidence in support of our initial impression that the assumption of countable additivity owes very little to genuine probabilistic considerations; in other words, that it is more a mathematical embellishment than a necessary property of probability. Many other anomalies and peculiarities (one might even say—in a psychological rather than a formal sense—contradictions) strengthen the same impression. The fact that 'equal probabilities' are perfectly acceptable in the finite and continuous (points of an interval) cases, but are not allowed in the countable case, can be explained only by drawing attention to our habit of applying, in particular cases, the most widely used tools (in the countable case we are accustomed to summing series!), rather than adhering to the principle of coherence. The very fact that one treats the finite and uncountable cases differently from the countable case (axioms *Zb* and *Kb*) is sufficient to show that more thought is given to the mathematical structure than to the logical problem—for which the meaningful distinction, if any, would seem to be that between the finite and the infinite (of whatever kind).

Bearing this in mind, the line we have followed here seems to represent, independently of the reasoning we have put forward—which, hopefully, is more persuasive—the most natural way of connecting together attitudes which are, at least in part, inspired by fragmentary and irreconcilable points of view.

Finally, we should mention a concept and a result which have come to be considered as a justification for the systematic use of σ-additivity. The basic

† One can say even more: those of the same order must posses a convergent sum; the different 'orders', arranged in decreasing order, must form a 'well-ordered' sequence, so that there always exists a 'maximal order'.

idea is the possibility of stretching the interpretation in such a way as to be able to attribute the 'missing' probability in the partition to new fictitious entities in order that everything adds up properly. In some cases, in order to salvage countable additivity, it is even claimed that the new entities are not fictitious, but real. I remember having seen something of this kind in a paper (by Kingman, I think) which involved a probability distribution for discrete processes concentrated in the neighbourhood of a limit case where the process would become continuous.† This was taken as an indication of the necessity of including the continuous limit cases among the possible cases, in order to be able to foist upon them the probability missing from the sum.

Using this kind of argument, one could say that if the possible cases were the rationals, and if to each of them is attributed zero probability, then we have demonstrated that the real numbers must also *exist*, and be possible, because they are required as the indispensable support for the probability as a whole ($= 1$).

The more general kinds of considerations we have alluded to are more abstract, although, in the final analysis, they reduce to the same type of argument as is involved in the addition of fictitious entities. In mathematics, this kind of argument or procedure is well-known to be fruitful (like, for example, the addition of new points in order to compactify a space), but, in our case, events must be events, and not abstractions, if the theory is to preserve a concrete meaning; that is to say 'has some meaning', and this, in the formulations we have mentioned, does not happen.

In fact, it is necessary to have recourse to 'ultrafilters' (and this gives only a theoretical possibility of obtaining the desired result). In any case, this would only hold in a field which has been modified with respect to the original one, and the latter is the only one that we are interested in. I have never seen any application to the study of actual cases (and it seems impossible that it should constitute a simplification, rather than an unnecessary complication, introduced for the purpose of permitting yet another unnecessary complication, i.e. σ-additivity). It seems to me that the only result has been to encourage people even further to consider just those cases where σ-additivity holds directly, and to ignore the others because they can, in theory, be transformed in such a way as to turn out to enjoy, fictitiously, the property which, in the field one ought to be considering, does not actually hold.

In practice, there are quite different purposes for which consideration of ultrafilters can be useful. In particular, for studying '*agglutinated* probabilities' (i.e. probabilities that cannot be subdivided), which can arise in

---

† *Translators' note.* J. F. C. Kingman, 'Additive set functions and the theory of probability', *Proc. Cam. Phil. Soc.*, **63** (1967), pp. 767–775.

distributions.† Think of the case (which we have already mentioned) of distributions on an ultrafilter: an ultrafilter is a family of events (sets) to which one and only one element of a partition can belong, and we attribute probability 1 to the events belonging to it (and, therefore, probability 0 to the others).

The consideration of *filters* can also be useful if one wishes to analyse further the possibility of dividing up the *missing* probability. In general, given a partition into events E_h, with $\sum p_h < 1$, it is sufficient to consider another event B (or a partition, $B_1, B_2, \ldots, B_n$), and to form the partition BE_h (or the partitions $B_1E_h, B_2E_h, \ldots, B_nE_h$) of B (or of B_1, of $B_2, \ldots$, of B_n). The missing probability $1 - \sum p_h$ can then be divided up between the filters generated by B and by $\tilde{B}$ (or by $B_1, B_2, \ldots, B_n$). Think, in particular, of the mass adherent to a point (in a distribution on the real line).

The probability adherent to the left (or to the right) may be further divided up by considering filters; for example, in the case of the sequence of sets I_n of rationals between $x - 1/n$ and x, one obtains the probability adherent from the left on the rationals, or on the irrationals, etc. Of course, just as knowledge of $F(x)$ is not sufficient to separate possible adherent masses from the concentrated ones, it is even less sufficient for these subdivisions, which have to be established on the basis of other considerations.

18.4. *Concerning what is 'reasonable'*. It would be very difficult to reach any conclusion, or to make any constructive progress by attempting to conduct a discussion of this topic with supporters of opposing points of view. Each would first of all attempt to challenge the 'reasonableness' of the assumptions of the others, judging them to be too 'theoretical', lacking any concrete value, and based on the assumption of an absolutely unrealistic degree of precision.

There would be no difficulty for anyone in criticizing the formulations of others, and no doubt anyone making such criticisms against the formulation we have adopted here would find good reasons for so doing. The complications we have considered, however, do not arise, unless one wishes to isolate cases which are in a certain sense 'pathological'. For problems which are 'sensible from a practical point of view', one not only avoids these complications, but also those imposed everywhere *a priori* by the assumption of σ-additivity. The latter are harmful because they go beyond what is required in simple cases and, moreover, are over-restrictive in the complicated cases.

† Cf. B. de Finetti, 'La struttura delle distribuzioni in un insieme astratto qualsiasi', *Giorn. Ist. Ital. Attuari*, **XVIII** (1955), pp. 1–14. An English translation of this paper, 'The structure of distributions on abstract spaces', forms Chapter 7 of B. de Finetti, *Probability, Induction and Statistics*, Wiley (1972).

Our criticism of countable additivity on the grounds that it precludes one attributing probability to all events (for example, the extension of Jordan–Peano measure to all the sets of the interval [0, 1]) is in no way intended as implying that the extension to Lebesgue measure is considered insufficient, or that one actually wishes to go further. On the contrary, it means that we consider it to be usually quite sufficient to confine attention to Jordan–Peano measure, but that, if one wishes to go further, the extension should be neither predetermined, nor ruled out in any way. In other words, Lebesgue measure is just one of the infinite number of extensions to larger families of sets, and one should be free to choose any of these if one wishes to make such an extension. Should anyone opt for countable additivity as a matter of preference, there is no objection (just as, in a practical case, it is open to one to choose a distribution possessing a continuous density, rather than a less 'regular' one, without feeling that one is forced to make such a choice by virtue of some law of probability). Any other extension (to all sets) is equally legitimate (in principle: so far as its usefulness is concerned, it is not clear whether the Lebesgue extension should be regarded as useful, once it has been made clear that the consequences one derives from it are not consequences of the initial evaluation by virtue of the 'law' of countable additivity, but rather that they derive from the arbitrary choice of a particular one of the possible extensions of the given evaluation).

On the other hand, a similar criticism can be made at a much more basic and fundamental level. In many approaches, one establishes *a priori* that the probability $\mathbf{P}(E)$ has to be given for all the events E of some given family obeying some given conditions; for example, forming a field (in the above case a Borel field) which is considered fixed once and for all. One can then go on to consider only those problems which belong within that field (considered as a single, closed system, and often referred to as a 'probability space'). It is not then possible to evaluate the probabilities of two events A and B without doing the same for the product AB. But it could well be that sometimes one either has to, or wishes to, proceed (albeit temporarily) without the knowledge or evaluation of $\mathbf{P}(AB)$; or that the family of events initially considered (and 'arbitrary') does not contain all the products. The conclusions hold for all events and random quantities linearly dependent on those events one starts from. Indeed, we could start from random quantities, X, for which the previsions, $\mathbf{P}(X)$, were evaluated: the particular fact of whether all, or just a few, or none, of them are events is in itself irrelevant. The case of events seems simpler and more intuitive only because it is more familiar, as well as being more schematic, and capable of more varied representations (set-theoretic, for example). One may then examine for each problem (with no limitations of any kind) the implications of the evaluations already assumed made, and one can complete them by means of any further evaluations that are required to answer the questions of interest.

There is no need to make use of, or mention, probabilities conditional on events of zero probability (or to compare zero probabilities, as non-Archimedean quantities) except when this might be useful for more careful consideration of delicate situations—where it is otherwise easy to adopt a cursory attitude.

As is illustrated in the cases used as examples, cases which are representative of the general situation, the approach we adopt consists in keeping the treatment at the simplest and most concrete level, adhering to the practical meaning, rejecting assumptions which are not supported by compelling arguments (like that of replacing finite additivity by σ-additivity), rejecting the once-and-for-all fixing of closed structures, and, instead, in always open-mindedly allowing the possibility of extending the probability field to be studied, as and when required.

In this sense, we obtain the maximum simplification. There are, however, certain circumstances in which complications do, in fact, arise. This happens when the assumption of σ-additivity leads to simple conclusions (well-known from measure theory), which either no longer hold if we abandon the assumption, or require more careful and less 'intuitive' formulations. What should we reply to someone who objects to complications of this kind?

Our reply is that such complications are inherent in the fact that when we speak of probability (or prevision) we are referring to functions **P** which may well be finitely additive (instead of being assumed, *a priori*, by virtue of an 'axiom', to be σ-additive over the field under consideration; or, and it amounts to the same thing, that the term 'event' can only be applied to members of some σ-field, restricted in such a way that **P** is σ-additive over it).

There are, therefore, two possibilities. On the one hand, it may be that we wish to be able to make a statement which holds only under the assumption of σ-additivity; in this case, one can state it as it is, making explicit the assumption that **P** be σ-additive (over some given field, or, often, just over some particular partition). There are no complications, apart from that of stating the assumption, and this has the advantage over the abandoned axiom in that it only requires the assumption of the latter over the minimal field for which it is required. The approach is rather like forcing a person to declare that a property holds for continuous functions, or for functions continuous in a given interval, or at a particular point, when the person is accustomed to stating it as valid for all functions (leaving it to be understood that he is only referring to functions which are continuous everywhere).

Alternatively, it may be that one wishes the statement to be valid without the restrictive hypothesis under which those things which held under σ-additivity continue to be true. In this case, matters become more complicated (unless, for reasons of simplicity, one prefers an inaccurate statement). To put it concisely, there is always the possibility of choice: either stick to the assumption of σ-additivity (no longer considered as an axiom), making it clear that one is doing so, or state things in the form necessary for them to turn out to be true independently of this assumption.

It will suffice to recall various of the cases we have already examined (the reader can, if necessary, refer back to the extensive discussion given in the text); the possibility of masses adherent to a point (instead of concentrated at it), and, in particular, concentrated at infinity (improper distributions); the indeterminacy of $\mathbf{P}(X)$ with respect to distributional knowledge in the case of unbounded distributions; the bogus formulation of the 'strong law of large numbers', and related topics.

In many cases, simple, minor modifications of the kind put forward are sufficient to ensure the validity of a statement, independently of the axiom of countable additivity. Moreover—although this is a matter of taste—they serve, because of their 'finitistic' character, to give a more concrete air to things.

All previous considerations can be regarded as variations on a single fundamental theme: the desirability of basing oneself on axioms which are the weakest (i.e. least restrictive) from a mathematical point of view, because they are from a logical point of view the most securely based (i.e. the least disputable), and which lead to results and statements which are the most secure (i.e. the least disputable).

Let us consider again the introduction of $\hat{\mathbf{P}}(X)$ (Chapter 6, 6.5.7). The mathematical definition is unexceptionable, and this, by the standards normally adopted, suffices to render $\hat{\mathbf{P}}(X)$ acceptable *by definition* as the value of $\mathbf{P}(X)$. In contrast, we put forward arguments whose purpose was to establish the existence of possible reasons for considering this choice as being, in addition, a 'reasonable' one (and this, in a certain sense, is even more important). We were careful, however, not to identify $\mathbf{P}(X)$ with $\hat{\mathbf{P}}(X)$. In fact, $\hat{\mathbf{P}}(X)$ is always just *one* of the possible values for the extension of $\mathbf{P}$ outside the field within which it is uniquely defined by the F (although, in a certain sense, it is the most 'reasonable' extension).

This example, and the discussion arising from it, serves also as an illustration of the kind of attitude that results from the choice of a 'conceptual' approach as opposed to a 'formal' one, in the sense already considered. All the ideas and results are drawn from the *meaning* that lies behind the axioms, and not from the mathematical conventions. In contrast to the tendency towards uniquely determining the extension of certain notions by means of special forms of passage to the limit, our effort consists in not admitting, even inadvertently, any restriction which is not the result of simple finitistic inequalities, and which—one might say—goes beyond the idea of the 'method of exhaustion'.

It is not a question of weighing up, *a priori*, one's preferences for this or that mathematical approach, but, on the contrary, of emphasizing the need to choose, in any application, the tools most suited to the nature and meaning of the problem. The nature and meaning must not be distorted or disguised in order to introduce tools of a more or less elegant, sophisticated, or 'fashionable' kind.

18.5. *Countable additivity as continuity.* We return to the topic of countable additivity once again, this time in a different (although equivalent and suggestive) guise. We shall examine some further questions, and provide further discussion.

The condition of countable additivity for events (as considered so far) can be expressed in an even more meaningful form as a 'continuity' condition. This is the condition which appears among the axioms given by Kolmogorov (and other authors) in the following form (which we shall call 'axiom' Kb'):

if $E_1, E_2, \ldots, E_n, \ldots$ is a sequence of events, each of which is contained in the preceding one, and whose product is empty (i.e. there are no 'elementary outcomes' common to all the E_n), then $\mathbf{P}(E_n) \to 0$ as $n \to \infty$.

We can see immediately that this condition is equivalent to countable additivity. Let us write

$$E_1 = (E_1 - E_2) + (E_2 - E_3) + \ldots + (E_{n-1} - E_n) + E_n;$$

all the terms in brackets are events by virtue of the inclusion hypothesis, and the probability of E_1 is the sum of the probabilities of the $(E_h - E_{h+1})$ up to some point, plus the remainder. If the latter tends to zero, as axiom Kb' requires, the probability is given by the sum of the series, and countable additivity holds. The argument can be turned around straightforwardly: starting from a sequence $C_1, C_2, \ldots, C_n, \ldots$ of incompatible events, and setting $E_n = C_n + C_{n+1} + \ldots$, we reduce to the preceding case (with $C_n = (E_n - E_{n+1})$); in order that the series of the $\mathbf{P}(C_n)$ converges, the remainder, i.e. $\mathbf{P}(E_n)$, must tend to zero.

In fact, it is easily seen that Kb' leads, in general, to a further property, even more meaningful, and showing more clearly the appropriateness of the term 'continuity'. Note that for any sequence of events, or random quantities, we can consider the *lower limit* and the *upper limit* (and also, if these coincide, we can consider the *limit*, their common value), just as in analysis. The fact of whether the values of the sequence are known or not (random) is irrelevant. In particular, in the case of events, $E' = \liminf E_n$ and $E'' = \limsup E_n$ are the events which consist in the fact that a finite number of the E_n are *false* and an infinite number are *true*, respectively (i.e. a finite number take the value 0 and an infinite number take the value 1, respectively). To say that $E_n \to E$ (i.e. that E' and E'' *coincide*), or that the limit E of the sequence E_n exists, is to say that *necessarily*, in the case under consideration, from some N onwards the events E_n are either all true or all false (in other words, it is impossible for infinite sequences of both true and false events to occur).

Well, then: in the case of countable additivity one has

$$\mathbf{P}(E') = \mathbf{P}(\liminf E_n) \leqslant \liminf \mathbf{P}(E_n) \leqslant \limsup \mathbf{P}(E_n) \leqslant \mathbf{P}(\limsup E_n) = \mathbf{P}(E'');$$

in particular, if the limit E exists, $\lim \mathbf{P}(E_n) = \mathbf{P}(\lim E_n) = \mathbf{P}(E)$.

It remains to check that the same condition holds more generally when we have a sequence of random quantities X_n rather than events.

The property

$$\mathbf{P}(X_n) \to 0 \quad \text{if } X_n \to 0$$

is valid (under the assumption of countable additivity) *if the random quantities X_n are uniformly bounded*. Suppose, in fact, that, for all n, $|X_n| < K$; then, for any (small) $\varepsilon > 0$, we have

$$|\mathbf{P}(X_n)| \leqslant \mathbf{P}(|X_n|) < \varepsilon + K\,.\,\mathbf{P}(|X_n| > \varepsilon)$$

(because $|X_n| < \varepsilon + K\,.\,(|X_n| > \varepsilon) = \varepsilon$ if $|X_n| < \varepsilon$, and $= \varepsilon + K$ otherwise). But if $X_n \to 0$ we also have $(|X_n| > \varepsilon) \to 0$, and this means—recall the above—that we cannot have an infinite number of $|X_n| > \varepsilon$, and hence (assuming Kb') $\mathbf{P}(|X_n| > \varepsilon) \to 0$. It follows that $\lim |\mathbf{P}(X_n)| < \varepsilon$, and, since ε is arbitrary, that

$$\lim \mathbf{P}(X_n) = \lim |\mathbf{P}(X_n)| = 0.$$

Since events are uniformly bounded ($|E_n| \leqslant 1$) the property we have established is equivalent to Kb' (i.e. to countable additivity). If we remove the condition of uniform boundedness, the property does not hold (even if countable additivity holds). Suppose we take a countable partition into events E_n to which we assign non-zero probabilities p_n with sum $= 1$, and let us consider the sequence of random quantities $X_n = E_n/p_n$ (which are not uniformly bounded). We obtain $\mathbf{P}(X_n) = p_n/p_n = 1 \to 1 \neq 0$, although $X_n \to 0$ (all the X_n but one are, in fact, $= 0$). By slightly modifying the example, putting $X_n = E_n/p_n^\alpha$ for instance, we obtain $\mathbf{P}(X_n) = p_n^{1-\alpha}$, and we see, therefore, that the property $\mathbf{P}(X_n) \to 0$ holds for $\alpha < 1$, whereas, if $\alpha > 1$, we have $\mathbf{P}(X_n) \to \infty$ (although, for the same reason as before, we still have $X_n \to 0$).

The extension of the property to the limit is similar. Assuming Kb, we have

$$\mathbf{P}(X') = \mathbf{P}(\liminf X_n) \leqslant \liminf \mathbf{P}(X_n) \leqslant \limsup \mathbf{P}(X_n) \leqslant \mathbf{P}(\limsup X_n) = \mathbf{P}(X'')$$

if the X_n are uniformly bounded. We shall give the proof in the case of the upper limit (the other case is clearly symmetric), and our proof includes the case of events (where a proof was not given).

Putting $X''_n = \sup X_h$ (for $h \geqslant n$), $X'' = \lim\sup X_n = \inf X''_n$, we obtain $X''_n = X_n + (X''_n - X_n) = X'' + (X''_n - X'')$, where $(X''_n - X_n)$ and $(X''_n - X'')$ are non-negative. We therefore have, in every case,

$$\mathbf{P}(X_n) \leqslant \mathbf{P}(X'') + \mathbf{P}(X''_n - X''),$$

and, if $\mathbf{P}(X''_n - X'') \to 0$, we shall have $\lim\sup \mathbf{P}(X_n) \leqslant \mathbf{P}(X'')$. But

$$X''_n - X'' \to 0,$$

by definition, and, if we assume the X_n to be uniformly bounded (which implies, *a fortiori*, that the $X''_n - X''$ are), then, assuming countable additivity, the condition will be satisfied.

In particular, if the sequence of the X_n converges (definitely) to a limit (in general random), $X = \lim X_n$, we can, if we assume countable additivity plus the uniform boundedness of the X_n, state that $\mathbf{P}(X) = \lim \mathbf{P}(X_n)$. The case of a series $\sum X_h$ reduces to the preceding case if we consider the partial sums, $Y_n = \sum X_h$ $(h \leqslant n)$. If we call Y' and Y'' the infimum and supremum of the sums, we have

$$\mathbf{P}(Y') \leqslant \inf \sum \mathbf{P}(X_h) \leqslant \sup \sum \mathbf{P}(X_h) \leqslant \mathbf{P}(Y''),$$

and, in particular, we have $\mathbf{P}(Y) = \mathbf{P}(\sum X_h) = \sum \mathbf{P}(X_h)$ if $Y' = Y'' = Y$ (i.e. the series is definitely convergent), under the condition that the remainders are uniformly bounded (and always, of course, with the assumption of countable additivity).†

We have said that we do not intend to consider countable additivity as an axiom; for us, it is a property which may appear more or less interesting, and which will hold over certain partitions and not over others. Interpreting the condition as one of continuity, we can reformulate this fact in a more meaningful way by saying that it will hold *over certain linear spaces* and not over others.

This approach has the merit of spotlighting the real essence of the problem; the fact that the property of countable or finite additivity, i.e. of continuity or the absence of continuity, concerns the behaviour of the function $\mathbf{P}$ over a linear space $\mathscr{L}$. To give a complete account of the behaviour of $\mathbf{P}$ in terms of continuity involves, therefore, distinguishing which linear spaces $\mathscr{L}$ belong

† The convergence of the series $\sum \mathbf{P}(|X_h|)$ (together with the assumption of countable additivity) is a sufficient condition to establish that $\sum X_h$ has probability $=1$ of being convergent, and that (putting therefore, arbitrarily, $Y = Y'$, or $Y = Y''$, or $Y = Y' = Y''$, if they coincide, and otherwise $Y = 0$, etc.) one has $\hat{\mathbf{P}}(Y) = \sum \mathbf{P}(X_h)$ (N.B.: $\hat{\mathbf{P}}$ not $\mathbf{P}$). Under this hypothesis, in fact, for arbitrary choices of positive ε and λ, there exists an N such that, for any q, we have $\sum \mathbf{P}(|X_h|)(N \leqslant h \leqslant N+q) < \lambda\varepsilon$, i.e. $\mathbf{P}\{\sum |X_h|(N \leqslant h \leqslant N+q)\} < \lambda\varepsilon$, and, *a fortiori* (denoting the preceding summation by $\{\sum_{N,q}\}$ for short), $\mathbf{P}(\sum_{N,q} > \lambda) < \varepsilon$ (because, if X is certainly positive, $(X > \lambda) \leqslant X/\lambda$). If the axiom of continuity holds, as we have assumed, the limit-event $(\sum_N > \lambda) = \lim(\sum_{N,q} > \lambda)$ (as $q \to \infty$) also has probability $\leqslant \varepsilon$, and, *a fortiori*, it follows that the fact that the series diverges (in which case the remainder $\sum_N$ will be ∞) has probability $\leqslant \varepsilon$, and hence (since ε is arbitrary) *zero*.

to the complex Λ_p of linear spaces over which $\mathbf{P}$ is continuous, and which do not.

More precisely, in line with what we have said previously, and with the condition of coherence, we shall say that $\mathbf{P}$ is *coherent and continuous* on $\mathscr{L}$ if no random quantity X of the form

$$X = k_1(X_1 - \mathbf{P}(X_1)) + k_2(X_2 - \mathbf{P}(X_2)) + \ldots + k_n(X_n - \mathbf{P}(X_n)) + \ldots$$

turns out to be uniformly positive (where the k_h are any real numbers, and the X_h belong to $\mathscr{L}$), not only for sums involving a finite number of terms (as is required for coherence), but also for series (convergent, and with uniformly bounded remainders).

It is clear that if $\mathscr{L}$ belongs to Λ_p then so does every linear space contained in it, and so does the closure, $\bar{\mathscr{L}}$, formed by all the random quantities that can be obtained from $\mathscr{L}$ by means of the passage to the limit in the sense given:

$$(X_n \to X, |X_n - X| < K).$$

If $\mathscr{L}_1$ and $\mathscr{L}_2$ belong, then so does $\mathscr{L}_1 + \mathscr{L}_2$ (the linear space of sums $X_1 + X_2$, $X_1 \in \mathscr{L}_1$ and $X_2 \in \mathscr{L}_2$): this holds for any finite number of spaces $\mathscr{L}_h$, *but not for an infinite number.*† This indicates that the most 'natural' hypothesis is not true; a hypothesis which corresponds most closely to the standard point of view because it leads to a distinction between those events and random quantities which belong to a certain system of 'probabilizable' entities, and those which do not. This is the hypothesis that the complex Λ_p consists of all and only those linear spaces which belong to some given linear space $\mathscr{L}^*$, which, in this case, would have acquired the meaning of 'total field of continuity'.

19 QUESTIONS CONCERNING QUALITATIVE FORMULATIONS

19.1. There are many senses in which the words qualitative probability have been used, some of them very different from each other. To attempt to list them and classify them would be both tedious and pointless, but something must be said in order to point out the necessity of not confusing things which do differ, and of not being put off by apparent absurdities. Among the latter, for example, we include the fact that one might expect to encounter

† Consider a countable partition of events $E_{hk} (h, k = 1, 2, \ldots, n, \ldots)$. Let us denote by $E_h = \sum_k E_{hk}$ the sum of events whose first subscript is h, and suppose we attribute the values $p_{hk} = \mathbf{P}(E_{hk})$ and $p_h = \mathbf{P}(E_h)$ in such a way that $\sum_k p_{hk} = p_h$ (for each h), but $\sum_h p_h < 1$ (i.e. $\sum_{hk} p_{hk} < 1$). On the linear spaces $\mathscr{L}_h$ defined by the E_{hk} and E_h, $\mathbf{P}$ is continuous, and hence is also continuous on every linear space $\mathscr{L}$ determined by a finite number of $\mathscr{L}_h$. This no longer holds, however, if we consider the space $\mathscr{L}$ determined by the whole infinite collection of $\mathscr{L}_h$.

rather vague considerations, but can, in fact, find oneself forced into hair-splitting detail, obliging one to apply, in all cases, the methods of comparison introduced for zero probabilities.

Our day to day judgements are on the whole rather vague, and we usually limit ourselves to just a few verbal gradations (quite probable, or very, very much, not much, very little,...), or to percentage approximations (50%, 75%, 90%, 99%,...). In comparisons between two events, the probabilities will be said to be 'roughly equal' if the dominance of one over the other does not appear to be obvious. At this level, however, there is not even the possibility of arguing in mathematical terms.

Sometimes, one thinks of vagueness in the sense of 'indeterminacy' (for example, between precise numerical bounds); we have already referred to this, and we shall return to it later. At other times, one is willing to compare (let us assume exact comparison, in order not to get lost in too many subcases) the probabilities of events, but without using numerical probabilities. A physician might have a quite precise opinion, in a comparative sense, concerning the probabilities that a small number of patients will overcome their present disease, but without knowing what to do if he were required to compare them with the probability of obtaining something other than a '6' on the role of a die (or, more explicitly, were he required to state whether they were more or less than $\frac{5}{6}$; i.e. 83·3%). Sometimes, this inability to compare them with numbers is attributed to innate peculiarities of the events in question (rather than to contingent reasons, such as lack of practice; cf. Borel's review of Keynes' treatise†), or to the fact of not having at one's disposal (or not wishing to have) devices such as dice, urns, and so on. In this case, if comparability is assumed exact, as in the comparison of intervals where one is led to say that a closed interval (i.e. end-points included) is greater than one of equal length but open (i.e. end-points excluded), it is clear that a non-Archimedean scale results (and this is the absurdity we referred to at the beginning).

Other considerations arise when indeterminacy has a precise meaning; when, on the basis of some data, one can establish only that a probability p belongs to an interval $p' \leqslant p \leqslant p''$. That we are not dealing with an essential indeterminacy is a point that we have stressed. Nevertheless, there is something to be said here, and a few points will have to be made in connection with the discussion (in Sections 5–7) relating to the verifiability of events and measurement of quantities.

19.2. *Axiomatic formulations in qualitative form.* In all the methods of approach we have so far looked at, we have introduced, straightaway,

† The article is reprinted in Borel's *Traité* (as note 2 in issue III of Vol. IV); an English translation is given in H. E. Kyburg and H. E. Smokler, *Studies in Subjective Probability*, Wiley, New York (1964).

numerical values for probabilities under the intuitive guise of prices, and as parameters required for optimal decision-making. In so doing, we referred (albeit indirectly) to percentages of white balls, or to number of successes, etc. This is certainly the most direct way of learning how to express one's own opinions, and how to formulate the mathematical conditions which they must satisfy (and in terms of which they can be manipulated in a probabilistic argument).

There are occasions, on the other hand, when it seems preferable to start from a purely ordinal relation—i.e. a qualitative one—which either replaces the quantitative notion (should one consider it to be meaningless, or, anyway, if one simply wishes to avoid it), or is used as a first step towards its definition. For example, given two commodities (or two economic alternatives) A and B, one can ask which is preferable (or whether they are equally preferable) before defining utility (or perhaps even rejecting the very idea of measurable utility); and the same can be said for temperature, the pitch of a note, the length of intervals, etc.

One could proceed in a similar manner for probabilities, too. In fact (if one accepts the subjective point of view), one can apply precisely the same notion of preference as we mentioned for utility. Instead of two commodities A and B, one compares one and the same gain (let us say 1 lira) conditional on the occurrence of event A, or event B. Our preference (apart from reservations concerning 'distorting factors'; cf. Section 13 above) will be for the event judged more probable (or, if the two events are judged to have the same probability, we will be indifferent).

This approach has been studied, and, provided one does not insist on splitting hairs, leads quickly and naturally to the usual conclusions (although in a form less directly applicable to the general case). The properties one needs to take as axioms are simple and intuitive (the standard order properties, plus the qualitative equivalent of additivity): given that E' and E'' are incompatible with E, then $E \vee E'$ is more or less probable than $E \vee E''$, or equally probable, according as E' is more or less probable than E'', or equally probable; in other words, logical sums preserve order.† All the same, the 'qualitative' comparison inevitably turns out to be far too precise (indeed, far too sophisticated), from a theoretical point of view, for what is required for the quantitative (numerical) evaluation; in any case, it is not conveniently translatable into such an evaluation (unless one considers the possibility of constructing special scales of comparison).

The complication derives from the fact that, in a qualitative sense, a possible event (no matter what probability p one attributes to it, even

† Cf. B. de Finetti, 'Sul significato soggettivo della probabilitá', *Fundamenta Mathematicae*, **17**, Warsaw (1931); an improvement in the argument given in the notes of my course on the Calculus of Probability, University of Padua, 1937–38, was made by Professor A. Gennaro who succeeded me in presenting the course.

$p = 0$) is obviously 'more probable' than an impossible event. Similarly, by adding to an event E a possible event A, incompatible with it, even if of zero probability, one obtains an event $E + A$, which is 'more probable' than E. It follows that, having other events E' with (numerical) probabilities equal to that of E, $\mathbf{P}(E') = \mathbf{P}(E)$, the qualitative comparison would have to establish for each one whether it had the same probability as E, or $E + A$, or greater than the first and less than the second, or greater than both, or less than both. Even worse; consider an arbitrary sequence of events $A_1, A_2, \ldots, A_h, \ldots$, all of zero probability, mutually incompatible, and incompatible with E, and another sequence of events $B_1, B_2, \ldots, B_h, \ldots$, all of zero probability, mutually incompatible, and contained in E: setting

$$E_0 = E, \qquad E_h = E + A_1 + A_2 + \ldots + A_h,$$
$$E_{-h} = E - B_1 - B_2 - \ldots - B_h, \qquad (h > 0),$$

one obtains an increasing ($E_h \subset E_k$ for $h < k$) and doubly unbounded sequence of events E_h ($h = 0, \pm 1, \pm 2, \ldots, \pm n, \ldots$), all with probability $\mathbf{P}(E_h) = \mathbf{P}(E)$. Any comparison of an E' (also having probability $\mathbf{P}(E') = \mathbf{P}(E)$) with the E_h should make precise which (if any) of the E_h have the same probability as E'; or, otherwise, in which of the intervals E_h, E_{h+1}, it finds itself; or if it precedes, or follows, all the E_h, for h between $\pm\infty$.

The necessity, now explained, of this much more refined comparison, has led us to use the phrase 'having the *same* probability' for two events which, in the ordering, belong to the same 'equivalence class', rather than 'equally probable', which we use when we refer to the equality of their numerical probabilities.

The situation is that which would present itself (in a less serious way) in a comparison between intervals, if intervals of equal length were to be called 'equally long' only if they both contained either 0, 1 or 2 of their end-points (otherwise, the one containing more end-points would be called 'longer'). One arrives at a closer analogy by extending the example to sets which are unions of a finite number of intervals, and (the sum of the lengths being equal) calling 'longer' the set for which the difference between the number of closed and open components is greatest (the intervals containing only one end-point are not counted; any isolated points are counted as closed intervals; these conventions are necessary if we are to have additivity, as in the case of probability).

Using such partitions into intervals as an image for our probabilistic partitions, one sees, for example, that, if the certain event is thought of as represented by a closed interval of length 1, it is impossible to divide it into two intervals (or, more generally, into n) which have the same probability (there are $n + 1$ end-points, one too many). This difficulty cannot be overcome by appealing to partitions into sums of intervals: it is always a question of dividing up the length 1 (into intervals with one end-point), plus one point. Shifting an end-point from one of the intervals to another one creates a disparity between them, but, in total, there always remains one end-point too many.

Conversely, if one does not consider that one has (included in the field of events to be compared) events which are suitable for furnishing a scale of comparison (for example, drawing balls numbered 1 to n, and judged to have the same probabilities, from an urn, with n arbitrarily large†), then the inequalities arising can provide totally inadequate information about the numerical values of the probabilities. Given, for example, a partition into three (incompatible) events A, B, C (assumed in order of decreasing probability), and that the only remaining comparison open to us is between A and $B + C$, this will tell us whether the probability of A is greater than or less than $\frac{1}{2}$. In the former case, we know only that $\mathbf{P}(A)$ lies between $\frac{1}{2}$ and 1, $\mathbf{P}(B)$ between 0 and $\frac{1}{2}$, $\mathbf{P}(C)$ between 0 and $\frac{1}{4}$; in the latter case, we know that $\mathbf{P}(A)$ lies between $\frac{1}{3}$ and $\frac{1}{2}$, $\mathbf{P}(B)$ between $\frac{1}{4}$ and $\frac{1}{2}$, $\mathbf{P}(C)$ between 0 and $\frac{1}{3}$ (cf. Figure A.2). And it cannot be said that things necessarily improve if we

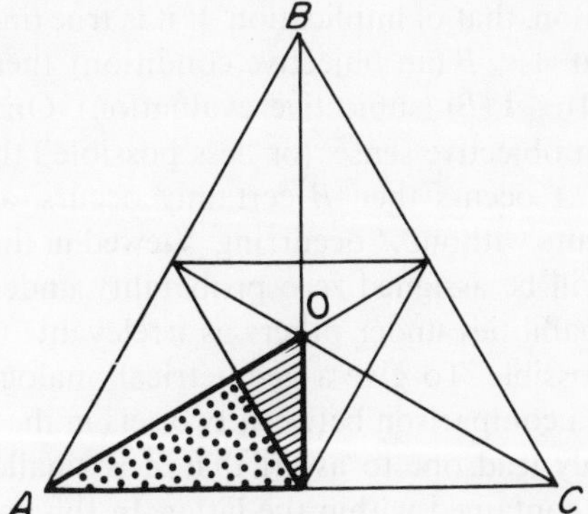

Figure A.2 Areas distinguished by means of the comparisons of probabilities for the events of a partition and their sums. The above refers to the case $n = 3$.

Dotted area: $A > B > C$, $A > B + C$.

Shaded area: $A > B > C$, $A < B + C$.

For $n = 4$ (tetrahedron), and $n > 4$ (simplex in higher-dimensional space), the mode of subdivision is similar

† The inconvenience of having to postulate the existence of partitions into events having the same probabilities has been overcome in L. J. Savage, *The Foundations of Statistics* (Chapter 3, 3: 'Quantitative personal probability') by means of a weaker assumption: for any N, one can construct a partition into N parts such that no union of n parts is more probable than one of $n + 1$ parts (for any $n < N$).

consider more than three events. If, for example, the most probable of them is more probable than the union of the others, one can only say that its probability lies between $\frac{1}{2}$ and 1; the others, therefore, in aggregate, can have probability close to $\frac{1}{2}$, or arbitrarily close to zero, or even zero.†

We could avoid complications of this kind by assigning to the comparison 'A is more probable than B', a meaning equivalent to $\mathbf{P}(A) > \mathbf{P}(B)$, and, in particular, calling a possible event of zero probability 'equal in probability to the impossible event'. In order to do this, it would be necessary to introduce the Archimedean property; in other words, to characterize those events 'more probable than the impossible one' as those with a positive numerical probability by means of a condition like the following: 'there exists a finite N such that, in every partition into N events, at least one is less probable than the given event' (whose probability is then $\geqslant 1/N$). But why resort to this distortion of an arithmetic condition instead of proceeding directly, given that one's desired goal is, in fact, the arithmetic notion?

On the other hand, the 'sophisticated', non-Archimedean, criterion corresponds exactly to the purely logical meaning that one would like to give to the comparison of probabilities in the particular case in which it relates to an objective condition, that of implication. If it is true (indeed, if it is certain; i.e. if we know) that $A \subset B$ (an objective condition), then coherence obliges us to evaluate $\mathbf{P}(A) \leqslant \mathbf{P}(B)$ (subjective evaluation). One can say that A is 'less probable in an objective sense' (or 'less possible') than B, if and only if $A \subset B$, because if A occurs then B certainly occurs, and, moreover, it is possible that B occurs without A occurring. Viewed in this light, the fact that the event $B - A$ will be assigned zero probability under some evaluations, and non-zero probabilities under others, is irrelevant: the important thing is that $B - A$ is possible. To give a geometrical analogy in this case, also, one could say that a comparison between two sets in the absence of a notion of a metric can only lead one to assert that A is smaller than B when the former is properly contained within the latter. In this case, and only in this case, will it be true that $m(A) \leqslant m(B)$ for *whatever* measure m might be introduced (the exact form being $<$ or $=$, according to whether the measure in question attributes to $B - A$ a positive or zero value).

Given disparate events E' and E'', it seems too much to hope that by comparing their probabilities one can decide if $\mathbf{P}(E') = \mathbf{P}(E'')$ exactly, rather than whether there is a difference of about 10^{-6}, or 10^{-1000}, and so on. Should one wish to square the reasonableness of the procedure with the

† In the case of $n > 3$ parts (and in the absence of a 'scale of comparison') it becomes complicated to even establish the compatibility of a system of inequalities (between sums of events of the partition). For $n = 4$, there is a simple sufficient condition (as I showed in my paper 'La logica del plausibile secondo la concezione di Pólya', *Atti. Riun. S.I.P.S.* 1949 (1951)). Contrary to what I had supposed, however, this does not hold for $n > 4$, as was shown by C. H. Kraft, J. Pratt and A. Seidenberg, 'Intuitive probabilities on finite sets', *Annals of Mathematical Statistics*, **XXX** (1959).

logical scruples mentioned above (i.e. the seemingly obvious fact that, for the same price, one prefers to have an extra possibility of winning, even though the extra possibility has probability zero), one could perhaps consider an intermediate order relation. One might define (for example):

$A < B$: 'A less probable than B' if $\mathbf{P}(A) < \mathbf{P}(B)$, or if $\mathbf{P}(A) = \mathbf{P}(B)$ and $A \subset B$;

$A \sim B$: 'A not comparable with B' if neither $A < B$ nor $B < A$; i.e. if $\mathbf{P}(A) = \mathbf{P}(B)$ and we have neither $A \subset B$ nor $B \subset A$.

Instead of following this rather abstract comparison of the various possibilities, it is more useful to ask oneself whether the exigencies of the problem themselves indicate the appropriate form, which could then profitably be adopted. This approach leads to something quite similar to the above, but, in the case where $\mathbf{P}(A) = \mathbf{P}(B)$, comparability will be given by a condition less restrictive than demanding that either $A \subset B$ or $B \subset A$ (i.e. either $A - AB = 0$, $\mathbf{P}(B - AB) = 0$, or vice-versa). Specifically, the condition consists of $\mathbf{P}(A - AB) = \mathbf{P}(B - AB) = 0$, together with the comparability of these two zero probabilities, along the lines suggested in Chapter 4 and developed in Section 18.2 of this Appendix.

It is a question of comparing the conditional probabilities

$$\mathbf{P}(A - AB|A + B - AB) \quad \text{and} \quad \mathbf{P}(B - AB|A + B - AB)$$

(whose sum $= 1$), saying that A is more or less probable than B according as the first expression is greater than the second or is less. Note that in this way one can deal with every case within the one formulation; it does not matter whether $\mathbf{P}(A)$ and $\mathbf{P}(B)$ are different, or are equal, or whether, if they are zero, one has to proceed to the comparison of residuals. By definition, we have that A and B are not comparable if $\mathbf{P}(A) = \mathbf{P}(B)$ and the residuals $A - AB$ and $B - AB$ have equal probabilities (which is automatic if they are not zero).

Note also that it would be wrong to claim that events which are not comparable have the same probabilities. If A is more probable than B on account of the zero probability of $A - AB$ being greater than that of $B - AB$, and if C is an event such that $\mathbf{P}(C) = \mathbf{P}(A) = \mathbf{P}(B)$ but $\mathbf{P}(AC) < \mathbf{P}(C)$, then both A and B turn out not to be comparable with C; but to say 'having the same probability as C' would imply that they had the same probability as each other, and this is false by hypothesis.†

† The only case in which the two events E' and E'' could be said 'to have the same probability' is that in which they consisted, respectively, of m' out of n', and m'' out of n'', events of two partitions into events having the same probabilities, where $m'/n' = m''/n''$. This remark should not be taken to mean that we wish these futile and absurd complications to be taken seriously, but, on the contrary, that we wish to remove them, without ignoring, however, the issue of what can or cannot be expressed in a correct way.

19.3. *Do 'imprecise probabilities' exist?* The question as it stands is rather ill-defined, and we must first of all make precise what we mean. In actual fact, there is no doubt that quantities can neither be measured, nor thought of as really defined, with the absolute precision demanded by mathematical abstraction (can we say whether the number in question is algebraic or transendental? Or are we capable of giving millions of significant figures, or even a few dozen?). A subjective evaluation, like that involved in expressing a probability, attracts this criticism to an even greater degree (but this is no reason for regarding the problem differently in this case, as somehow being more essentially rooted in the concepts involved). The same is true for 'objective probability': the person putting his faith in 'objective probabilities' is in precisely the same situation, except insofar as he is restricting himself to cases in which everyone (he himself, or even a subjectivist) has at his disposal criteria and information which make the judgement easier.

In this sense, it should be sufficient to say that all probabilities, like all quantities, are in practice imprecise, and that in every problem involving probability one should provide, just as one does for other measurements, evaluations whose precision is adequate in relation to the importance of the consequences that may follow. In any case, one should take into account that there is always this margin of error (for instance, it might be worth repeating the calculations with several slightly different values).

The question posed originally, however, really concerns a different issue, one which has been raised by several authors (each of whom, it seems to me, imparts a different shade of meaning to the problem). It concerns the possibility of cases in which one is not able to speak of a single value p for a given probability, but rather of two values, p' and p'', which bound an area of indeterminacy, $p' \leqslant p \leqslant p''$, possessing some essential significance.

The idea can be traced back to Keynes (cf. the remark in the last section concerning Borel's review), and was later taken up by B. O. Koopman and I. J. Good, developed considerably by C. A. B. Smith,† and more recently by other authors, like Ellsberg and Dempster.

Several different situations may lead one to express oneself in terms of an imprecise evaluation.

An example of this occurs when one wishes to distinguish various hypotheses, and attributes different probabilities $\mathbf{P}(E|H_i)$ to an E, depending on the various hypotheses H_i; if one then ignores the hypotheses, one can only conclude that the probability lies between the maximum and the minimum.

† These questions are examined in detail in Sections 26 and 27 of the paper by B. de Finetti and L. J. Savage that we have frequently referred to; particular reference is made to the (then very recent) paper of C. A. B. Smith, and to the interesting discussion to which it gave rise at a Royal Statistical Society meeting, with contributions from Barnard, Cox, Lindley, Finney, Armitage, Pike, Kerridge, Bartlett and (in the form of a written contribution) Anscombe. The reader will find there many other points which we have not found room for here.

We have already dealt with this case in Chapter 4, Section 4.8, especially in 4.8.3 and 4.8.5. The probability is what it is on the basis of the information that one has. It is clear that with additional information the probability could take on all conceivable values, finally reaching, and then remaining at, either 1 or 0, when it is finally known whether it is true or false. If we are dealing with hypotheses H_i about which we expect soon to have some information, then it may be reasonable to wait until this information is available, rather than making a provisional evaluation by taking a weighted average with respect to the probabilities which, in the meantime, are attributed to the H_i. It would be naïve, however, to assert that $\mathbf{P}(E)$ will take on a value lying somewhere between the $\mathbf{P}(E|H_i)$. There is an infinite number of partitions into hypotheses, and the information which comes along might be anything at all (for instance, it may confirm that out of the H_i, H'_j, H''_l those with $i = 3, j = 1, l = 7$ are true); the $\mathbf{P}(E|H_3H'_1H''_7)$ could vary anywhere between 0 and 1, even though the $\mathbf{P}(E|H_i)$ are all very close to one another (and even if they are equal; this case is only without interest insofar as the question would then, of course, never have been raised).

A second example occurs when one has not given sufficient thought to the matter, and hence possesses only a vague idea of the evaluation one wishes to make. A special case of this occurs when one has expressed the evaluation in terms of a formula (e.g. $p = \mathrm{e}^{-a}a^n/n!$, with $a = 5813$ and $n = 12$), but, having not yet carried out the numerical calculations, one has only a rough idea of the order of magnitude. In the final analysis, however, nothing has changed. One either carries out the calculations, or one is obliged to take as the probability the prevision of the result according to one's own, more or less haphazard, crude estimation: there is no better solution.

The idea of translating the imprecision into bounds, $p' \leqslant p \leqslant p''$, even in the weaker sense proposed by Good (who regards p' and p'' not as absurd, rigid bounds, capable of 'making the imprecision precise', but merely as indications of maxima), is inadequate if one wishes to take it to the limit in the sense in which it serves to give an idea of the imprecision with which every quantity is known or can be considered. One should think of the imprecision in the choice of the function $\mathbf{P}$ (extended, for example, to some neighbourhood $\mathscr{P}^*$ of a given $\mathbf{P}_0$ in the space $\mathscr{P}$). The imprecision for individual events and random quantities would, as a consequence, be determined not as isolated features, but with the certain or uncertain connections deriving from logical or probabilistic relations.

19.4. *What one can do in practice.* A similar kind of discussion can be given concerning what one can actually do in practice; the main purpose of this, however, is to make clear that the issues involved here are rather different, and do not give rise to any difficulties or anxieties.

The (theoretical) possibility of attributing a *precise probability* to all the events appears to be an indispensable requirement if one considers probability as a notion which applies to events *per se*, independently (cf. Section 3) of the existence and nature of any properties (e.g. topological) of the fields to which the definition of certain events can be referred. On the other hand, this does not imply that these probabilities are determined, as unique extensions of those conferring to a subfield (extensions which may or may not provide a sharpening of bounds; sometimes, as a special case, they may provide a unique value), nor does it imply that we are obliged to complete the evaluation, nor even to worry about it. Indeed, one can even call a halt well before this (earlier than usual) if there is no real interest in proceeding further: this is even more the case if the hypotheses upon which one would base oneself in proceeding further appear rather artificial and devoid of any realistic meaning.

If, in the context of the above (Section 19.3), we stop thinking in terms of a certain subset $\mathscr{P}^*$ of previsions $\mathbf{P} \in \mathscr{P}^*$, among which we are uncertain as to which one to choose, and we consider instead that we are dealing with the set of all the $\mathscr{P}$ which extend a given $\mathbf{P}$ in $\mathscr{L}$, then the bounds of indeterminacy mentioned above (at the end of Section 15) would follow. But it is not a question of imprecision. The fact that, in the case we are considering, it is only possible to say of a $\mathbf{P}(E)$ that it lies between $\mathbf{P}^-(E)$ and $\mathbf{P}^+(E)$, does not imply that certain events, like E, have an indeterminate probability: it merely implies that the probability is not uniquely defined by the initial data that one has considered. As an analogy, it would, in the same way, be nonsense to say that 'a rectangle having a perimeter of 12 m has an indeterminate diagonal'; it is 'determinate', in the sense that it is what it is, but one has to measure it, or measure a side, or something else, in order to obtain sufficient information to be able to 'determine' it (in the sense of obtaining, by means of a calculation, its well-determined value, notwithstanding the fact that knowledge of the perimeter alone is not sufficient).

There is a context in which one might refer to indeterminacy, but only in a precise, technical sense. This would arise if a certain individual were familiar with the evaluations that another individual had made within $\mathscr{L}$, and wished to establish what *the latter* should do outside that ambit in order to remain coherent. This makes it clear, however, that the indeterminacy is not a property of the events, but rather that it lies in the fact that an outsider cannot remove it, since he cannot replace the individual who is interested in the, as yet unprejudiced, evaluation.

Another aspect of the problem links up with the discussion (Sections 5–11) concerning the 'verifiability' of events (or the 'realizability' of measurements).

If, for a given problem, in a given situation, an event is not, in practice, verifiable, then any discussion about its probability is mere idle talk. For this reason, leaving aside the question of whether or not one accepts the necessity of admitting countable additivity as an axiom, it seems that if one is discussing the probability that $X \in I$, where I is non-measurable in the Jordan–Peano sense, and there is an interval in which both I and its complement $\tilde{I}$ are everywhere dense, then no measurement—not even with the assumption of 'unbounded precision' (Section 7)—can decide, on the basis of an observation x lying in that interval, whether or not the exact value lies in the interval. The same difficulty remains even under weaker assumptions: for example, if there exist points of both I and $\tilde{I}$ whose distance from x is smaller than the margin of measurement error.

In other fields, too, for realistic applications it seems much more useful to use methods which avoid too rigorous an assumption of precision. For example, when one speaks of 'convergence' it seems preferable, when considering results to have asymptotic validity, to confine attention to those which are valid for a large, but finite, number of cases (without taking seriously the notion of considering an infinite number of them).

20 CONCLUSIONS

Can one, having now come to the end, draw some conclusions?

I have in mind, of course, the critical questions which we have examined in this Appendix (some of which we anticipated in the text, to the extent that the topic in question required). In other words, I am referring to questions of a predominantly technical nature—if the word technical is adequate to characterize the difference between these matters and those concerning the meaning and formulation of the entire (subjective and Bayesian) theory; matters which occupied us throughout the text (Chapters 1–12).

Some kind of summary is required, if only to avoid the possibility of the reader being left with a feeling of confusion or bewilderment. The latter is a distinct possibility, because of the apparent contrast between our tendency on the one hand to simplify things, refusing to go beyond the level of practical applications, and on the other towards throwing ourselves headlong into hair-splitting and complicated analyses (which are not only far removed from any foreseeable application, but even strain the limits of good sense).

Why then, someone will surely ask, not be content with the 'happy medium' provided by the standard approach? This consists in proceeding to the point where countable additivity make everything work beautifully, and then stopping when the miracle ceases.

Because—I would answer—so far as I am concerned it is by no means a 'happy medium', but rather a case of 'two wrongs not making a right'. In my opinion, anything in the formulation which proceeds beyond what the Jordan–Peano–Riemann machinery provides is irrelevant for practical purposes, and unjustifiable on theoretical and conceptual grounds.

The two kinds of discussion to which we referred above, although apparently in contrast to one another, are intended, converging from opposite directions, to demonstrate one and the same point: *one can do without complications* (and this is perhaps the wisest course of action), but, should one decide to embark upon them, *one must do so whole-heartedly, in a constructive manner, even though this may prove troublesome.*

I may be wrong. My criticisms will not have been in vain, however, if in order to refute them someone comes forward and explains and justifies, in a sensible and meaningful way, those things which, up until now, have merely been 'Adhockeries for mathematical convenience'.

Index

Reference is made to the important ideas and terms, and to the authors cited. A detailed bibliography is *not* provided; the interested reader is referred to those given in the following texts:

W. Feller, *An Introduction to Probability Theory and its Applications*, Vol. 1, 1958, and Vol. 2, 1966, Wiley, New York;

B. de Finetti, *Probability, Induction and Statistics*, 1972, Wiley, New York;

H. E. Kyburg and H. E. Smokler, *Studies in Subjective Probability*, 1964, Wiley, New York;

P. Lévy, *Théorie de l'Addition des Variables Aléatoires*, 1954, Gauthier-Villars, Paris.